Dimitrios Karamichos (Ed.)

Ocular Tissue Engineering

MDPI

This book is a reprint of the Special Issue that appeared in the online, open access journal, *Journal of Functional Biomaterials* (ISSN 2079-4983) from 2015–2016 (available at: http://www.mdpi.com/journal/jfb/special_issues/ocular-tissue-eng).

Guest Editor
Dimitrios Karamichos
University of Oklahoma Health Sciences Center,
USA

Editorial Office
MDPI AG
Klybeckstrasse 64
Basel, Switzerland

Publisher
Shu-Kun Lin

Senior Assistant Editor
Zhiqiao Dong

1. Edition 2016

MDPI • Basel • Beijing • Wuhan • Barcelona

ISBN 978-3-03842-201-3 (Hbk)
ISBN 978-3-03842-202-0 (PDF)

Table of Contents

Chapter 1: An Editorial

Chapter 2: Ocular Disease and Future Biomaterials

Chapter 3: Ocular Nanotechnology and Tissue Engineering

List of Contributors

Raphaelle Alzonne Queensland Eye Institute, 140 Melbourne Street, South Brisbane, Queensland 4101, Australia.

Traian V. Chirila Queensland Eye Institute, 140 Melbourne Street, South Brisbane, Queensland 4101, Australia; Australian Institute for Bioengineering and Nanotechnology, Faculty of Medicine and Biomedical Sciences, Science and Engineering Faculty, Queensland University of Technology, Brisbane, Queensland 4001, Australia; Faculty of Science, University of Western Australia, Crawley, Western Australia 6009, Australia.

Che J. Connon Institute of Genetic Medicine, Newcastle University, International Centre for Life, Central Parkway, Newcastle upon Tyne NE1 3BZ, UK.

Darlene A. Dartt Schepens Eye Research Institute, Massachusetts Eye and Ear/Harvard Medical School, Boston, MA 02114, USA.

Cula N. Dautriche State University of New York (SUNY) Polytechnic Institute, Colleges of Nanoscale Science and Engineering, 257 Fuller Road, Albany, NY 12203, USA.

Rebecca A. Dawson Queensland Eye Institute, South Brisbane, Queensland 4101, Australia; School of Biomedical Sciences, Faculty of Health, Queensland University of Technology, Brisbane, Queensland 4001, Australia.

Grant A. Edwards Australian Institute for Bioengineering and Nanotechnology, University of Queensland, St Lucia, Queensland 4072, Australia.

Jon Roger Eidet Department of Ophthalmology, Oslo University Hospital, Oslo 0424, Norway.

Maria Fideliz de la Paz El centro de Oftalmología Barraquer, Universitari Barraquer/Universitat Autonoma de Barcelona, Barcelona 08021, Spain.

Ricardo M. Gouveia Institute of Genetic Medicine, Newcastle University, International Centre for Life, Central Parkway, Newcastle upon Tyne NE1 3BZ, UK.

Damien G. Harkin Queensland Eye Institute, South Brisbane, Queensland 4101, Australia; Institute of Health and Biomedical Innovation and School of Biomedical Sciences, Faculty of Health, Queensland University of Technology, Brisbane, Queensland 4001, Australia.

Masatoshi Hirayama Department of Ophthalmology, Keio University School of Medicine, Shinjuku-ku, Tokyo 160-8582, Japan.

Thomas A. Hogerheyde Queensland Eye Institute, South Brisbane, Queensland 4101, Australia; Institute of Health and Biomedical Innovation, School of Biomedical Sciences, Faculty of Health, Queensland University of Technology, Brisbane, Queensland 4001, Australia.

Dimitrios Karamichos Department of Cell Biology, University of Oklahoma Health Sciences Center, Oklahoma City, OK 73104, USA; Department of Ophthalmology, Dean McGee Eye Institute, University of Oklahoma Health Sciences Center, Oklahoma City, OK 73104, USA.

Qalb-E-Saleem Khan Department of Medical Biology, Faculty of Health Sciences, University of Tromsø, Tromsø 9037, Norway.

Darren J. Lee Department of Ophthalmology/Dean McGee Eye Institute, University of Oklahoma Health Sciences Center, 608 Stanton L. Young Blvd, DMEI PA404, Oklahoma City, OK 73104, USA.

Desiree' Lyon Department of Ophthalmology/Dean McGee Eye Institute, University of Oklahoma Health Sciences Center, Oklahoma City, OK 73104, USA.

Tina B. McKay Department of Cell Biology, University of Oklahoma Health Sciences Center, Oklahoma City, OK 73104, USA.

Martina Miotto Institute of Genetic Medicine, Newcastle University, International Centre for Life, Central Parkway, Newcastle upon Tyne NE1 3BZ, UK.

Jesintha Navaratnam Department of Ophthalmology, Oslo University Hospital, Postbox 4950 Nydalen, Oslo 0424, Norway.

Ole Kristoffer Olstad Department of Medical Biochemistry, Oslo University Hospital, Oslo 0407, Norway.

Kristoffer Ommundsen Department of Medical Biochemistry, Oslo University Hospital, Kirkeveien 166, Oslo 0407, Norway.

Lara Pasovic Department of Medical Biochemistry, Oslo University Hospital, Oslo 0407, Norway; Faculty of Medicine, University of Oslo, Oslo 0372, Norway.

Shrestha Priyadarsini Department of Ophthalmology/Dean McGee Eye Institute, University of Oklahoma Health Sciences Center, Oklahoma City, OK 73104, USA.

Sten Ræder The Norwegian Dry Eye Clinic, 0159 Oslo, Norway.

Ammaji Rajala Dean A. McGee Eye Institute, Oklahoma City, OK 73104, USA; Department of Ophthalmology, College of Medicine, University of Oklahoma, Oklahoma City, OK 73014, USA.

Raju V. S. Rajala Dean A. McGee Eye Institute, Oklahoma City, OK 73104, USA; Department of Ophthalmology, College of Medicine, University of Oklahoma, Oklahoma City, OK 73014, USA; Department of Physiology and Harold Hamm Diabetes Center, University of Oklahoma Health Sciences Center, Oklahoma City, OK 73014, USA.

Vinagolu K. Rajasekhar Memorial Sloan Kettering Cancer Center, Rockefeller Research Building, Room 1163, 430 East 67th Street/1275 York Avenue, New York, NY 10065, USA.

Neil A. Richardson Queensland Eye Institute, 140 Melbourne Street, South Brisbane, Queensland 4101, Australia; School of Biomedical Sciences and Institute of Health & Biomedical Innovation, Queensland University of Technology, 2 George Street, Brisbane, Queensland 4001, Australia.

Panagiotis Salvanos Department of Ophthalmology, Drammen Hospital, Vestre Viken Hospital Trust, Drammen 3004, Norway; Faculty of Medicine, University of Oslo, Oslo 0372, Norway.

Akhee Sarkar-Nag Department of Ophthalmology/Dean McGee Eye Institute, University of Oklahoma Health Sciences Center, Oklahoma City, OK 73104, USA.

Amer Sehic Department of Oral Biology, Faculty of Dentistry, University of Oslo, Sognsvannsveien 10, Oslo 0372, Norway; Department of Ophthalmology, Drammen Hospital, Vestre Viken Hospital Trust, Drammen 3004, Norway.

Audra M. A. Shadforth Queensland Eye Institute, South Brisbane, Queensland 4101, Australia; School of Biomedical Sciences, Faculty of Health, Queensland University of Technology, Brisbane, Queensland 4001, Australia.

Aboulghassem Shahdadfar Department of Ophthalmology, Oslo University Hospital, Postbox 4950 Nydalen, Oslo 0424, Norway.

Susan T. Sharfstein State University of New York (SUNY) Polytechnic Institute, Colleges of Nanoscale Science and Engineering, 257 Fuller Road, Albany, NY 12203, USA.

Shuko Suzuki Queensland Eye Institute, South Brisbane, Queensland 4101, Australia; Queensland Eye Institute, 140 Melbourne Street, South Brisbane, Queensland 4101, Australia.

Yangzi Tian State University of New York (SUNY) Polytechnic Institute, Colleges of Nanoscale Science and Engineering, 257 Fuller Road, Albany, NY 12203, USA.

Kazuo Tsubota Department of Ophthalmology, Keio University School of Medicine, Shinjuku-ku, Tokyo 160-8582, Japan.

Takashi Tsuji Laboratory of Organ Regeneration, RIKEN Center for Developmental Biology, Kobe, Hyogo 650-0047, Japan; Organ Technologies Inc., Chiyoda-ku, Tokyo 101-0048, Japan.

Øygunn Aass Utheim Department of Oral Biology, Faculty of Dentistry, University of Oslo, Sognsvannsveien 10, Oslo 0372, Norway; Department of Medical Biochemistry, Oslo University Hospital, Kirkeveien 166, Oslo 0407, Norway.

Tor Paaske Utheim Department of Medical Biochemistry, Oslo University Hospital, Oslo 0407, Norway; Department of Ophthalmology, Drammen Hospital, Vestre Viken Hospital Trust, Drammen 3004, Norway; Department of Oral Biology, Faculty of Dentistry, University of Oslo, Oslo 0372, Norway; The Norwegian Dry Eye Clinic, 0159 Oslo, Norway; Faculty of Health Sciences, University College of South East Norway, Kongsberg 3603, Norway; Department of Ophthalmology, Drammen Hospital, Vestre Viken Hospital Trust, Drammen 3004, Norway.

Yuhong Wang Dean A. McGee Eye Institute, Oklahoma City, OK 73104, USA; Department of Ophthalmology, College of Medicine, University of Oklahoma, Oklahoma City, OK 73014, USA.

Yubing Xie State University of New York (SUNY) Polytechnic Institute, Colleges of Nanoscale Science and Engineering, 257 Fuller Road, Albany, NY 12203, USA.

About the Guest Editor

Dimitrios Karamichos is a corneal scientist and holds a Ph.D. in Tissue Engineering/Molecular Biology from the University of London, London, UK. He has dedicated his research career to wound healing and corneal diseases. Dr. Karamichos is currently an Assistant Professor of Ophthalmology and Cell Biology at the University of Oklahoma Health Sciences Center, Dean McGee Eye Institute, Oklahoma, USA. He is also the Vice President of the Graduate College of the University of Oklahoma Health Sciences Center and the President of the Society for Neuroscience, Oklahoma Chapter. Dr. Karamichos's research interests include corneal wound healing and fibrosis, keratoconus, corneal diabetes, and drug delivery. He is the principal investigator on numerous grants funded by the National Eye Institute/National Institute of Health.

Preface to "Ocular Tissue Engineering"

Tissue engineering is a rapidly growing area, and complex three-dimensional tissue substitutes are emerging. Although cells are routinely cultured outside the body, current research shows that tissue engineered constructs can be used as replacement tissues for damaged or diseased human organs. This book is an outgrowth of a Special Issue of the Journal of Functional Biomaterials (JFB) devoted to Ocular Tissue Engineering and contains both original research and review articles. Each of the articles included here provides an up-to-date analysis and cutting edge technology in this fast growing field. Biomaterials and nanotechnology in cellular processes, as well as in ocular disease, are highlighted. We sincerely hope that readers will enjoy these articles and be inspired by the ideas presented.

Dimitrios Karamichos
Guest Editor

Chapter 1:
An Editorial

Ocular Tissue Engineering: Current and Future Directions

D. Karamichos

Reprinted from *J. Funct. Biomater.* Cite as: Karamichos, D. Ocular Tissue Engineering: Current and Future Directions. *J. Funct. Biomater.* **2015**, *6*, 77–80.

Tissue engineering (TE) is a concept that was first emerged in the early 1990s to provide solutions to severe injured tissues and/or organs [1]. The dream was to be able to restore and replace the damaged tissue with an engineered version which would ultimately help overcome problems such as donor shortages, graft rejections, and inflammatory responses following transplantation. While an incredible amount of progress has been made, suggesting that TE concept is viable, we are still not able to overcome major obstacles. In TE, there are two main strategies that researchers have adopted: (1) cell-based, where cells are been manipulated to create their own environment before transplanted to the host, and (2) scaffold-based, where an extracellular matrix is created to mimic *in vivo* structures. TE approaches for ocular tissues are available and have indeed come a long way, over the last decades; however more clinically relevant ocular tissue substitutes are needed. Figure 1 highlights the importance of TE in ocular applications and indicates the avenues available based on each tissue.

In cornea, TE approaches are vital in order to maintain the transparent barrier between the eye and the environment. Of the three corneal layers (epithelium, stroma, and endothelium) probably the most difficult one to replace is the stroma. Stroma is a thick, transparent middle layer, consisting of regularly arranged collagen fibers along with sparsely distributed resident cells commonly known as keratocytes. The corneal stroma consists of approximately 200 collagen fibril layers and account for up to 90% of the total corneal thickness. Corneal transplantation is currently the only surgical procedure for replacing damaged or diseased corneas. Damaged cornea is replaced by donated corneal tissue in its entirety (penetrating keratoplasty) or in part (lamellar keratoplasty). While the surgical procedure has been somewhat successful, major problems remain including donor corneas shortage, risks of infection, and graft rejection. In an attempt for an alternative avenue, several studies have reported successful cultivation of corneal stroma, in combination with corneal epithelium and endothelium, however the long-term *in vivo* data and clinical applications are still lacking [1]. The corneal epithelium has been targeted by scientists and a variety of TE applications using both cell and scaffold-based approaches have been developed [2–6]. Studies reporting the successful transplantation of mucosal epithelial cells [5,6] as well as limbal stem cells [2] are promising. Tissue grafts such amniotic membranes [3,4] have also been reported and used in humans. While these have been assessed in clinical setting, long-term studies are still needed in order to safely assess the benefits.

In lens, despite the limited number of studies developing TE solutions, there is a clear need for cataract surgeries alternatives. Currently, lens opacification or else known as cataracts are treated surgically by removing the lens and replacing it with artificial intraocular lenses (IOL) [1,7]. Most of the people receiving cataract surgery will need to come back for a second surgery due to the posterior capsule opacification (PCO). PCO occurs because lens epithelial cells remaining after cataract surgery have grown on the capsule causing it to become hazy and opaque [1,7,8]. Development of alternatives is almost nonexistent and urgently needed. One of the few TE approaches was reported by Tsionis *et al.* [9] where a human retinal PE cell line cultured in Matrigel was differentiated in lentoids and lens-like structures. Nevertheless, therapies based on this technique or others are far away and it remains unknown if TE is the future for lens related clinical problems.

In retina, both cell and substrate-based TE approaches have been reported mainly in animal models. Homologous retinal pigment epithelium (RPE) cells have been transplanted in the subretinal space with no visual benefits to the patients [10,11]. On the other hand autologous RPE transplantation resulted in clinically significant improvement of vision; however the limited number of healthy cells that can be isolated from the patient is a huge problem [12,13]. The concept of the use of polymers for retinal TE is rather new and has only been emerged in the last decade or so. As reviewed by Trese and co-authors [14] the ideal polymer for retinal transplantation should be thinner than 50 μm, porous, biodegradable, and have the correct Young's modulus. Several polymers fulfill this criteria including but not limited to poly(lactic-co-glycolic acid) (PLGA), poly(lactic acid (PLLA), poly(glucerol-sebacate) (PGS), and poly(caprolactone) (PCL) [14,15]. However, only a few studies have shown promising results using these or other polymers for TE retinal applications. The combination of PLLA-PLGA polymer reported by Thomson and co-authors [16] showed good RPE cellular viability, adhesion and proliferation for the course of the month long study. However, the main limitation of this study was the use of cell lines instead of primary cells which are known to be different in terms of their behavior. The general consensus is that embryonic stem cells (ESC) and induced pluripotent stem (iPS) cells are a better choice since they more closely resemble actual RPE. This, however, remains to be seen. Regardless of the cell source, technical challenges still remain before cell-substrate based therapies can be successful.

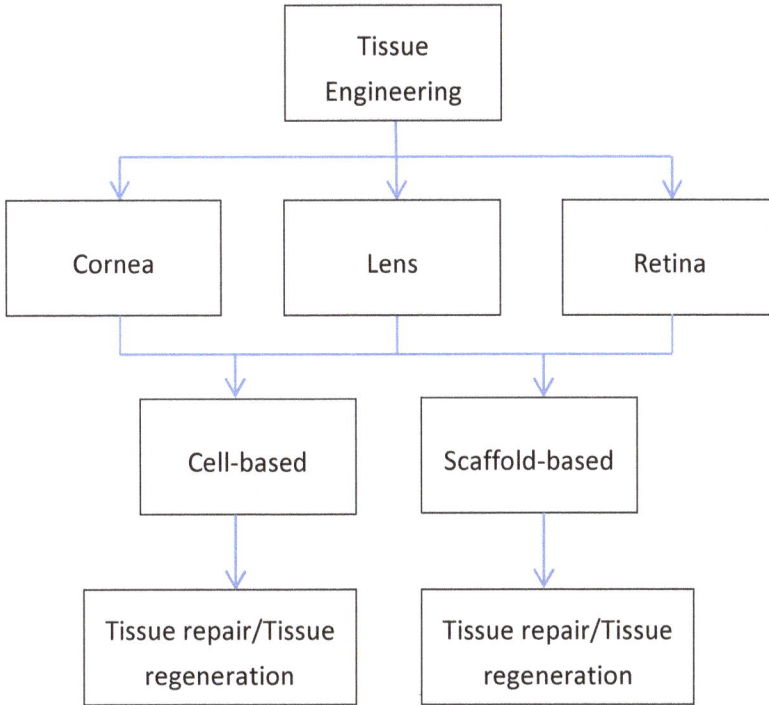

Figure 1. Schematic diagram highlighting the importance of tissue engineering (TE) approaches in ocular tissues: cornea, lens, and retina.

In conclusion, the human eye with the different structures, cell types, and tissues is an ideal candidate for TE approaches. The eye structures and the inadequate to-date therapies make this a very attractive tissue for TE. This is well understood within the scientific community and that is why significant discoveries and knowledge advancements have been made. Perhaps the one tissue with the most success is the corneal epithelium. There is no reason why the other structures cannot be regenerated or reconstructed using TE techniques. The challenge here is to be able to get the scientists, engineers, and clinicians to work together in order to tackle today's challenges and give our patients the best possible treatment.

Conflicts of Interest: The author declares no conflict of interest.

Acknowledgments: The author would like to thank the National Eye Institute of the National Institutes of Health for the support (EY023568 and EY020886). The content is solely the responsibility of the author and does not necessarily represent the official views of the National Institutes of Health.

References

1. Sommer, F.; Brandl, F.; Gopferich, A. Ocular tissue engineering. *Adv. Exp. Med. Biol.* **2006**, *585*, 413–429.

2. Pellegrini, G.; Traverso, C.E.; Franzi, A.T.; Zingirian, M.; Cancedda, R.; De Luca, M. Long-term restoration of damaged corneal surfaces with autologous cultivated corneal epithelium. *Lancet* **1997**, *349*, 990–993.

3. Tsai, R.J.; Li, L.M.; Chen, J.K. Reconstruction of damaged corneas by transplantation of autologous limbal epithelial cells. *New. Engl. J. Med.* **2000**, *343*, 86–93.

4. Schechter, B.A.; Rand, W.J.; Nagler, R.S.; Estrin, I.; Arnold, S.S.; Villate, N.; Velazquez, G.E. Corneal melt after amniotic membrane transplant. *Cornea* **2005**, *24*, 106–107.

5. Kinoshita, S.; Nakamura, T. Development of cultivated mucosal epithelial sheet transplantation for ocular surface reconstruction. *Artif. Organs* **2004**, *28*, 22–27.

6. Kinoshita, S.; Koizumi, N.; Nakamura, T. Transplantable cultivated mucosal epithelial sheet for ocular surface reconstruction. *Exp. Eye Res.* **2004**, *78*, 483–491.

7. Francis, P.J.; Berry, V.; Moore, A.T.; Bhattacharya, S. Lens biology: Development and human cataractogenesis. *Trends Genet.* **1999**, *15*, 191–196.

8. Apple, D.J.; Solomon, K.D.; Tetz, M.R.; Assia, E.I.; Holland, E.Y.; Legler, U.F.; Tsai, J.C.; Castaneda, V.E.; Hoggatt, J.P.; Kostick, A.M. Posterior capsule opacification. *Surv. Ophthalmol.* **1992**, *37*, 73–116.

9. Tsonis, P.A.; Jang, W.; Del Rio-Tsonis, K.; Eguchi, G. A unique aged human retinal pigmented epithelial cell line useful for studying lens differentiation *in vitro*. *Int. J. Dev. Biol.* **2001**, *45*, 753–758.

10. Algvere, P.V.; Berglin, L.; Gouras, P.; Sheng, Y.; Kopp, E.D. Transplantation of RPE in age-related macular degeneration: Observations in disciform lesions and dry RPE atrophy. *Graefes. Arch. Clin. Exp. Ophthalmol.* **1997**, *235*, 149–158.

11. Algvere, P.V.; Gouras, P.; Dafgard Kopp, E. Long-term outcome of RPE allografts in non-immunosuppressed patients with AMD. *Eur. J. Ophthalmol.* **1999**, *9*, 217–230.

12. Binder, S.; Krebs, I.; Hilgers, R.D.; Abri, A.; Stolba, U.; Assadoulina, A.; Kellner, L.; Stanzel, B.V.; Jahn, C.; Feichtinger, H. Outcome of transplantation of autologous retinal pigment epithelium in age-related macular degeneration: A prospective trial. *Invest. Ophthalmol. Vis. Sci.* **2004**, *45*, 4151–4160.

13. Binder, S.; Stolba, U.; Krebs, I.; Kellner, L.; Jahn, C.; Feichtinger, H.; Povelka, M.; Frohner, U.; Kruger, A.; Hilgers, R.D.; *et al.* Transplantation of autologous retinal pigment epithelium in eyes with foveal neovascularization resulting from age-related macular degeneration: A pilot study. *Am. J. Ophthalmol.* **2002**, *133*, 215–225.

14. Trese, M.; Regatieri, C.V.; Young, M.J. Advances in retinal tissue engineering. *Materials* **2012**, *5*, 108–120.

15. Langer, R. Biomaterials in drug delivery and tissue engineering: One laboratory's experience. *Acc. Chem. Res.* **2000**, *33*, 94–101.

16. Thomson, H.A.; Treharne, A.J.; Walker, P.; Grossel, M.C.; Lotery, A.J. Optimisation of polymer scaffolds for retinal pigment epithelium (RPE) cell transplantation. *Br. J. Ophthalmol.* **2011**, *95*, 563–568.

Chapter 2:
Ocular Disease and Future Biomaterials

Intraocular Implants for the Treatment of Autoimmune Uveitis

Darren J. Lee

Abstract: Uveitis is the third leading cause of blindness in developed countries. Currently, the most widely used treatment of non-infectious uveitis is corticosteroids. Posterior uveitis and macular edema can be treated with intraocular injection of corticosteroids, however, this is problematic in chronic cases because of the need for repeat injections. Another option is systemic immunosuppressive therapies that have their own undesirable side effects. These systemic therapies result in a widespread suppression of the entire immune system, leaving the patient susceptible to infection. Therefore, an effective localized treatment option is preferred. With the recent advances in bioengineering, biodegradable polymers that allow for a slow sustained-release of a medication. These advances have culminated in drug delivery implants that are food and drug administration (FDA) approved for the treatment of non-infectious uveitis. In this review, we discuss the types of ocular implants available and some of the polymers used, implants used for the treatment of non-infectious uveitis, and bioengineered alternatives that are on the horizon.

Reprinted from *J. Funct. Biomater.* Cite as: Lee, D.J. Intraocular Implants for the Treatment of Autoimmune Uveitis. *J. Funct. Biomater.* **2015**, *6*, 650–666.

1. Introduction

Uveitis is the third leading cause of blindness in developed countries [1–3], with an incidence of 52–93 cases per 100,000 persons per year and a prevalence of 115 cases per 100,000 persons [4,5]. Uveitis can affect different parts of the eye and the affected part distinguishes the different types of uveitis. Anterior uveitis involves the iris, cornea, and ciliary body, intermediate uveitis involves the vitreous and pars plana, and posterior uveitis is an inflammation of the retina [6,7]. Posterior uveitis can be devastating to vision and is difficult to diagnose and treat [8], whereas anterior uveitis is more common and can be easier to diagnose and treat [9]. Following an initial episode of anterior uveitis the recurrence rate is 36% for three or more episodes within five years [10]. Suppression of the inflammation can be achieved through the use of topical corticosteroids, but is an ineffective long-term solution because the steroids also cause cataracts and glaucoma [11–13]. The administration of systemic steroids is more effective than topical steroids for posterior uveitis and has a lower incidence of elevated intraocular pressure and cataracts [14]. However, systemic steroids can have much more serious side effects, such as weight gain, hyperglycemia, osteoporosis, gastrointestinal ulceration, intense mood changes, depression, and violent aggressive behavior [15].

The current treatment paradigm for autoimmune uveitis focuses on the use of non-steroidal immunosuppressive therapies to keep the ocular inflammation suppressed [1]. These therapies include anti-metabolites, such as methotrexate [2,16,17]; calcineurin inhibitors, such as cyclosporine A [2,18]; DNA alkylating agents, such as cyclophosphamide and chlorambucil [2,18]; and biologics, such as Adalimumab and Infliximab [19–22]. Most of these therapies are systemic treatments that are not specific for the eye so can cause systemic side effects. Therefore, a localized treatment approach has the advantage of avoiding systemic complications. Intravitreal injection of methotrexate or sirolimus is a localized treatment approach for chronic uveitis, but does not always lead to sustained suppression, so is not a viable long-term solution in every case [17,23,24]. Moreover, frequent invasive intravitreal injections can lead to retinal detachment, hemorrhage, or endophthalmitis. There is clearly a need for a localized treatment approach for autoimmune uveitis and retinal diseases. In this review, we will discuss the different types of FDA approved intraocular implants, some of the polymers involved in their composition, and what other implants are on the horizon.

2. Biodegradable Implants

An implantable long-lasting sustained release of medication is advantageous for treatment of ocular disease because it allows for effective delivery to the posterior region of the eye and eliminates the need for frequent intraocular injections of the medication. This sustained release can be achieved by imbedding a biodegradable polymer with the medication so as the polymer is degraded the medication is released. The process of converting a hydrophobic polymer into a water-soluble material is termed erosion. Erosion can occur through either surface-erosion or bulk-erosion. The type of erosion that occurs is dependent on the permeability of the polymer to water and the rate at which erosion of the polymer occurs. If the polymer erodes slower than water can penetrate into the core, the surface continues to erode as the molecular weight decreases and is termed surface-erosion [25]. In bulk-eroding polymers the water penetrates into the polymer faster than degradation of the polymer resulting in degradation of the entire material [25]. Because of the penetration of water into the polymer during bulk-erosion, interaction of the water with the medication can result in the destruction of the medication before it can be released. Another problem with bulk erosion polymers is auto-catalysis, in which the core is degraded quickly and once a pore is eroded the drug will be suddenly released—resulting in a sudden increase in drug release [25,26]. Since the type of polymers influences the surface or bulk erosion properties, the choice of materials is important to consider for the application. The bulk-erosion polymer would be useful for tissue engineering where a permeable membrane is useful for hydrolytic diffusion [27]. In the case of a desired sustained delivery, surface-eroding polymers would be more appropriate because of the stable drug release [28]. Biodegradable implants are advantageous because they do not require removal when the drug has been exhausted [25].

2.1. Bulk-Eroding Polymers

Polyglycolide or poly(glycolic acid) (PGA) has been used as a degradable suture since 1970 [29]. Glycolic acid is readily incorporated by cells through the citric acid cycle [30]. However, it can affect the pH and elicit an inflammatory response [31–33], so can be problematic for ocular implants.

Polylactide or poly(lactic acid) (PLA) has four different chiral forms [34]. Depending on the chirality, PLA can take 1–5 years or more to completely degrade [35]. Because of the extended degradation time research on PLA has been limited as such, PLA alone has not been explored as a potential drug delivery system [36–39]. Instead, PLA can be combined with PGA to form poly(lactide-co-glycolide) (PLGA) in order to accelerate the degradation time. Moreover, by changing the ratio of PGA to PLA the degradation time can be adjusted from 1–2 months to 5–6 months [40]. PLGA has been used for sutures since 1974 [41], and has been used for the delivery of proteins [42–45], anti-inflammatory drugs [46,47], and siRNA [48–50]. Unfortunately, two drawbacks to PLGA is that PLGA degradation products can both lower the pH and bulk-erosion can result in a sudden increase of the drug and destruction of the medication due to water that has diffused into the matrix [51–53].

Polycarpolactones (PCL) have been used as a contraceptive that delivers levonorgestrel for over a year and has been on the market for over 25 years [54]. Because degradation of micro-particles and nano-particles occurs over 2–3 years, blending with other polymers, such as PLGA is done to accelerate erosion [49,55,56].

2.2. Surface-Eroding Polymers

Polyanhydrides are a class of polymers that contain two carbonyl groups joined by an ether bond [34]. Hydrolysis of this bond results in two carboxylic acids that are readily metabolized, so can lower the pH. Polyanhydrides are unique because the polymer backbone has a direct relationship with the degradation rate that can vary by more than six orders of magnitude [56,57]. Combining aromatic and aliphatic diacids can slow degradation and extend drug delivery from days to years [28].

Polyacetals have two ether bonds connected to the same carbon (germinal) and maintain a milder pH because they do not degrade into carboxylic acids [58]. Some degrade when entering the lysosome at pH 4–5 [58] and have been used to deliver siRNA, DNA, and proteins for acute inflammation [59–66]. Despite

11

these properties, polyacetals have found limited used because they do not meet the mechanical strength needs of most implant applications.

Polycarbonates are very stable and consist of two geminal ether bonds and a carbonyl bond. It is thought that enzymatic degradation is responsible for surface erosion of this polymer [67]. Poly(trimethylene carbonate) (PTMC) is the most characterized polycarbonate [68]. In order to increase the mechanical and degradation properties, PTMC is copolymerized with PLA, PCL, polyether, or poly(L-glutamic acid) [69–74]. The copolymer used can give the polycarbonate drastically different mechanical properties. A gel that rapidly degrades can be made with polycarbonate and dihydroxacetone for use in clotting [75]. In contrast, significant mechanical strength and slow degradation can be achieved if the copolymer is a tyrosine-derivative, this can be used in tissue engineering of bone or muscle [76–84].

3. Non-Biodegradable Implants

The use of biodegradable implants for ocular disease has been limited because intravitreal injection of PLGA microspheres can cloud vision [85] and movement of the implant into the anterior chamber or in front of the retina can be a complication [8,86]. Some non-biodegradable implants can be anchored to the sclera for easy removal and to prevent the implant from mobilizing into an inconvenient position [87]. The variable drug release kinetics associated with biodegradable polymers [8,26] can be avoided by coating the polymer with a non-biodegradable polymer or through storage of the drug in a reservoir encased in a non-biodegradable polymer [8]. The coating can be porous or have a small hole to allow for a small area of diffusion, these typically have an initial burst that is followed by a consistent release of the medication [8,25,88,89]. A depleted non-biodegradable implant has a risk of irritating the tissue or eliciting an inflammatory response, so requires a second procedure to have the implant removed [25,90].

Silicon and ethylene-vinyl acetate copolymer (EVA) are used as hydrophobic membranes for non-biodegradable implants [8,91,92]. Poly(vinyl alcohol) (PVA) is more hydrophilic, so is more permeable to a wider range of drugs [88]. Polyimide has been used for a variety of applications from photovoltaic cells to biomedical implants [8,25,93]. Poly(methyl methacrylate) (PMMA) is a clear plastic that can be used for drug delivery [94]. Phosphatidylglycerol is a negatively charged phospholipid that can be used as a vehicle for drug delivery [95].

4. FDA Approved Implants for Uveitis

Ozurdex® is sold by Allergan and has been FDA approved to treat macular edema secondary to branch or central retinal vein occlusion and non-infectious posterior uveitis. Dexamethosone is released from a PLGA matrix for up to four months (Table 1) and because it is completely biodegradable it does not need to be removed at a later date [87]. No patients that received the Ozurdex® implant required intra-ocular pressure (IOP) lowering medications or surgery, and only 17% of patients experienced an increase in IOP of 10 mmHg or more [96–98]. Migration of the Ozurdex® implant to the anterior chamber has been reported [86,99]. However, it is possible that this complication can be avoided with careful screening of patients for post-lensectomy-vitrectomy aphakia [100,101].

Retisert® has been developed by pSivdea Corp. (Watertown, MA, USA) and is a non-biodegradable implant that delivers fluocinolone acetonide. It has been FDA approved for noninfectious posterior uveitis. The PVA and silicon coating allows for consistent release of fluocinolone for up to 30 months (Table 1) [87,101]. The Retisert® implant has been associated with the development of cataracts [102] and an increase in IOP of more that 10 mmHg in more than 60% of patients [102–104].

Table 1. FDA approved intraocular implants.

Implant	Medication	Method of Implantation	Size	Release Time	Reference
Ozurdex	Dexamethosone	22 gauge designer applicator	rod-shaped, 0.46 mm diameter, 6 mm long	up to 6 months	[25,101]
Retisert	Fluocinolone acetonide	surgical insertion	1.5 mm diameter, 6 mm long, 2 mm wide	up to 2.5 years	[25,30,101,104]
Iluvien	Fluocinolone acetonide	injection with 25 gauge needle	3.5 mm × 0.37 mm tube	up to 3 years	[8,25,87,101]
Vitrasert	Gancyclovir	surgical insertion	4–5 sclerotomy	up to 8 months	[25,101,105]
Surodex *	Dexamethosone	25 gauge needle, placed during cataract surgery	1 mm × 0.5 mm	7–10 days	[25,101]

* Surodex has been approved for use in China and Singapore.

5. FDA Approved Implants for Ocular Disease

Iluvien™ is a non-biodegradable implant that delivers fluocinolone acetate that is in a PVA matrix within a polyimide tube. The tube has membrane caps on each end to allow for diffusion of water into the matrix. Iluvien has been FDA approved for diabetic macular edema, and delivers medication for up to 36 months (Table 1) [8,87,101].

Vitrasert® contains gancyclovir in a PVA matrix with a non-biodegradable EVA coating. Gancyclovir is delivered for 5–8 months (Table 1) and is effective in the treatment of CMV [101,105,106].

Surodex® is a poly(lactic-glycolic acid) device approved for use in China and Singapore that is used to control post-cataract surgery inflammation. This is inserted in the posterior or anterior chamber during cataract surgery and dexamethosone is delivered for up to 10 days (Table 1) [101,107–109].

6. Implants in Development

While the above-discussed implants have been effective for the treatment of uveitis and other ocular diseases, these devices still have additional challenges. For instance, if elevated intraocular pressure control is a concern, these corticosteroid delivery devices should be carefully considered [97,100,102–104]. In this case topical steroid use is preferred over injectable steroids because the topical steroids can be discontinued if an increase in pressure occurs. Therefore, a tunable implant that allows for control over the delivery of medication is advantageous to avoid unnecessary delivery of the drug. This type of device would also be advantageous because uveitis is a relapsing and remitting disease and it would be attractive to be able to control the drug concentration depending on the disease state [110]. We also discuss the development of non-steroidal immunosuppressive drug delivery devices.

6.1. Non-Steroid Implants

In order to avoid the side effects associated with sustained corticosteroid use it would be advantageous to have a non-steroidal implant to deliver a localized immunosuppressant to the eye. Methotrexate has been used safely and effectively as a systemic treatment for noninfectious uveitis for years [16]. It can also be injected into the vitreous as a localized treatment for uveitis [17,23,24]. A nanogel of PEGylated poly ethyleneimine containing methotrexate has shown to be effective in reducing joint inflammation in a murine model of arthritis [111]. Cyclosporine A is a calcineurin inhibitor that has been used as a systemic treatment for uveitis [2,18]. PCL and PGLC nanoparticles are being developed as a vehicle to deliver cyclosporine A as an injection into the subconjunctiva or vitreous [112,113]. There has been some investigation into tethering neutralizing antibodies such as, anti-TNF, to polymers [112,114]. If these non-steroidal immunosuppressive medications could be adapted as an ocular implant they would provide an excellent alternative to the current sustained corticosteroid delivery devices.

Another interesting device in development is NT-501, through Neurotech, Inc. NT-501 contains polyethylene terephthalate yarn that is loaded with retinal pigmented epithelial cells (RPE). The polyethylene terephthalate yarn is contained within a polysulfone [8,88] membrane that is sutured with a titanium loop to the scleral wall. The semipermeable membrane allows for the diffusion of nutrients into the device to sustain the RPE cells and diffusion of RPE products out into the vitreous. Neurotech is developing this device for the treatment of retinitis pigmentosa, so the RPE is genetically engineered to produce recombinant CTNF or VEGF neutralizing antibodies or both [25,115–117]. If the RPE cells could be genetically engineered to secrete steroidal or non-steroidal immunosuppressive drugs, this could be another excellent implant for the treatment of uveitis.

6.2. Tunable Implants

A healthy eye has a clear cornea and lens that allows for the passage of light to the retina [118]. This property can be exploited to deliver a specific wavelength of light to an implant in the vitreous. Polymers can be formulated to include light sensitive components that allow for a permanent or temporary change in the chemistry or structure of the polymer matrix to either trigger drug release or prevent drug release [8,119]. Temporary photo-activated changes are achieved with chromophores that allow for drug release only in the presence of the photo-stimulus because when that specific wavelength of light is removed the chromophore returns to the stable state [8,120]. In contrast, an irreversible change occurs with pyrene derivatives when the light is removed [120]. Another type of photo tunable system is to convert the light into thermal energy. This can be achieved by coating the polymer with a nontransparent metal that converts the light into thermal energy [121]. The thermal energy then breaks down the polymer to create permeable pores or an orifice for the drug to diffuse out. This photo-activated technology is limited because wavelengths of light more than 900 nm cannot penetrate the eye, and wavelengths too short can cause damage to ocular structures [122]. In addition, many of the chromophores are too toxic to use in biological systems [8].

Another noninvasive method to achieve precise control of drug release from a polymer matrix is with magnetic fields. Magnetically modulated systems for drug release utilize a matrix or reservoir-based design. The matrix systems consist of magnetic particles imbedded in the polymer matrix. Upon exposure to a magnetic field, the magnetic particles vibrate in the pores to increase the pore size and allow for a greater rate of drug release [123]. The reservoir-based device contains one or more magnetic components that allows for modulating the diffusion of the drug with an external magnetic field [124]. Repeated usage of these devices results in a reduced magnetic response [125], so long term usage is not practical. Moreover, these devices would be problematic if computerized tomography (CT) and magnetic resonance imaging (MRI). There is potential for this technology, but further research is necessary before it can be implemented in the clinic.

Conductive polymers have both polymer and metal properties [8,126]. Since 1977, conducting polymers have been studied for many biomedical applications, in particular for the electrically tunable property for drug delivery [127,128]. Electric stimulation alters the redox state of the polymer to affect the charge, volume, permeability and hydrophobicity. Because the volume of the polymer can be altered it is possible to contain the drug in a reservoir and upon electrical stimulation the volume change causes the drug to be released from the polymer [129]. These polymers are biocompatible, non-toxic, and allow for fine control of drug release [130]. They are also non-biodegradable and can be altered to biodegrade, but at the expense of lowering the conductivity, and drug release capacity [131]. The disadvantage of conductive polymers is that the power source requires a bulky battery and wires to actuate the device.

The Micro Electro Mechanical System (MEMS) is an implant that achieves actuation through temperature, electrical stimulus, magnetic field, or osmotic pressure [124,132–134]. These devices consist of a reservoir and an actuator to mechanically push the medication out of the reservoir upon proper stimulation. Ideally, these devices have the advantage of being able to refill the reservoir [8] and have lower power requirements, so a wireless signal could be employed [124,135]. Unfortunately, MEMS are still in the developmental stages as the reservoir is small and the lifetime is less than a year [25]. However, a new model is in clinical trials that could last up to five years [25]. There is also a magnetic stimuli responsive MEMS implant being investigated to deliver medication to the posterior of the eye [124]. However, the long-term feasibility of this magnetic device is still to be determined as discussed with the magnetically modulated systems.

7. Summary

The current treatment paradigm for chronic noninfectious uveitis is to suppress the inflammation with localized or systemic immunosuppressive medications for a period of time that is sufficient for the patient to be slowly weaned off of the medications [1]. Presumably, during the time that the uveitis is suppressed the patient establishes a regulatory immune response to provide a resistance to relapse [136–139]. It is also probable that some aspects of ocular immune privilege are re-established during this period of immunosuppression [140].

If medication is providing systemic immunosuppression the patient will require careful monitoring to ensure systemic side effects do not occur [15]. Systemic side effects may be severe enough that termination of a treatment may be necessary and will often be related to a relapse. This is where localized treatments are ideally suited for uveitis patients. We have discussed several ocular implants for the treatment of uveitis and other ocular disease. Another advantage of ocular implants is that they are effective in delivering drugs to the retina [8].

The implants available for the treatment of ocular inflammation are either biodegradable or non-biodegradable. Biodegradable implants have the advantage of only requiring one procedure to install the implant, the disadvantage is that the implant can move and the release rate of the medication is not consistent [8,25]. The non-biodegradable implants allow for a continuous release rate and some can be secured to the sclera to prevent movement away from the implant site and for ease of removal [8,25,87]. Both types of implants have their advantages and disadvantages so until additional advancement can eliminate the disadvantages the ophthalmologist will need to evaluate the implants carefully before choosing which is more appropriate for the particular patient.

Current ocular implants available for the treatment of ocular inflammation deliver dexamethasone or fluocinolone acetate, both of which can trigger an increase in IOP [96,97,100,102–104,141]. This can be problematic, especially if a patient has steroid induced glaucoma. Fortunately, there are additional implants in development that deliver non-steroidal medication as an alternative to the current implants available. Polymers imbedded with non-steroidal immunosuppressive drugs, such as cyclosporine A and anti-TNF are in development and could provide an excellent alternative to the current sustained release corticosteroid delivery devices available for the treatment of uveitis and other ocular disease [88]. The development of tunable ocular implants is desirable because the ability to control the drug release can help to reduce side effects and can extend the availability of drug in the implant, thereby extending the life of the device [8]. Another interesting device is NT-501, currently in clinical testing by Neurotech, Inc. NT-501 is in development for retinitis pigmentosa, but if found to be a feasible long-term treatment, it could be adapted to function for ocular inflammatory disease as well.

The availability of implantable corticosteroid delivery devices has improved the outcome for noninfectious uveitis patients, particularly those with posterior uveitis [100,141]. This represents an exciting area of research and the success of the current devices available has bolstered an interest in the development of additional implants for the treatment of uveitis and other ocular diseases. In the next 4–5 years, we should see the translation of many new drug delivery devices into early stage implantable systems and we hope that many more ocular implants that are currently in clinical trials will become available for the treatment of ocular inflammatory diseases.

Acknowledgments: This work was funded in part through an unrestricted Research to Prevent Blindness grant.

Conflicts of Interest: The authors declare no conflict of interest.

References

1. Siddique, S.S.; Shah, R.; Suelves, A.M.; Foster, C.S. Road to remission: A comprehensive review of therapy in uveitis. *Expert Opin. Investig. Drugs* **2011**, *20*, 1497–1515.
2. Durrani, K.; Zakka, F.R.; Ahmed, M.; Memon, M.; Siddique, S.S.; Foster, C.S. Systemic therapy with conventional and novel immunomodulatory agents for ocular inflammatory disease. *Surv. Ophthalmol.* **2011**, *56*, 474–510.
3. Rothova, A.; Suttorp-van Schulten, M.S.; Frits Treffers, W.; Kijlstra, A. Causes and frequency of blindness in patients with intraocular inflammatory disease. *Br. J. Ophthalmol.* **1996**, *80*, 332–336.

4. Darrell, R.W.; Wagener, H.P.; Kurland, L.T. Epidemiology of uveitis. Incidence and prevalence in a small urban community. *Arch. Ophthalmol.* **1962**, *68*, 502–514.
5. Gritz, D.C.; Wong, I.G. Incidence and prevalence of uveitis in northern california; the northern california epidemiology of uveitis study. *Ophthalmology* **2004**, *111*, 491–500.
6. Jabs, D.A.; Nussenblatt, R.B.; Rosenbaum, J.T. Standardization of Uveitis Nomenclature Working Group. Standardization of uveitis nomenclature for reporting clinical data. Results of the First International Workshop. *Am. J. Ophthalmol.* **2005**, *140*, 509–516.
7. Prete, M.; Dammacco, R.; Fatone, M.C.; Racanelli, V. Autoimmune uveitis: Clinical, pathogenetic, and therapeutic features. *Clin. Exp. Med.* **2015**.
8. Yasin, M.N.; Svirskis, D.; Seyfoddin, A.; Rupenthal, I.D. Implants for drug delivery to the posterior segment of the eye: A focus on stimuli-responsive and tunable release systems. *J. Control. Release* **2014**, *196*, 208–221.
9. Barisani-Asenbauer, T.; Maca, S.M.; Mejdoubi, L.; Emminger, W.; Machold, K.; Auer, H. Uveitis—a rare disease often associated with systemic diseases and infections—a systematic review of 2619 patients. *Orphanet J. Rare Dis.* **2012**, *7*.
10. Natkunarajah, M.; Kaptoge, S.; Edelsten, C. Risks of relapse in patients with acute anterior uveitis. *Br. J. Ophthalmol.* **2007**, *91*, 330–334.
11. Pleyer, U.; Ursell, P.G.; Rama, P. Intraocular pressure effects of common topical steroids for post-cataract inflammation: Are they all the same? *Ophthalmol. Ther.* **2013**, *2*, 55–72.
12. Clark, A.F.; Steely, H.T.; Dickerson, J.E., Jr.; English-Wright, S.; Stropki, K.; McCartney, M.D.; Jacobson, N.; Shepard, A.R.; Clark, J.I.; Matsushima, H.; *et al.* Glucocorticoid induction of the glaucoma gene myoc in human and monkey trabecular meshwork cells and tissues. *Investig. Ophthalmol. Vis. Sci.* **2001**, *42*, 1769–1780.
13. Nerome, Y.; Imanaka, H.; Nonaka, Y.; Takei, S.; Kawano, Y. Frequent methylprednisone pulse therapy is a risk factor for steroid cataracts in children. *Pediatr. Int.* **2008**, *50*, 541–545.
14. Friedman, D.S.; Holbrook, J.T.; Ansari, H.; Alexander, J.; Burke, A.; Reed, S.B.; Katz, J.; Thorne, J.E.; Lightman, S.L.; Kempen, J.H.; *et al.* Risk of elevated intraocular pressure and glaucoma in patients with uveitis: Results of the multicenter uveitis steroid treatment trial. *Ophthalmology* **2013**, *120*, 1571–1579.
15. Buchman, A.L. Side effects of corticosteroid therapy. *J. Clin. Gastroenterol.* **2001**, *33*, 289–294.
16. Gangaputra, S.; Newcomb, C.W.; Liesegang, T.L.; Kacmaz, R.O.; Jabs, D.A.; Levy-Clarke, G.A.; Nussenblatt, R.B.; Rosenbaum, J.T.; Suhler, E.B.; Thorne, J.E.; *et al.* Methotrexate for ocular inflammatory diseases. *Ophthalmology* **2009**, *116*, 2188–2198.
17. Taylor, S.R.; Habot-Wilner, Z.; Pacheco, P.; Lightman, S.L. Intraocular methotrexate in the treatment of uveitis and uveitic cystoid macular edema. *Ophthalmology* **2009**, *116*, 797–801.
18. Kruh, J.; Foster, C.S. Corticosteroid-sparing agents: Conventional systemic immunosuppressants. *Dev. Ophthalmol.* **2012**, *51*, 29–46.

19. Zannin, M.E.; Birolo, C.; Gerloni, V.M.; Miserocchi, E.; Pontikaki, I.; Paroli, M.P.; Bracaglia, C.; Shardlow, A.; Parentin, F.; Cimaz, R.; *et al.* Safety and efficacy of infliximab and adalimumab for refractory uveitis in juvenile idiopathic arthritis: 1-year followup data from the italian registry. *J. Rheumatol.* **2013**, *40*, 74–79.

20. Ardoin, S.P.; Kredich, D.; Rabinovich, E.; Schanberg, L.E.; Jaffe, G.J. Infliximab to treat chronic noninfectious uveitis in children: Retrospective case series with long-term follow-up. *Am. J. Ophthalmol.* **2007**, *144*, 844–849.

21. Doycheva, D.; Zierhut, M.; Blumenstock, G.; Stuebiger, N.; Januschowski, K.; Voykov, B.; Deuter, C. Immunomodulatory therapy with tumour necrosis factor alpha inhibitors in children with antinuclear antibody-associated chronic anterior uveitis: Long-term results. *Br. J. Ophthalmol.* **2014**, *98*, 523–528.

22. Ramanan, A.V.; Dick, A.D.; Benton, D.; Compeyrot-Lacassagne, S.; Dawoud, D.; Hardwick, B.; Hickey, H.; Hughes, D.; Jones, A.; Woo, P.; *et al.* A randomised controlled trial of the clinical effectiveness, safety and cost-effectiveness of adalimumab in combination with methotrexate for the treatment of juvenile idiopathic arthritis associated uveitis (sycamore trial). *Trials* **2014**, *15*.

23. Mikhail, M.; Sallam, A. Novel intraocular therapy in non-infectious uveitis of the posterior segment of the eye. *Med. Hypothesis Discov. Innov. Ophthalmol.* **2013**, *2*, 113–120.

24. Taylor, S.R.; Banker, A.; Schlaen, A.; Couto, C.; Matthe, E.; Joshi, L.; Menezo, V.; Nguyen, E.; Tomkins-Netzer, O.; Bar, A.; *et al.* Intraocular methotrexate can induce extended remission in some patients in noninfectious uveitis. *Retina* **2013**, *33*, 2149–2154.

25. Kuno, N.; Fujii, S. Biodegradable intraocular therapies for retinal disorders: Progress to date. *Drugs Aging* **2010**, *27*, 117–134.

26. Vert, M.; Li, S.M.; Garreau, H. Attempts to map the structure and degradation characteristics of aliphatic polyesters derived from lactic and glycolic acids. *J. Biomater. Sci. Polym. Ed.* **1994**, *6*, 639–649.

27. Wen, X.; Tresco, P.A. Fabrication and characterization of permeable degradable poly(DL-lactide-co-glycolide) (PLGA) hollow fiber phase inversion membranes for use as nerve tract guidance channels. *Biomaterials* **2006**, *27*, 3800–3809.

28. Determan, A.S.; Trewyn, B.G.; Lin, V.S.; Nilsen-Hamilton, M.; Narasimhan, B. Encapsulation, stabilization, and release of bsa-fitc from polyanhydride microspheres. *J. Control. Release* **2004**, *100*, 97–109.

29. Katz, A.R.; Turner, R.J. Evaluation of tensile and absorption properties of polyglycolic acid sutures. *Surg. Gynecol. Obstet.* **1970**, *131*, 701–716.

30. Gunatillake, P.; Mayadunne, R.; Adhikari, R. Recent developments in biodegradable synthetic polymers. *Biotechnol. Annu. Rev.* **2006**, *12*, 301–347.

31. Ceonzo, K.; Gaynor, A.; Shaffer, L.; Kojima, K.; Vacanti, C.A.; Stahl, G.L. Polyglycolic acid-induced inflammation: Role of hydrolysis and resulting complement activation. *Tissue Eng.* **2006**, *12*, 301–308.

32. Pihlajamaki, H.; Salminen, S.; Laitinen, O.; Tynninen, O.; Bostman, O. Tissue response to polyglycolide, polydioxanone, polylevolactide, and metallic pins in cancellous bone: An experimental study on rabbits. *J. Orthop. Res.* **2006**, *24*, 1597–1606.

33. Otto, J.; Binnebosel, M.; Pietsch, S.; Anurov, M.; Titkova, S.; Ottinger, A.P.; Jansen, M.; Rosch, R.; Kammer, D.; Klinge, U. Large-pore pds mesh compared to small-pore pg mesh. *J. Investig. Surg.* **2010**, *23*, 190–196.

34. Ulery, B.D.; Nair, L.S.; Laurencin, C.T. Biomedical applications of biodegradable polymers. *J. Polym. Sci. B Polym. Phys.* **2011**, *49*, 832–864.

35. Suuronen, R.; Pohjonen, T.; Hietanen, J.; Lindqvist, C. A 5-year *in vitro* and *in vivo* study of the biodegradation of polylactide plates. *J. Oral Maxillofac. Surg.* **1998**, *56*, 604–614.

36. Zielhuis, S.W.; Nijsen, J.F.; Seppenwoolde, J.H.; Bakker, C.J.; Krijger, G.C.; Dullens, H.F.; Zonnenberg, B.A.; van Rijk, P.P.; Hennink, W.E.; van het Schip, A.D. Long-term toxicity of holmium-loaded poly(l-lactic acid) microspheres in rats. *Biomaterials* **2007**, *28*, 4591–4599.

37. Lu, J.; Jackson, J.K.; Gleave, M.E.; Burt, H.M. The preparation and characterization of anti-VEGFR2 conjugated, paclitaxel-loaded plla or plga microspheres for the systemic targeting of human prostate tumors. *Cancer Chemother. Pharmacol.* **2008**, *61*, 997–1005.

38. Chen, A.Z.; Li, Y.; Chau, F.T.; Lau, T.Y.; Hu, J.Y.; Zhao, Z.; Mok, D.K. Microencapsulation of puerarin nanoparticles by poly(L-lactide) in a supercritical CO_2 process. *Acta. Biomater.* **2009**, *5*, 2913–2919.

39. Lensen, D.; van Breukelen, K.; Vriezema, D.M.; van Hest, J.C. Preparation of biodegradable liquid core plla microcapsules and hollow plla microcapsules using microfluidics. *Macromol. Biosci.* **2010**, *10*, 475–480.

40. Middleton, J.C.; Tipton, A.J. Synthetic biodegradable polymers as orthopedic devices. *Biomaterials* **2000**, *21*, 2335–2346.

41. Conn, J., Jr.; Oyasu, R.; Welsh, M.; Beal, J.M. Vicryl (polyglactin 910) synthetic absorbable sutures. *Am. J. Surg.* **1974**, *128*, 19–23.

42. Quintilio, W.; Takata, C.S.; Sant'Anna, O.A.; da Costa, M.H.; Raw, I. Evaluation of a diphtheria and tetanus plga microencapsulated vaccine formulation without stabilizers. *Curr. Drug Deliv.* **2009**, *6*, 297–204.

43. Jiang, W.; Schwendeman, S.P. Stabilization of tetanus toxoid encapsulated in PLGA microspheres. *Mol. Pharm.* **2008**, *5*, 808–817.

44. Thomas, C.; Gupta, V.; Ahsan, F. Influence of surface charge of PLGA particles of recombinant hepatitis b surface antigen in enhancing systemic and mucosal immune responses. *Int. J. Pharm.* **2009**, *379*, 41–50.

45. Thomas, C.; Gupta, V.; Ahsan, F. Particle size influences the immune response produced by hepatitis b vaccine formulated in inhalable particles. *Pharm. Res.* **2010**, *27*, 905–919.

46. Zolnik, B.S.; Burgess, D.J. Evaluation of *in vivo—In vitro* release of dexamethasone from PLGA microspheres. *J. Control. Release* **2008**, *127*, 137–145.

47. Eperon, S.; Bossy-Nobs, L.; Petropoulos, I.K.; Gurny, R.; Guex-Crosier, Y. A biodegradable drug delivery system for the treatment of postoperative inflammation. *Int. J. Pharm.* **2008**, *352*, 240–247.

48. Murata, N.; Takashima, Y.; Toyoshima, K.; Yamamoto, M.; Okada, H. Anti-tumor effects of anti-VEGF sirna encapsulated with plga microspheres in mice. *J. Control. Release* **2008**, *126*, 246–254.

49. Singh, A.; Nie, H.; Ghosn, B.; Qin, H.; Kwak, L.W.; Roy, K. Efficient modulation of T-cell response by dual-mode, single-carrier delivery of cytokine-targeted sirna and DNA vaccine to antigen-presenting cells. *Mol. Ther.* **2008**, *16*, 2011–2021.

50. Patil, Y.; Panyam, J. Polymeric nanoparticles for sirna delivery and gene silencing. *Int. J. Pharm.* **2009**, *367*, 195–203.

51. Fu, K.; Pack, D.W.; Klibanov, A.M.; Langer, R. Visual evidence of acidic environment within degrading poly(lactic-co-glycolic acid) (PLGA) microspheres. *Pharm. Res.* **2000**, *17*, 100–106.

52. Ding, A.G.; Schwendeman, S.P. Acidic microclimate ph distribution in plga microspheres monitored by confocal laser scanning microscopy. *Pharm. Res.* **2008**, *25*, 2041–2052.

53. Ionescu, L.C.; Lee, G.C.; Sennett, B.J.; Burdick, J.A.; Mauck, R.L. An anisotropic nanofiber/microsphere composite with controlled release of biomolecules for fibrous tissue engineering. *Biomaterials* **2010**, *31*, 4113–4120.

54. Darney, P.D.; Monroe, S.E.; Klaisle, C.M.; Alvarado, A. Clinical evaluation of the capronor contraceptive implant: Preliminary report. *Am. J. Obstet. Gynecol.* **1989**, *160*, 1292–1295.

55. Mundargi, R.C.; Srirangarajan, S.; Agnihotri, S.A.; Patil, S.A.; Ravindra, S.; Setty, S.B.; Aminabhavi, T.M. Development and evaluation of novel biodegradable microspheres based on poly(D,L-lactide-co-glycolide) and poly(epsilon-caprolactone) for controlled delivery of doxycycline in the treatment of human periodontal pocket: *In vitro* and *in vivo* studies. *J. Control. Release* **2007**, *119*, 59–68.

56. Singh, J.; Pandit, S.; Bramwell, V.W.; Alpar, H.O. Diphtheria toxoid loaded poly-(epsilon-caprolactone) nanoparticles as mucosal vaccine delivery systems. *Methods* **2006**, *38*, 96–105.

57. Leong, K.W.; Brott, B.C.; Langer, R. Bioerodible polyanhydrides as drug-carrier matrices. I: Characterization, degradation, and release characteristics. *J. Biomed. Mater. Res.* **1985**, *19*, 941–955.

58. Heffernan, M.J.; Murthy, N. Polyketal nanoparticles: A new ph-sensitive biodegradable drug delivery vehicle. *Bioconjug. Chem.* **2005**, *16*, 1340–1342.

59. Lee, S.; Yang, S.C.; Heffernan, M.J.; Taylor, W.R.; Murthy, N. Polyketal microparticles: A new delivery vehicle for superoxide dismutase. *Bioconjug. Chem.* **2007**, *18*, 4–7.

60. Lee, S.; Yang, S.C.; Kao, C.Y.; Pierce, R.H.; Murthy, N. Solid polymeric microparticles enhance the delivery of sirna to macrophages *in vivo*. *Nucleic Acids Res.* **2009**, *37*.

61. Goh, S.L.; Murthy, N.; Xu, M.; Frechet, J.M. Cross-linked microparticles as carriers for the delivery of plasmid DNA for vaccine development. *Bioconjug. Chem.* **2004**, *15*, 467–474.

62. Murthy, N.; Xu, M.; Schuck, S.; Kunisawa, J.; Shastri, N.; Frechet, J.M. A macromolecular delivery vehicle for protein-based vaccines: Acid-degradable protein-loaded microgels. *Proc. Natl. Acad. Sci. USA* **2003**, *100*, 4995–5000.

63. Standley, S.M.; Kwon, Y.J.; Murthy, N.; Kunisawa, J.; Shastri, N.; Guillaudeu, S.J.; Lau, L.; Frechet, J.M. Acid-degradable particles for protein-based vaccines: Enhanced survival rate for tumor-challenged mice using ovalbumin model. *Bioconjug. Chem.* **2004**, *15*, 1281–1288.

64. Yang, S.C.; Bhide, M.; Crispe, I.N.; Pierce, R.H.; Murthy, N. Polyketal copolymers: A new acid-sensitive delivery vehicle for treating acute inflammatory diseases. *Bioconjug. Chem.* **2008**, *19*, 1164–1169.

65. Heffernan, M.J.; Kasturi, S.P.; Yang, S.C.; Pulendran, B.; Murthy, N. The stimulation of CD8+ T cells by dendritic cells pulsed with polyketal microparticles containing ion-paired protein antigen and poly(inosinic acid)-poly(cytidylic acid). *Biomaterials* **2009**, *30*, 910–918.

66. Seshadri, G.; Sy, J.C.; Brown, M.; Dikalov, S.; Yang, S.C.; Murthy, N.; Davis, M.E. The delivery of superoxide dismutase encapsulated in polyketal microparticles to rat myocardium and protection from myocardial ischemia-reperfusion injury. *Biomaterials* **2010**, *31*, 1372–1379.

67. Zhang, Z.; Kuijer, R.; Bulstra, S.K.; Grijpma, D.W.; Feijen, J. The *in vivo* and *in vitro* degradation behavior of poly(trimethylene carbonate). *Biomaterials* **2006**, *27*, 1741–1748.

68. Pego, A.P.; van Luyn, M.J.; Brouwer, L.A.; van Wachem, P.B.; Poot, A.A.; Grijpma, D.W.; Feijen, J. *In vivo* behavior of poly(1,3-trimethylene carbonate) and copolymers of 1,3-trimethylene carbonate with D,L-lactide or epsilon-caprolactone: Degradation and tissue response. *J. Biomed. Mater. Res. A* **2003**, *67*, 1044–1054.

69. Zurita, R.; Puiggali, J.; Rodriguez-Galan, A. Loading and release of ibuprofen in multi- and monofilament surgical sutures. *Macromol. Biosci.* **2006**, *6*, 767–775.

70. Chen, W.; Meng, F.; Li, F.; Ji, S.J.; Zhong, Z. pH-responsive biodegradable micelles based on acid-labile polycarbonate hydrophobe: Synthesis and triggered drug release. *Biomacromolecules* **2009**, *10*, 1727–1735.

71. Chen, W.; Meng, F.; Cheng, R.; Zhong, Z. Ph-sensitive degradable polymersomes for triggered release of anticancer drugs: A comparative study with micelles. *J. Control. Release* **2010**, *142*, 40–46.

72. Kim, S.H.; Tan, J.P.; Nederberg, F.; Fukushima, K.; Colson, J.; Yang, C.; Nelson, A.; Yang, Y.Y.; Hedrick, J.L. Hydrogen bonding-enhanced micelle assemblies for drug delivery. *Biomaterials* **2010**, *31*, 8063–8071.

73. Sanson, C.; Schatz, C.; le Meins, J.F.; Brulet, A.; Soum, A.; Lecommandoux, S. Biocompatible and biodegradable poly(trimethylene carbonate)-b-poly(l-glutamic acid) polymersomes: Size control and stability. *Langmuir* **2010**, *26*, 2751–2760.

74. Sanson, C.; Schatz, C.; le Meins, J.F.; Soum, A.; Thevenot, J.; Garanger, E.; Lecommandoux, S. A simple method to achieve high doxorubicin loading in biodegradable polymersomes. *J. Control. Release* **2010**, *147*, 428–435.

75. Henderson, P.W.; Kadouch, D.J.; Singh, S.P.; Zawaneh, P.N.; Weiser, J.; Yazdi, S.; Weinstein, A.; Krotscheck, U.; Wechsler, B.; Putnam, D.; *et al.* A rapidly resorbable hemostatic biomaterial based on dihydroxyacetone. *J. Biomed. Mater. Res. A* **2010**, *93*, 776–782.

76. Ertel, S.I.; Kohn, J. Evaluation of a series of tyrosine-derived polycarbonates as degradable biomaterials. *J. Biomed. Mater. Res.* **1994**, *28*, 919–930.

77. Asikainen, A.J.; Noponen, J.; Mesimaki, K.; Laitinen, O.; Peltola, J.; Pelto, M.; Kellomaki, M.; Ashammakhi, N.; Lindqvist, C.; Suuronen, R. Tyrosine derived polycarbonate membrane is useful for guided bone regeneration in rabbit mandibular defects. *J. Mater. Sci. Mater. Med.* **2005**, *16*, 753–758.

78. Asikainen, A.J.; Noponen, J.; Lindqvist, C.; Pelto, M.; Kellomaki, M.; Juuti, H.; Pihlajamaki, H.; Suuronen, R. Tyrosine-derived polycarbonate membrane in treating mandibular bone defects. An experimental study. *J. R. Soc. Interface* **2006**, *3*, 629–635.

79. Bailey, L.O.; Becker, M.L.; Stephens, J.S.; Gallant, N.D.; Mahoney, C.M.; Washburn, N.R.; Rege, A.; Kohn, J.; Amis, E.J. Cellular response to phase-separated blends of tyrosine-derived polycarbonates. *J. Biomed. Mater. Res. A* **2006**, *76*, 491–502.

80. Briggs, T.; Treiser, M.D.; Holmes, P.F.; Kohn, J.; Moghe, P.V.; Arinzeh, T.L. Osteogenic differentiation of human mesenchymal stem cells on poly(ethylene glycol)-variant biomaterials. *J. Biomed. Mater. Res. A* **2009**, *91*, 975–984.

81. Johnson, P.A.; Luk, A.; Demtchouk, A.; Patel, H.; Sung, H.J.; Treiser, M.D.; Gordonov, S.; Sheihet, L.; Bolikal, D.; Kohn, J.; *et al.* Interplay of anionic charge, poly(ethylene glycol), and iodinated tyrosine incorporation within tyrosine-derived polycarbonates: Effects on vascular smooth muscle cell adhesion, proliferation, and motility. *J. Biomed. Mater. Res. A* **2010**, *93*, 505–514.

82. Meechaisue, C.; Dubin, R.; Supaphol, P.; Hoven, V.P.; Kohn, J. Electrospun mat of tyrosine-derived polycarbonate fibers for potential use as tissue scaffolding material. *J. Biomater. Sci. Polym. Ed.* **2006**, *17*, 1039–1056.

83. Sung, H.J.; Sakala Labazzo, K.M.; Bolikal, D.; Weiner, M.J.; Zimnisky, R.; Kohn, J. Angiogenic competency of biodegradable hydrogels fabricated from polyethylene glycol-crosslinked tyrosine-derived polycarbonates. *Eur. Cell Mater.* **2008**, *15*, 77–87.

84. Costache, M.C.; Qu, H.; Ducheyne, P.; Devore, D.I. Polymer-xerogel composites for controlled release wound dressings. *Biomaterials* **2010**, *31*, 6336–6343.

85. Herrero-Vanrell, R.; Refojo, M.F. Biodegradable microspheres for vitreoretinal drug delivery. *Adv. Drug Deliv. Rev.* **2001**, *52*, 5–16.

86. Pardo-Lopez, D.; Frances-Munoz, E.; Gallego-Pinazo, R.; Diaz-Llopis, M. Anterior chamber migration of dexametasone intravitreal implant (ozurdex(r)). *Graefes Arch. Clin. Exp. Ophthalmol.* **2012**, *250*, 1703–1704.

87. Morrison, P.W.; Khutoryanskiy, V.V. Advances in ophthalmic drug delivery. *Ther. Deliv.* **2014**, *5*, 1297–1315.

88. Bourges, J.L.; Bloquel, C.; Thomas, A.; Froussart, F.; Bochot, A.; Azan, F.; Gurny, R.; BenEzra, D.; Behar-Cohen, F. Intraocular implants for extended drug delivery: Therapeutic applications. *Adv. Drug Deliv. Rev.* **2006**, *58*, 1182–1202.

89. Liechty, W.B.; Kryscio, D.R.; Slaughter, B.V.; Peppas, N.A. Polymers for drug delivery systems. *Annu. Rev. Chem. Biomol. Eng.* **2010**, *1*, 149–173.

90. Choonara, Y.E.; Pillay, V.; Danckwerts, M.P.; Carmichael, T.R.; du Toit, L.C. A review of implantable intravitreal drug delivery technologies for the treatment of posterior segment eye diseases. *J. Pharm. Sci.* **2010**, *99*, 2219–2239.

91. Malcolm, R.K.; Woolfson, A.D.; Toner, C.F.; Morrow, R.J.; McCullagh, S.D. Long-term, controlled release of the hiv microbicide tmc120 from silicone elastomer vaginal rings. *J. Antimicrob. Chemother.* **2005**, *56*, 954–956.

92. Van Laarhoven, J.A.; Kruft, M.A.; Vromans, H. Effect of supersaturation and crystallization phenomena on the release properties of a controlled release device based on eva copolymer. *J. Control. Release* **2002**, *82*, 309–317.

93. Lin, J.Y.; Wang, W.Y.; Lin, Y.T.; Chou, S.W. Ni3S2/Ni-P bilayer coated on polyimide as a Pt- and TCO-free flexible counter electrode for dye-sensitized solar cells. *ACS Appl. Mater. Interfaces* **2014**, *6*, 3357–3364.

94. Bettencourt, A.; Almeida, A.J. Poly(methyl methacrylate) particulate carriers in drug delivery. *J. Microencapsul.* **2012**, *29*, 353–367.

95. Amin, K.; Wasan, K.M.; Albrecht, R.M.; Heath, T.D. Cell association of liposomes with high fluid anionic phospholipid content is mediated specifically by LDL and its receptor, LDLr. *J. Pharm. Sci.* **2002**, *91*, 1233–1244.

96. Lobo, A.M.; Sobrin, L.; Papaliodis, G.N. Drug delivery options for the treatment of ocular inflammation. *Semin. Ophthalmol.* **2010**, *25*, 283–288.

97. Kuppermann, B.D.; Blumenkranz, M.S.; Haller, J.A.; Williams, G.A.; Weinberg, D.V.; Chou, C.; Whitcup, S.M.; Dexamethasone, D.D.S.P.I.I.S.G. Randomized controlled study of an intravitreous dexamethasone drug delivery system in patients with persistent macular edema. *Arch. Ophthalmol.* **2007**, *125*, 309–317.

98. Williams, G.A.; Haller, J.A.; Kuppermann, B.D.; Blumenkranz, M.S.; Weinberg, D.V.; Chou, C.; Whitcup, S.M.; Dexamethasone, D.D.S.P.I.I.S.G. Dexamethasone posterior-segment drug delivery system in the treatment of macular edema resulting from uveitis or irvine-gass syndrome. *Am. J. Ophthalmol.* **2009**, *147*, 1048–1054.

99. Bansal, R.; Bansal, P.; Kulkarni, P.; Gupta, V.; Sharma, A.; Gupta, A. Wandering ozurdex((r)) implant. *J. Ophthalmic Inflamm. Infect.* **2012**, *2*, 1–5.

100. Arcinue, C.A.; Ceron, O.M.; Foster, C.S. A comparison between the fluocinolone acetonide (retisert) and dexamethasone (ozurdex) intravitreal implants in uveitis. *J. Ocul. Pharmacol. Ther.* **2013**, *29*, 501–507.

101. Wang, J.; Jiang, A.; Joshi, M.; Christoforidis, J. Drug delivery implants in the treatment of vitreous inflammation. *Mediat. Inflamm.* **2013**, *2013*.

102. Callanan, D.G.; Jaffe, G.J.; Martin, D.F.; Pearson, P.A.; Comstock, T.L. Treatment of posterior uveitis with a fluocinolone acetonide implant: Three-year clinical trial results. *Arch. Ophthalmol.* **2008**, *126*, 1191–1201.

103. Chieh, J.J.; Carlson, A.N.; Jaffe, G.J. Combined fluocinolone acetonide intraocular delivery system insertion, phacoemulsification, and intraocular lens implantation for severe uveitis. *Am. J. Ophthalmol.* **2008**, *146*, 589–594.

104. Jaffe, G.J.; McCallum, R.M.; Branchaud, B.; Skalak, C.; Butuner, Z.; Ashton, P. Long-term follow-up results of a pilot trial of a fluocinolone acetonide implant to treat posterior uveitis. *Ophthalmology* **2005**, *112*, 1192–1198.

105. Del Amo, E.M.; Urtti, A. Current and future ophthalmic drug delivery systems. A shift to the posterior segment. *Drug Discov. Today* **2008**, *13*, 135–143.

106. Conway, B.R. Recent patents on ocular drug delivery systems. *Recent Pat. Drug Deliv. Formul.* **2008**, *2*, 1–8.

107. Tan, D.T.; Chee, S.P.; Lim, L.; Lim, A.S. Randomized clinical trial of a new dexamethasone delivery system (surodex) for treatment of post-cataract surgery inflammation. *Ophthalmology* **1999**, *106*, 223–231.

108. Gulati, V.; Pahuja, S.; Fan, S.; Toris, C.B. An experimental steroid responsive model of ocular inflammation in rabbits using an SLT frequency doubled Q switched Nd:YAG laser. *J. Ocul. Pharmacol. Ther.* **2013**, *29*, 663–669.

109. Cheng, Y.; Xu, Z.; Ma, M.; Xu, T. Dendrimers as drug carriers: Applications in different routes of drug administration. *J. Pharm. Sci.* **2008**, *97*, 123–143.

110. Ong, F.S.; Kuo, J.Z.; Wu, W.C.; Cheng, C.Y.; Blackwell, W.L.; Taylor, B.L.; Grody, W.W.; Rotter, J.I.; Lai, C.C.; Wong, T.Y. Personalized medicine in ophthalmology: From pharmacogenetic biomarkers to therapeutic and dosage optimization. *J. Pers. Med.* **2013**, *3*, 40–69.

111. Abolmaali, S.; Tamaddon, A.; Kamali-Sarvestani, E.; Ashraf, M.; Dinarvand, R. Stealth nanogels of histinylated poly ethyleneimine for sustained delivery of methotrexate in collagen-induced arthritis model. *Pharm. Res.* **2015**.

112. Pehlivan, S.B.; Yavuz, B.; Calamak, S.; Ulubayram, K.; Kaffashi, A.; Vural, I.; Cakmak, H.B.; Durgun, M.E.; Denkbas, E.B.; Unlu, N. Preparation and *in vitro/in vivo* evaluation of cyclosporin a-loaded nanodecorated ocular implants for subconjunctival application. *J. Pharm. Sci.* **2015**, *104*, 1709–1720.

113. Dong, X.; Shi, W.; Yuan, G.; Xie, L.; Wang, S.; Lin, P. Intravitreal implantation of the biodegradable cyclosporin a drug delivery system for experimental chronic uveitis. *Graefes Arch. Clin. Exp. Ophthalmol.* **2006**, *244*, 492–497.

114. Friedrich, E.E.; Azofiefa, A.; Fisch, E.; Washburn, N.R. Local delivery of antitumor necrosis factor-alpha through conjugation to hyaluronic acid: Dosing strategies and early healing effects in a rat burn model. *J. Burn Care Res.* **2015**, *36*, e90–e101.

115. Tao, W.; Wen, R.; Goddard, M.B.; Sherman, S.D.; O'Rourke, P.J.; Stabila, P.F.; Bell, W.J.; Dean, B.J.; Kauper, K.A.; Budz, V.A.; *et al.* Encapsulated cell-based delivery of cntf reduces photoreceptor degeneration in animal models of retinitis pigmentosa. *Investig. Ophthalmol. Vis. Sci.* **2002**, *43*, 3292–3298.

116. Bush, R.A.; Lei, B.; Tao, W.; Raz, D.; Chan, C.C.; Cox, T.A.; Santos-Muffley, M.; Sieving, P.A. Encapsulated cell-based intraocular delivery of ciliary neurotrophic factor in normal rabbit: Dose-dependent effects on erg and retinal histology. *Investig. Ophthalmol. Vis. Sci.* **2004**, *45*, 2420–2430.

117. Thanos, C.G.; Bell, W.J.; O'Rourke, P.; Kauper, K.; Sherman, S.; Stabila, P.; Tao, W. Sustained secretion of ciliary neurotrophic factor to the vitreous, using the encapsulated cell therapy-based nt-501 intraocular device. *Tissue Eng.* **2004**, *10*, 1617–1622.

118. Christie, J.G.; Kompella, U.B. Ophthalmic light sensitive nanocarrier systems. *Drug Discov. Today* **2008**, *13*, 124–134.

119. Wells, L.A.; Sheardown, H. Photoresponsive polymers for ocular drug delivery. In *Ocular Drug Delivery Systems*; Thassu, D., Chader, G.J., Eds.; CRC Press: Boca Raton, FL, USA, 2012; pp. 383–400.

120. Rijcken, C.J.; Soga, O.; Hennink, W.E.; van Nostrum, C.F. Triggered destabilisation of polymeric micelles and vesicles by changing polymers polarity: An attractive tool for drug delivery. *J. Control. Release* **2007**, *120*, 131–148.

121. Hirsch, L.R.; Stafford, R.J.; Bankson, J.A.; Sershen, S.R.; Rivera, B.; Price, R.E.; Hazle, J.D.; Halas, N.J.; West, J.L. Nanoshell-mediated near-infrared thermal therapy of tumors under magnetic resonance guidance. *Proc. Natl. Acad. Sci. USA* **2003**, *100*, 13549–13554.

122. Juzenas, P.; Juzeniene, A.; Kaalhus, O.; Iani, V.; Moan, J. Noninvasive fluorescence excitation spectroscopy during application of 5-aminolevulinic acid *in vivo*. *Photochem. Photobiol. Sci.* **2002**, *1*, 745–748.

123. Kost, J.; Noecker, R.; Kunica, E.; Langer, R. Magnetically controlled release systems: Effect of polymer composition. *J. Biomed. Mater. Res.* **1985**, *19*, 935–940.

124. Pirmoradi, F.N.; Jackson, J.K.; Burt, H.M.; Chiao, M. On-demand controlled release of docetaxel from a battery-less mems drug delivery device. *Lab. Chip.* **2011**, *11*, 2744–2752.

125. Kost, J.; Langer, R. Responsive polymeric delivery systems. *Adv. Drug Deliv. Rev.* **2001**, *46*, 125–148.

126. Balint, R.; Cassidy, N.J.; Cartmell, S.H. Conductive polymers: Towards a smart biomaterial for tissue engineering. *Acta. Biomater.* **2014**, *10*, 2341–2353.

127. Geetha, S.; Rao, C.R.; Vijayan, M.; Trivedi, D.C. Biosensing and drug delivery by polypyrrole. *Anal. Chim. Acta.* **2006**, *568*, 119–125.

128. Svirskis, D.; Sharma, M.; Yu, Y.; Garg, S. Electrically switchable polypyrrole film for the tunable release of progesterone. *Ther. Deliv.* **2013**, *4*, 307–313.

129. Valdes-Ramirez, G.; Windmiller, J.R.; Claussen, J.C.; Martinez, A.G.; Kuralay, F.; Zhou, M.; Zhou, N.; Polsky, R.; Miller, P.R.; Narayan, R.; *et al.* Multiplexed and switchable release of distinct fluids from microneedle platforms via conducting polymer nanoactuators for potential drug delivery. *Sens. Actuators B Chem.* **2012**, *161*, 1018–1024.

130. Svirskis, D.; Travas-Sejdic, J.; Rodgers, A.; Garg, S. Electrochemically controlled drug delivery based on intrinsically conducting polymers. *J. Control. Release* **2010**, *146*, 6–15.

131. Rivers, T.J.; Hudson, T.W.; Schmidt, C.E. Synthesis of a novel, biodegradable electrically conducting polymer for biomedical applications. *Adv. Funct.Mater.* **2002**, *12*, 33–37.

132. Yu-Chuan, S.; Lin, L. A water-powered micro drug delivery system. *J. Microelectromech. Syst.* **2004**, *13*, 75–82.

133. Zengerle, R.; Ulrich, J.; Kluge, S.; Richter, M.; Richter, A. A bidirectional silicon micropump. *Sens. Actuators A Phys.* **1995**, *50*, 81–86.

134. Jeong, O.C.; Yang, S.S. Fabrication and test of a thermopneumatic micropump with a corrugated p+ diaphragm. *Sens. Actuators A Phys.* **2000**, *83*, 249–255.

135. Li, P.-Y.; Shih, J.; Lo, R.; Saati, S.; Agrawal, R.; Humayun, M.S.; Tai, Y.-C.; Meng, E. An electrochemical intraocular drug delivery device. *Sens. Actuators A Phys.* **2008**, *143*, 41–48.

136. Lee, D.J.; Taylor, A.W. Following EAU recovery there is an associated MC5r-dependent APC induction of regulatory immunity in the spleen. *Investig. Ophthalmol. Vis. Sci.* **2011**, *52*, 8862–8867.

137. Lee, D.J.; Taylor, A.W. Both mc5r and a2ar are required for protective regulatory immunity in the spleen of post-experimental autoimmune uveitis in mice. *J. Immunol.* **2013**, *191*, 4103–4111.

138. Lee, D.J.; Taylor, A.W. Recovery from experimental autoimmune uveitis promotes induction of antiuveitic inducible tregs. *J. Leukoc. Biol.* **2015**, *97*, 1101–1109.

139. Taylor, A.W.; Kitaichi, N.; Biros, D. Melanocortin 5 receptor and ocular immunity. *Cell. Mol. Biol.* **2006**, *52*, 53–59.

140. Ohta, K.; Wiggert, B.; Taylor, A.W.; Streilein, J.W. Effects of experimental ocular inflammation on ocular immune privilege. *Investig. Ophthalmol. Vis. Sci.* **1999**, *40*, 2010–2018.

141. Cabrera, M.; Yeh, S.; Albini, T.A. Sustained-release corticosteroid options. *J. Ophthalmol.* **2014**, *2014*.

A Closer Look at Schlemm's Canal Cell Physiology: Implications for Biomimetics

Cula N. Dautriche, Yangzi Tian, Yubing Xie and Susan T. Sharfstein

Abstract: Among ocular pathologies, glaucoma is the second leading cause of progressive vision loss, expected to affect 80 million people worldwide by 2020. A primary cause of glaucoma appears to be damage to the conventional outflow tract. Conventional outflow tissues, a composite of the trabecular meshwork and the Schlemm's canal, regulate and maintain homeostatic responses to intraocular pressure. In glaucoma, filtration of aqueous humor into the Schlemm's canal is hindered, leading to an increase in intraocular pressure and subsequent damage to the optic nerve, with progressive vision loss. The Schlemm's canal encompasses a unique endothelium. Recent advances in culturing and manipulating Schlemm's canal cells have elucidated several aspects of their physiology, including ultrastructure, cell-specific marker expression, and biomechanical properties. This review highlights these advances and discusses implications for engineering a 3D, biomimetic, *in vitro* model of the Schlemm's canal endothelium to further advance glaucoma research, including drug testing and gene therapy screening.

Reprinted from *J. Funct. Biomater.* Cite as: Dautriche, C.N.; Tian, Y.; Xie, Y.; Sharfstein, S.T. A Closer Look at Schlemm's Canal Cell Physiology: Implications for Biomimetics. *J. Funct. Biomater.* **2015**, *6*, 963–985.

1. Introduction

The Schlemm's canal (SC), named after the German anatomist, Friedrich Schlemm and first identified in 1830 [1], is a unique, ring-shaped, endothelium-lined vessel that encircles the cornea [2,3] (Figure 1). Anatomically, it is situated directly against the juxtacanalicular (JCT) region of the trabecular meshwork (TM). As a consequence, one of its primary functions is to deliver aqueous humor into the collecting channels, following filtration through the TM. Because of its close apposition to the JCT, not all SC cells are created equal. As a result, the SC is divided into the inner and outer wall, each possessing endothelial cells that differ in morphology [4], cell-specific marker expression [5,6], specialized cellular organelles, and functions (Table 1). However, these differences may be due to the differences in biomechanical environment between the inner and outer wall, rather than any underlying biological or biochemical differences between the inner and outer wall endothelia. The inner wall has been more extensively studied, as the greatest resistance to aqueous humor outflow is generated in or close to the SC endothelium that lines the TM [7–11]. Excessive resistance leads to elevated intraocular pressure

(IOP), the leading modifiable risk factor for glaucoma. This review will focus on the development, anatomy, biology and physiology of SC inner wall endothelial cells as they are relevant to engineering the SC inner wall.

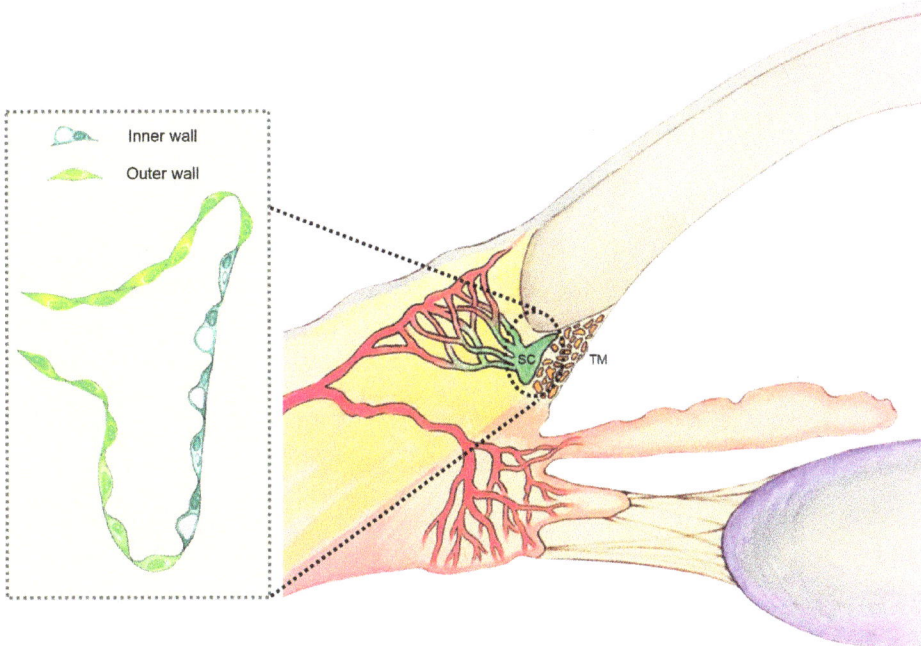

Figure 1. Schematic of the conventional outflow pathway. The left inset shows an expanded view of the Schlemm's canal's microanatomy detailing the cell morphology of the inner and outer wall.

The inner wall of the SC is a unique endothelium, specialized to maintain aqueous humor homeostasis, and IOP regulation in conjunction with the TM. Unlike the TM, controversy remains regarding the SC, especially in terms of development, its contribution to outflow resistance in normal and glaucomatous states, and its role in ocular immunity. This is, in part, due to the very limited amount of SC tissue available as well as the difficulty in isolating SC cells. Moreover, 2D culture of SC cells on tissue culture plastic results in dedifferentiation of the SC cells [12], limiting its utility as a clinical model. Advances in nanotechnology, particularly materials science, have permitted cultures of SC cells in more biomimetic environments, leading to significant advances in characterizing SC cell mechano-biology and physiology, which highlights the extremely dynamic nature of the inner wall [13,14].

Table 1. Characteristics of Schlemm's canal endothelial cells.

Property	Inner Wall	Outer Wall
Morphology	Cobblestone appearance [15] Discontinuous basement membrane [15,16]	Smooth and flat [5], continuous basement membrane [17]
Cell-specific marker	Zipper-like VE-cadherin [18]	Desmin Reactivity to Factor VIII-related antigen [19]
Subcellular structure	Giant vacuoles [20], paracellular pores [21]	Weibel-Palade bodies [17]
Function	Aqueous humor filtration IOP homeostasis [5,22,23]	Unknown

2. Embryological Origin: Of Lymph or Blood?

One of the greatest controversies that surrounds the unique endothelial nature of the SC is its embryological origin. Although earlier studies suggested a vascular origin [17,24–26], recent publications establish the SC as a lymphatic-like vessel. In humans, prenatal SC development begins at week 17 [5] and is completed by week 24 [27], whereas in mice, the SC development is postnatal [2]. The organogenesis of the SC is a stepwise process in which SC progenitors are first specified in the transscleral veins and bud off laterally to anastomose, with subsequent lumenization and development into the mature SC (Figure 2) [28–30]. Park and Aspelund and their respective coworkers elucidated key molecular mechanisms and characteristics of SC progenitors for their terminal differentiation into SC cells, while contrasting it with lymphatic cell development (Figure 2). Although in mice, SC development is postnatal, unlike the embryonic development of the lymphatic system, both processes involve migration of venous endothelial cell (VEC) progenitors that undergo precise, orchestrated changes in key markers for subsequent acquisition of lymphatic identity (Table 2) [28,31].

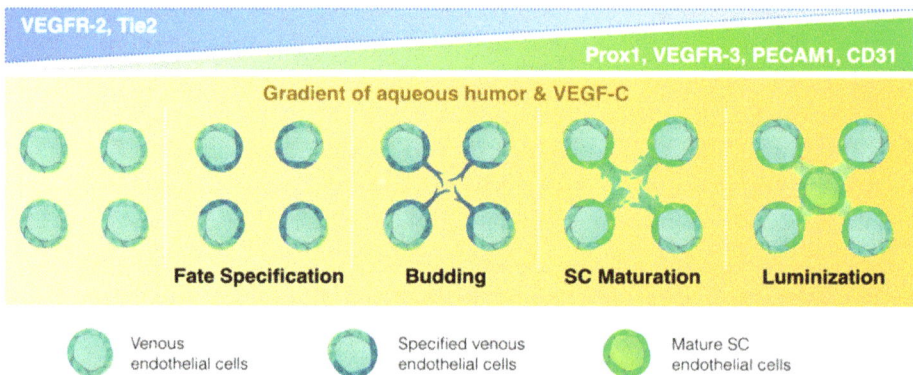

Figure 2. Organogenesis of the Schlemm's canal with a focus on key differential expression pattern along with essential soluble factors.

Table 2. Summary of key signaling necessary for Schlemm's canal or lymphatic development in mice.

Lineage	Development	Progenitors	Budding	Lumenization/Sac Formation	Separation from Venous Vasculature
Lymphatic	Embryonic	PROX1 [32], Sox18 [33], COUP-FII [34], VE-cadherin [35]	PDPN [36], VEFGR3 [37], CCBE1 [38], NRP2 [39], RAC1 [40], LYVE-1 [41]	NFATC1 [42], GATA2 [43,44], Calcr1 [45], Ramp2 [45], TIE1 [46]	Syk [47], SLP76 [47], Runx1 [48], PDPN [36], Meis1 [49], Clec2 [50,51], CXADR [52]
Schlemm's Canal	Postnatal	VEGFR-2, TIE 2 [2,28]	PROX1 [2,28,55]	VEFGR-3 [2,28]	PECAM1, VEFGR-3 [2,18,28]

The SC development may be classified into four stages, SC progenitor cell-fate specification, lateral sprouting, lumenization, and separation from venous vasculature. Unlike lymphatic progenitor cell-fate specification, the key regulatory molecules and specific markers needed to mediate SC progenitor cell-fate specification have not been clearly identified. Aspelund *et al.* demonstrated that vascular endothelial growth factor (VEGF)-C is necessary to initiate migration of VECs and lateral sprouting from the transscleral veins. Park and Aspelund and their respective coworkers elegantly demonstrated that SC progenitor cells are VECs that are positive for vascular endothelial growth factor receptor (VEGFR) 2 and the tunica interna endothelial cell kinase (TIE) 2. These SC progenitor cells subsequently acquired PROX1 expression for lumenization and VEGFR-3 for subsequent maturation into SC cells [2,28] (Figure 2). Truong *et al.* were the first to demonstrate high expression of the lymphatic transcription factor, PROX1, in SC endothelium, suggesting a closer similarity between SC endothelium and lymphatic endothelium [54]. Both aqueous humor and VEGF-C are required for proper SC development. VEGF-C (VEGFc+/LacZ) heterozygous mice exhibited delayed budding of SC endothelial cells from the venous system and retarded tubular fusion [2,28]. Meanwhile, reduction of aqueous humor resulted in endothelial-mesenchymal transition and loss of the lymphatic identity [28]. Thus, the SC in mice is a unique, specialized endothelium of vascular origin that undergoes partial lymphatic reprogramming during postnatal development to acquire a transient lymphatic identity required for maintaining its proper function in aqueous humor homeostasis [2,28,55]. Similar to lymphatic cells, SC cells experience flow from a basal to apical direction. While these studies were conducted in mice, expression of PROX1 by SC endothelial cells in humans, zebrafish, and mice indicates that the lymphatic-like identity of the SC is conserved in vertebrate evolution [2], and suggests that similar developmental pathways are likely to occur in humans, albeit prenatally rather than postnatally.

Despite their lymphatic nature and expression of several (though not all) lymphatic markers, SC cells do not appear to have lymphatic origins. Kizhatil *et al.*

recently detailed the organogenesis of the SC, which arises from the limbal vascular plexus (LVP) and radial vessels (RV) deep in the limbus that run in a direction perpendicular to the LVP. They coined the term canalogenesis to describe this process [53]. Canalogensis, the authors argued, is very similar to vascular development, emphasizing a more vascular origin or identity of the SC cells. However, there are important differences between angiogenesis and canalogenesis. In canalogenesis, following endothelial sprouting and tip cell formation, tip cells migrate into an intermediate zone between the LVP and RV to interact and adhere to each other, forming clusters of tip cells. The cells in these clusters divide, producing a chain of cells which acquire PROX1 expression for formation and remodeling into a tube, which is the SC. They further demonstrated that specification of the inner and outer wall of the SC is established during development with differential expression of key markers such FLT4 and PROX1.

Understanding the exact molecular footprint of SC organogenesis is at its infancy. However, these studies have radically advanced our knowledge of organogenesis of the SC. Still, important questions remain to be investigated to understand the critical contribution of the SC to aqueous humor homeostasis and glaucoma pathogenesis. For example, what determines the number of SC progenitors that will bud from the transscleral veins? What is the exact molecular footprint of SC progenitors? What triggers aqueous humor influx into the SC? What is the role of aqueous humor in the acquisition of the SC phenotype? What additional key regulators and signaling pathways are likely to participate in SC progenitor differentiation and maturation? What are the molecular events that facilitate separation from the venous system? What factors specify the cell fate of endothelial cells in the inner wall and the outer wall of the SC? Answers to these questions will facilitate establishment of platforms for manipulating SC progenitor cells to address the scarcity of SC cells available for research as well as further our understanding of human Schlemm's canal inner wall (HSCIW) cell biology and physiology.

3. Schlemm's Canal Anatomy

3.1. Macroarchitecture

The SC is located at the drainage or iridocorneal angle. The iridocorneal angle is lined by the TM, which overlies the SC. Together, they make up the conventional outflow tract and account for 50%–90% of aqueous humor outflow [22,54,56–58]. The SC is an endothelium-lined circular canal with branching of several aqueous channels. Until recently, the anatomy of the SC was characterized by histological stains, which estimated the SC cross-sectional area to be 1709 μm^2 [58]. Recent, live, 3D, non-invasive visualization has facilitated more detailed and physiological measurement of the SC. As a result, the cross-sectional area is now estimated to vary between 4064 to 7164 μm^2 [59–63], with many branched aqueous channels [64,65] (Figure 1).

3.2. Microarchitecture

The macroarchitecture of the SC dictates its microanatomy. The SC is lined by a continuous endothelium with tight junctions, which is divided into an outer and inner wall with regards to its relationship to the JCT (Figure 1). The SC endothelium that lies directly against the JCT is known as the inner wall and is the most celebrated and studied. The remaining endothelia comprise the outer wall. The endothelial cells of the inner wall differ from that of the outer wall in morphology, cell-specific markers and functions. In contrast to the outer wall, inner wall endothelial cells lie on a discontinuous basement membrane [15,66] and are specialized to handle flow in a basal-to-apical direction like lymphatic endothelium. The endothelial cells of the outer wall are differentiated from the inner wall endothelia by the presence of Weibel-Palade bodies [17], a positive desmin stain [67], and strong reactivity to Factor VIII-related antigen [17]. Because of their location against the JCT, the inner wall endothelial cells experience a unique biomechanical microenvironment that subjects them to a basal-apical pressure gradient. As a consequence, endothelial cells of the inner wall exhibit pores and giant vacuoles, as well as F-actin arrangements that are distinct from that of the outer wall [68]. Outer wall endothelial cells have stellate actin arrangements throughout much of the cell as compared to prominent peripheral F-actin bands observed in inner wall endothelial cells [68]. Giant vacuoles are not intracellular structures, but rather deformations of the inner wall to create a small potential space between the extracellular matrix (ECM) of the JCT and the inner wall [5]; whereas pores are inner wall structures with sizes between 0.6 and 3 μm [13] that mediate aqueous transport into the SC and may account for the SC contribution to aqueous outflow [21,69–72].

Two types of pores have been identified and characterized, I-pores (transcellular) and B-pores (paracellular) [21,73], which differ in location, sensitivity to strain and mechanisms of formation [13]. While B-pores result from local disassembly and widening of intercellular junctions, I-pores may be a result of fusion of the apical and basal cell membranes that may come into apposition as the cytoplasm thins under applied strain, with caveolae, vesicles, or "mini-pores" [13,74,75]. In addition, Braakman *et al.* recently illustrated aqueous outflow segmentation mediated by these pores, mainly B-pores [23]. Glaucomatous eyes exhibit decreased density of these pores, highlighting the vital role of the inner wall in aqueous humor homeostasis. Therefore, a goal of SC-targeted therapies might be to increase pore density and hence outflow, thus lowering IOP in glaucoma.

4. Characteristics of Human Schlemm's Canal Cells

The cobblestone appearance of the HSCIW cells is attributed to the significant biomechanical load experienced as well as the segmental flow of aqueous humor [5,76]. Segmental flow relates to the non-homogenous filtration of aqueous humor in the JCT, with greater flow occurring through certain portions of the TM and less through other portions, which has been attributed to the presence or absence of pores within the HSCIW [23,77,78]. The degree of biomechanical stress directly affects the morphology of the inner wall, as its endothelial cells are described as elongated and aligned to the longitudinal axis of the SC, with some flattened and some with dome-like outpouchings (giant vacuoles) [13,57]. Because of their unique development, SC cells share morphological characteristics and cell marker expression with both lymphatic and venous endothelial cells (Table 3). In conventional 2D culture, SC cells are characterized as a homogeneous and elongated monolayer. The characteristic monolayer exhibits a net transendothelial electrical resistance of 10 $\Omega \cdot cm^2$ or greater [79], an absence of myocilin induction by dexamethasone, and expression of vascular endothelial cadherin (VE-cadherin), integrin $\alpha 6$, and fibulin-2 [79,80]. *In vivo*, SC cells are positive for PROX1 (with much higher levels for HSCIW cells than outer wall cells), integrin $\alpha 9$, and CD31, but negative for the differentiated lymphatic markers LYVE-1 and podoplanin, as well as the blood vessel marker SMA [2,28,55,79,80]. The cytoskeleton of SC cells is enriched in both microfilaments and intermediate filaments, and has a prominent actin-enriched cell cortex [81]. Although traditional 2D culture systems allow for manipulation of the SC endothelial cells, SC cells in traditional culture systems usually lose essential signaling, both mechanical and biochemical, required for proper maintenance of their *in vivo* phenotypes [5,79], reducing the utility of information obtained from such systems. Therefore, 3D culture systems may promote *in vivo*-like SC morphology, marker expression and function.

Table 3. Comparison of Schlemm's canal, lymphatic, and vascular endothelial properties.

Molecular/Cellular Characteristics	Schlemm's Canal Endothelium [a]	Lymphatic Endothelium	Vascular Endothelium
Sox18	–	+ [33]	–
VEGFR-2	+ [2,28]	–	+ [82]
VEGFR-3	+ [2,28]	+ [38]	+ [83]
PROX1	+ [2,28,53]	+ [34]	–
CCL21	+ [2,28]	–	–
Itga9	+ [2]	–	–
Collagen IV	+ [2]	–	–
PECAM1	+ [18]	–	+ [84]
VE-cadherin	+ [18]	+ [35]	+ [85]
Endomucin	+ [53]	–	–
Foxc2	+ [2,28]	–	–
LYVE-1	–	+ [41]	–
Podoplanin	–	+ [36]	–
vWF	+ [17]	–	+ [86]
Wiebel-Palade bodies	+ [17]	–	+ [86]
Endothelial monolayer	continuous [5]	continuous [29,87]	continuous
Basement membrane	discontinuous [5]	discontinuous [29,87]	continuous
Basal-to-apical Flow	+ [5,88]	+ [87,89]	–

[a] Note that these studies did not distinguish between inner and outer wall endothelia.

We recently highlighted the importance of providing the proper 3D spatial and biochemical cues in engineering a 3D SC *in vitro* model [90]. We demonstrated that 3D culture of HSC cells on microfabricated scaffolds with well-defined physical and biochemical cues, rescued expression of key HSC markers, such as VE-cadherin and PECAM1, and mediated pore formation, crucial for the SC regulation of IOP. Whether the *in vivo* SC has been functionally or structurally replicated or even completely simulated remains to be determined with studies of physiological and structural responses to drugs and modulation of genes expression for genes such as VEGF-C.

5. Biomechanics

As a result of its location against the JCT, the HSCIW experiences a biomechanical microenvironment that is much closer to that of lymphatic endothelia than that of vascular endothelia. Similar to lymphatic endothelium, the HSCIW endothelia experience a basal to apical pressure gradient during aqueous outflow. Unlike lymphatic endothelium, the HSCIW endothelium is sealed by tight junctions, and thus, must support the basal-to-apical pressure drop between IOP and episcleral venous pressure, which tends to deform HSCIW cells off their supporting basement membrane, creating giant vacuoles [89,91–94]. As a result, HSCIW cells are highly contractile [95] with an estimated elastic modulus of 1–3 kPa, similar to, but somewhat larger than other endothelial cells [88,96]. In addition, as a consequence of the basal-to-apical flow, HSCIW cells also exhibit transcellular and paracellular pores to mediate aqueous humor transport [21,97–99]. Their unique location against the JCT subjects HSCIW endothelial cells to biomechanical signals from the JCT's ECM [100], causing modification of gene expression to accommodate changes in substrate stiffness. HSCIW cells stiffen in response to increasing substrate stiffness, with glaucomatous HSCIW cells being more sensitive to substrate stiffness and having a larger stiffening response [88]. HSCIW cells' ability to adapt to large deformations and respond to their microenvironment is reflected in a cytoskeletal arrangement enriched in actin microfilaments and intermediate filaments [81]. Clearly, the HSCIW's biomechanical microenvironment plays an important role in maintaining HSCIW cells phenotype and proper function.

6. Perspective on Schlemm's Canal Engineering

Although conventional 2D tissue culture is currently the primary system for evaluating and characterizing HSC cell properties, its limitations have severely impeded our understanding of trabecular outflow physiology as well as glaucoma pathology and drug screening. Currently, there is no glaucoma therapy that lowers IOP via mechanisms that target the physiology of HSCIW endothelial cells. This is, in part, due to our poor understanding of the pathology at the SC, particularly the HSCIW during glaucoma development as well as the lack of an *in vitro* system for 3D culture of these cells under flow conditions, which is necessary to capture their *in vivo* characteristics and obtain relevant clinical information. The remainder of this review will highlight the main challenges and opportunities in establishing cellular microenvironments for engineering 3D HSCIW constructs, including sources of HSCIW cells, biomaterials to mimic ECM, and soluble factors to direct and maintain functional HSCIW differentiation.

6.1. Criteria for a 3D in Vitro Model of the Schlemm's Canal Inner Wall

Until recently, HSC cell culture has been limited to traditional 2D culture or culture on microporous Transwell® inserts. These studies have resulted in a tremendous amount of information on cell biology, physiology, and biomechanics. More importantly, these studies have highlighted the limitations of current systems as well as the critical attributes that a 3D *in vitro* system should recreate to correlate well with the *in vivo* characteristics and physiologic cellular response. From these studies, it is clear that any *in vitro* 3D model of the HSCIW should perform the following functions:

(1) Express key cell-specific markers, necessary for the endothelial integrity and mechanosensing;

(2) Display both paracellular and transcellular pathways vital to aqueous outflow function of the SC;

(3)   Mimic the *in vivo* cellular micro architecture with respect to morphological features such as a cellular dimensions or surface area of cell-cell interactions within the cultured monolayer, the spatial distribution of subcellular organelles (vacuoles), the complexity of tight junctional strands;

(4)   Allow for ease of culture using phenotypically stable cell lines to facilitate high throughput screening. Thus, a well-characterized *in vitro* 3D model of the HSCIW would provide a system in which to study and understand the physiology, biomechanics, outflow functions, physiological drug responses as well as pathological processes in glaucoma.

6.2. Potential Sources of Human Schlemm's Canal Inner Wall Endothelial Cells

One of the biggest challenges facing SC inner wall engineering is the scarcity of these cells [101]. Selective isolation of HSC cells from the limited amount of corneoscleral remnants remains an art. To date, very few laboratories [6,79,102,103] have developed complex protocols for successful isolation and culture of these cells and have become the primary supply sources for HSC cells, which will certainly contain both inner and outer wall HSC cells. In addition, successful isolation of these cells depends on a variety of uncontrollable factors, such as the age of the donor, duration of storage of the tissue after surgical removal before cell isolation, *etc.* These challenges have dramatically hindered the availability of SC cells for research and speak to the need to identify new ways of obtaining these cells. Stem cell differentiation [104–106] is an attractive avenue to explore as an alternative way of obtaining HSC cells with possible selection for HSCIW cells. Although stem cell differentiation to generate TM cells has been successfully documented [107–110], there are no reports addressing SC differentiation from stem cells. Recent publications [2,28,55] on the organogenesis of the SC have highlighted key factors and signaling molecules (e.g., PROX1 [32], VEGFR-3) that are necessary for acquisition of the SC phenotype from transscleral veins. These recent studies suggest the possibility of using primordial endothelial cells and/or venous endothelial cells [111–114] as a strategy to obtain HSCIW cells through directed differentiation (Figure 3).

6.3. Biomaterials for 3D Culture of Human Schlemm's Canal Inner Wall Cells

In addition to the scarcity of HSC cells, challenges in conventional 2D culture of HSC cells have impeded our understanding of the functional contribution of the SC to outflow physiology and glaucoma pathology. HSC cells in conventional 2D culture are distinct from their *in vivo* counterpart as they lose expression of key *in vivo* cell-specific markers [18,79,80]. This dedifferentiation speaks to the need to engineer an *in vitro* model of the SC that can mimic the *in vivo* microenvironment, eventually capturing the 3D *in vivo* characteristics of these cells. Given that the SC inner wall layer is only a few microns thick [96,115], a top-down approach to engineering the inner wall may be the most feasible strategy. In the traditional top-down approach, exogenous biocompatible and mechanically competent scaffolds are fabricated for 3D culture of the cells, which are then allowed to populate the scaffold, deposit and remodel their ECM.

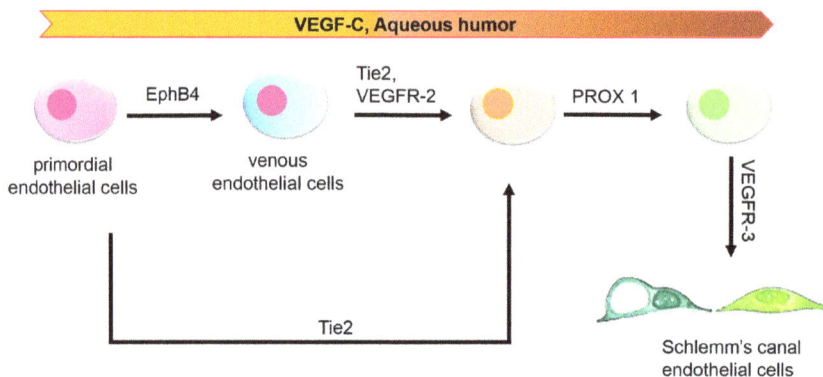

Figure 3. Potential strategies for stem cell differentiation into Schlemm's canal endothelial cells.

Scaffolds are fundamental to tissue engineering. Their functions include providing mechanical support, supporting ECM production and cell colonization, and waste-nutrient exchange [116–118]. As a consequence of its endothelial origin, scaffold materials and fabrication techniques being used in vascular tissue engineering can provide insight into engineering the SC [119–122] although different geometries will be required due to the unique nature of the SC canal, in particular, the polarization between in the inner and outer wall. Successful biomaterials for scaffolding for SC engineering might incorporate synthetic polymers, for their mechanical strength and well controlled porosity [123], and natural polymers, for their biochemical cues. In our previous work, we adapted the negative photoresist, SU-8, to provide the necessary topographical and mechanical cues while using the hydrogel Extracel™ to promote cell attachment and maintenance of SC cells differentiated functions [90].

Micro and nanofabrication techniques, such as lithography and electrospinning, are versatile fabrication techniques widely used in production of fibrous and porous scaffolds for vascular tissue engineering. Given the porous nature of the conventional outflow tract, scaffold considerations for SC engineering should be tailored to mimicking the pore structures (e.g., pore size, porosity), and in particular, extracellular, biochemical, and biomechanical microenvironments of the region directly against the SC, the JCT. Pore sizes and ECM fiber diameters in the JCT range from 2 to 15 μm [16]. Thus, scaffolds with various permutations of these properties might facilitate 3D culture of SC cells. In addition, biochemical cues play a paramount role in maintaining cell phenotype and function. The JCT ECM components, by virtue of their location against the HSCIW, might be providing key biochemical signaling for HSCIW cell growth and function. Thus, natural polymers of ECM components [8,124] like those found in the JCT, such as hyaluronic acid and collagen IV can be used to provide biochemical cues for HSCIW cell growth and function. Therefore, microfabricated scaffolds of synthetic polymers can be surface-coated or chemically modified with ECM components found at the JCT-SC border to provide key biochemical cues for successful HSCIW engineering. ECM components, such as hyaluronic acid [125–127] and collagens [128–130], have been used to support endothelial cell proliferation and function, indicating their potential to modify microfabricated scaffolds for HSCIW engineering. Additionally, the stratification of these ECM components at the JCT-SC interface highlights their potential as scaffolding materials or scaffold supplements for 3D HSCIW cell culture.

6.4. Soluble Factors for Directed Schlemm's Canal Cell Differentiation

Cellular differentiation is a result of coordinated dynamic expression of hundreds of genes and proteins in precise response to external signaling cues [106,131], which include soluble factors and spatio-physical cues from the ECM. Soluble factors mediate cellular differentiation by binding to cell surface receptors, thus activating downstream signaling [132]. Lineage specification of soluble factors, of the same stem cell type, differs depending on whether the stem cells were cultured in 3D or 2D configurations. For example, when induced to differentiate in restrictive ECM environments, adhesive, flattened human mesenchymal stem cells (hMSCs) in 2D preferentially adopt an osteogenic phenotype, whereas round hMSCs in 3D cultures preferentially undergo adipogenesis [133–135]. Thus, soluble factors and spatio-temporal cues in 3D culture are vital in providing the microenvironment necessary for differentiation, and may favor one lineage over another.

Park and Aspelund and their respective coworkers have elegantly elucidated key molecular footprints of venous endothelial cell differentiation to SC cells. Together, their data highlights the essential role of soluble factors such as VEGF-C, VEGF-D and aqueous humor for venous endothelial cells to acquire SC cell identity and for proper development of the SC. In addition, these data suggest the possibility of using these factors to mediate differentiation of induced pluripotent stem cells into SC cells [136]. Given the intimate relationship of the HSCIW to the JCT and that TM development precedes SC development, it is equally likely that soluble factors from the JCT cells may be vital to acquiring and maintaining the SC/HSCIW phenotype. Several groups have documented that cytokine (TNF-α, IL1-α, IL-β, and IL-8) release by TM cells mediates SC cell function in regulating aqueous humor outflow [100,137,138], highlighting the possible role of TM cells and their soluble factors in SC differentiation. Nitric oxide (NO) has been extensively studied for its role in modulating SC cell behavior to regulate aqueous outflow [3,103,139]. Because of the important role that NO plays in facilitating SC cell functions and endothelial junctional integrity, this warrants exploring the role of NO in SC development [140]. Furthermore, given the vascular origin of the SC and its lymphatic-like development and characteristics, it is important to consider soluble factors involved in vascular endothelial and lymphatic cell differentiation. Because of the unique biomechanical environment of the SC, direct addition of soluble factors to a 3D culture of the HSCIW cells may not be sufficient. Hence, controlled and sustained delivery of these soluble factors may be crucial for successful differentiation. Thus, other approaches such as nanoparticle-based delivery or conjugation to nanofibers that facilitate timed and spatial release should be consider for the delivery of soluble factors to induce SC cell differentiation and organogenesis [141–145].

6.5. Dynamic 3D Culture

The *in vivo* forces generated by aqueous humor flow play an important role in the SC organogenesis [2,28]. They are vital in maintaining the morphology and physiology of the HSCIW cells. Several groups have attempted to replicate the dynamic microenvironment of HSCIW cells *in vitro* through culture on microporous Transwell® membranes [100,146–148]. Together these studies have highlighted possible mechanisms for aqueous humor transport across the HSCIW, namely via formation of giant vacuoles and paracellular pores. They have further demonstrated the important role of the pressure gradient in modulating the barrier function of the HSCIW through regulation of junctional proteins. While these studies were able to capture the *in vivo* cell polarization and provided tremendous information regarding the biomechanics of the HSCIW, this system is not ideal to study aqueous outflow, or to perform continuous perfusion, medium exchange or gradient studies, particularly in response to key signaling factors like VEGF-C. These limitations are due to the nature of Transwell® membranes, which are track-etched and possess irregular pore structures or have low porosity (*i.e.*, 4%–20%), indicating poor topographical approximation [149] and limiting their performance for assessing physiologic parameters [150]. In addition, culture of endothelial cells on Transwell® membranes results in a less stringent endothelial barrier with the occurrence of irregular patterns of cell adhesion or "edge effect" [151], hampering the integrity of the endothelial monolayer. Thus, culture of HSCIW cells on Transwell® membranes to assess physiologic parameters uniquely associated with these cells may limit the clinical relevance obtained from such systems. Therefore, culture methods which incorporate a dynamic flow element might be instrumental not only for HSCIW cell differentiation, but for proper simulation and maintenance of the HSCIW phenotype *in vitro*. Dynamic flow system such as direct perfusion bioreactors and microfluidic devices that enable sophisticated control of the spatial, temporal profile of gradients as well as flow velocities are more suitable to culture of SC cells.

The *in vitro* system we previously described overcame some of the limitations of commercially available Transwell® membranes. Using photolithography techniques, we fabricated highly porous SU-8 membranes with pre-defined porosity, well-controlled, uniform pore size, shape, and beam width [90,152,153], demonstrating that well-defined SU-8 scaffolds support a more *in vivo*-like SC morphology, characterized by re-expression of key *in vivo* endothelial markers, PECAM1 and VE-cadherin, pore formation and outflow function. This system is a major step forward in the culture of SC cells, but it does not simulate the *in vivo* structure, in terms of fluidic or mechanical stress [90]. More studies are needed exploring 3D cultured cells in systems that can better simulate their *in vivo* dynamic microenvironment [154]. For instance, several studies have documented the importance of mimicking the dynamic microenvironment in lymphatic endothelial cell culture, demonstrating the critical role of interstitial flow in modulating lymphatic endothelial cell proliferation, migration and function [155,156]. In the case of cardiac cell differentiation, culture methods that incorporate a dynamic flow element improve cardiogenesis, beating percentage in size-controlled hESC-derived embryoid bodies (EBs), and cardiac gene expression at mRNA and protein levels in mESC-derived EBs [157–160] when compared to traditional static culture methods [161]. Therefore, in a similar fashion, replicating the dynamic microenvironment via continuous perfusion through SC cells might result in differences in cellular morphology, junctional complex expression and formation, and even outflow regulation. In addition, commercial Transwell® membranes are not ideal for co-culture, as they are too thick to allow for necessary cell-cell communications and their low porosity for appropriate rate of nutrients and paracrine signal exchange. Given that the HSCIW is in close apposition to the JCT, the most optimal culture system for both the SC cells and trabecular meshwork cells, is a co-culture on a membrane thin enough to allow for direct cell-cell communication, paracrine signals and a membrane strong enough to withstand appropriate flow velocities while under continuous perfusion. Thus, the fluid dynamics of the SC must be captured *in vitro*, for proper simulation of the *in vivo* tissue to obtain more clinically relevant responses, especially for drug studies.

7. Conclusions

The SC is a unique vascular endothelium with lymphatic-like characteristics that functions to mediate IOP outflow homeostasis together with the TM. The exact contribution of the SC is yet to be delineated. This is partly due to the lack of an *in vitro* model that can facilitate 3D culture of HSCIW cells, recapturing their *in vivo* phenotype and thus, obtaining more physiologically relevant information.

Research efforts targeting the engineering of the conventional outflow tract and its components are in their infancy. Applying nanotechnology for engineering the conventional outflow tract has great potential to mimic the nanoscale structure and outflow function of HSCIW cells. If successful, 3D culture of HSCIW cells will provide a valuable *in vitro* model that may revolutionize current thinking on the contribution of the SC to conventional outflow tract physiology and pathology and will hopefully translate into new drug modalities for glaucoma, targeting the SC.

Author Contributions: Cula N. Dautriche wrote and coordinated the overall preparation of this manuscript and tables. Yangzi Tian prepared all figures and index table. Overall editing, editorial direction and final editing of the manuscript were done by Yubing Xie and Susan T. Sharfstein.

Conflicts of Interest: The authors declare no conflict of interest.

Abbreviations

Calcr1	Calcitonin Receptor 1
CCBE1	Collagen and calcium-binding EGF domain-containing protein 1
CCL21	Chemokine (C-C motif) ligand 21
Clec2	C-type lectin-like receptor 2
COUP-FTII	Chicken ovalbumin upstream promoter-transcription factor 2
CXADR	Coxsackie virus and adenovirus receptor
Foxc2	Forkhead box protein C2
GATA2	GATA binding protein 2
Itga9	Integrin alpha-9
LYVE-1	Lymphatic vessel endothelial hyaluronan receptor
Meis1	Meis homeobox 1
Nfatc1	Nuclear factor of activated T-cells, cytoplasmic 1
NRP2	Neuropilin 2
PDPN	Podoplanin
PECAM1	Platelet endothelial cell adhesion molecule
PROX1	Prospero homeobox protein 1
RAC1	Ras-related C3 botulinum toxin substrate 1
Ramp2	Receptor activity modifying protein 2
Runx1	Runt-related transcription factor 1
SLP76	Lymphocyte cytosolic protein 2
Syk	Spleen tyrosine kinase
TIE1	Tunica interna endothelial cell kinase 1
TIE2	Tunica interna endothelial cell kinase 2
VE-cadherin	Vascular endothelial cadherin
VEFGR-2	Vascular endothelial growth factor 2
VEFGR-3	Vascular endothelial growth factor 3
vWF	Von Willebrand factor

References

1. Mansouri, K.; Shaarawy, T. Update on Schlemm's canal based procedures. *Middle East Afr. J. Ophthalmol.* **2015**, *22*, 38–44.

2. Aspelund, A.; Tammela, T.; Antila, S.; Nurmi, H.; Leppanen, V.M.; Zarkada, G.; Stanczuk, L.; Francois, M.; Makinen, T.; Saharinen, P.; *et al.* The Schlemm's canal is a VEGF-C/VEGFR-3-responsive lymphatic-like vessel. *J. Clin. Investig.* **2014**, *124*, 3975–3986.

3. Ashpole, N.E.; Overby, D.R.; Ethier, C.R.; Stamer, W.D. Shear stress-triggered nitric oxide release from Schlemm's canal cells. *Investig. Ophthalmol. Vis. Sci.* **2014**, *55*, 8067–8076.

4. Lutjen-Drecoll, E.; Rohen, J.W. [Endothelial studies of the Schlemm's canal using silver-impregnation technic]. *Albrecht Von Graefes Arch. Klin. Exp. Ophthalmol.* **1970**, *180*, 249–266.

5. Ethier, C.R. The inner wall of Schlemm's canal. *Exp. Eye Res.* **2002**, *74*, 161–172.

6. Karl, M.O.; Fleischhauer, J.C.; Stamer, W.D.; Peterson-Yantorno, K.; Mitchell, C.H.; Stone, R.A.; Civan, M.M. Differential P1-purinergic modulation of human Schlemm's canal inner-wall cells. *Am. J. Physiol. Cell Physiol.* **2005**, *288*, C784–C794.

7. Vranka, J.A.; Kelley, M.J.; Acott, T.S.; Keller, K.E. Extracellular matrix in the trabecular meshwork: Intraocular pressure regulation and dysregulation in glaucoma. *Exp. Eye Res.* **2015**, *133*, 112–125.

8. Acott, T.S.; Kelley, M.J. Extracellular matrix in the trabecular meshwork. *Exp. Eye Res.* **2008**, *86*, 543–561.

9. Maepea, O.; Bill, A. Pressures in the juxtacanalicular tissue and Schlemm's canal in monkeys. *Exp. Eye Res.* **1992**, *54*, 879–883.

10. Johnson, M. What controls aqueous humour outflow resistance? *Exp. Eye Res.* **2006**, *82*, 545–557.

11. Johnson, M.C.; Kamm, R.D. The role of Schlemm's canal in aqueous outflow from the human eye. *Investig. Ophthalmol. Vis. Sci.* **1983**, *24*, 320–325.

12. Pampaloni, F.; Reynaud, E.G.; Stelzer, E.H. The third dimension bridges the gap between cell culture and live tissue. *Nat. Rev. Mol. Cell Biol.* **2007**, *8*, 839–845.

13. Braakman, S.T.; Pedrigi, R.M.; Read, A.T.; Smith, J.A.E.; Stamer, W.D.; Ethier, C.R.; Overby, D.R. Biomechanical strain as a trigger for pore formation in Schlemm's canal endothelial cells. *Exp. Eye Res.* **2014**, *127*, 224–235.

14. Park, C.Y.; Zhou, E.H.; Tambe, D.; Chen, B.; Lavoie, T.; Dowell, M.; Simeonov, A.; Maloney, D.J.; Marinkovic, A.; Tschumperlin, D.J.; *et al.* High-throughput screening for modulators of cellular contractile force. *Integr. Biol.* **2015**.

15. Gong, H.; Tripathi, R.C.; Tripathi, B.J. Morphology of the aqueous outflow pathway. *Microsc. Res. Tech.* **1996**, *33*, 336–367.

16. Tamm, E.R. The trabecular meshwork outflow pathways: Structural and functional aspects. *Exp. Eye Res.* **2009**, *88*, 648–655.

17. Hamanaka, T.; Bill, A.; Ichinohasama, R.; Ishida, T. Aspects of the development of Schlemm's canal. *Exp. Eye Res.* **1992**, *55*, 479–488.

18. Heimark, R.L.; Kaochar, S.; Stamer, W.D. Human Schlemm's canal cells express the endothelial adherens proteins, VE-cadherin and PECAM-1. *Curr. Eye Res.* **2002**, *25*, 299–308.

19. Hamanaka, T.; Bill, A. Platelet aggregation on the endothelium of Schlemm's canal. *Exp. Eye Res.* **1994**, *59*, 249–256.

20. Johnstone, M.A.; Grant, W.M. Pressure-dependent changes in structures of the aqueous outflow system of human and monkey eyes. *Am. J. Ophthalmol.* **1973**, *75*, 365–383.

21. Ethier, C.R.; Coloma, F.M.; Sit, A.J.; Johnson, M. Two pore types in the inner-wall endothelium of Schlemm's canal. *Investig. Ophthalmol. Vis. Sci.* **1998**, *39*, 2041–2048.

22. Goel, M.; Picciani, R.G.; Lee, R.K.; Bhattacharya, S.K. Aqueous humor dynamics: A review. *Open Ophthalmol. J.* **2010**, *4*, 52–59.

23. Braakman, S.T.; Read, A.T.; Chan, D.W.; Ethier, C.R.; Overby, D.R. Colocalization of outflow segmentation and pores along the inner wall of Schlemm's canal. *Exp. Eye Res.* **2015**, *130*, 87–96.

24. Smelser, G.K.; Ozanics, V. The development of the trabecular meshwork in primate eyes. *Am. J. Ophthalmol.* **1971**, *71*, 366–385.

25. Foets, B.; van den Oord, J.; Engelmann, K.; Missotten, L. A comparative immunohistochemical study of human corneotrabecular tissue. *Graefes Arch. Clin. Exp. Ophthalmol.* **1992**, *230*, 269–274.

26. Wulle, K.G. Electron microscopic observations of the development of Schlemm's canal in the human eye. *Trans. Am. Acad. Ophthalmol. Otolaryngol.* **1968**, *72*, 765–773.

27. Ramirez, J.M.; Ramirez, A.I.; Salazar, J.J.; Rojas, B.; de Hoz, R.; Trivino, A. Schlemm's canal and the collector channels at different developmental stages in the human eye. *Cells Tissues Organs* **2004**, *178*, 180–185.

28. Park, D.Y.; Lee, J.; Park, I.; Choi, D.; Lee, S.; Song, S.; Hwang, Y.; Hong, K.Y.; Nakaoka, Y.; Makinen, T.; *et al.* Lymphatic regulator PROX1 determines Schlemm's canal integrity and identity. *J. Clin. Investig.* **2014**, *124*, 3960–3974.

29. Yang, Y.; Oliver, G. Development of the mammalian lymphatic vasculature. *J. Clin. Investig.* **2014**, *124*, 888–897.

30. Oliver, G.; Srinivasan, R.S. Endothelial cell plasticity: How to become and remain a lymphatic endothelial cell. *Development* **2010**, *137*, 363–372.

31. Hagerling, R.; Pollmann, C.; Andreas, M.; Schmidt, C.; Nurmi, H.; Adams, R.H.; Alitalo, K.; Andresen, V.; Schulte-Merker, S.; Kiefer, F. A novel multistep mechanism for initial lymphangiogenesis in mouse embryos based on ultramicroscopy. *EMBO J.* **2013**, *32*, 629–644.

32. Hong, Y.K.; Detmar, M. PROX1, master regulator of the lymphatic vasculature phenotype. *Cell Tissue Res.* **2003**, *314*, 85–92.

33. Francois, M.; Caprini, A.; Hosking, B.; Orsenigo, F.; Wilhelm, D.; Browne, C.; Paavonen, K.; Karnezis, T.; Shayan, R.; Downes, M.; *et al.* Sox18 induces development of the lymphatic vasculature in mice. *Nature* **2008**, *456*, 643–647.

34. Yamazaki, T.; Yoshimatsu, Y.; Morishita, Y.; Miyazono, K.; Watabe, T. Coup-TFII regulates the functions of PROX1 in lymphatic endothelial cells through direct interaction. *Genes Cells* **2009**, *14*, 425–434.

35. Yang, Y.; García-Verdugo, J.M.; Soriano-Navarro, M.; Srinivasan, R.S.; Scallan, J.P.; Singh, M.K.; Epstein, J.A.; Oliver, G. Lymphatic endothelial progenitors bud from the cardinal vein and intersomitic vessels in mammalian embryos. *Blood* **2012**, *120*, 2340–2348.

36. Breiteneder-Geleff, S.; Soleiman, A.; Kowalski, H.; Horvat, R.; Amann, G.; Kriehuber, E.; Diem, K.; Weninger, W.; Tschachler, E.; Alitalo, K.; *et al.* Angiosarcomas express mixed endothelial phenotypes of blood and lymphatic capillaries: Podoplanin as a specific marker for lymphatic endothelium. *Am. J. Pathol.* **1999**, *154*, 385–394.

37. Kaipainen, A.; Korhonen, J.; Mustonen, T.; van Hinsbergh, V.W.; Fang, G.H.; Dumont, D.; Breitman, M.; Alitalo, K. Expression of the FMS-like tyrosine kinase 4 gene becomes restricted to lymphatic endothelium during development. *Proc. Natl. Acad. Sci. USA* **1995**, *92*, 3566–3570.

38. Le Guen, L.; Karpanen, T.; Schulte, D.; Harris, N.C.; Koltowska, K.; Roukens, G.; Bower, N.I.; van Impel, A.; Stacker, S.A.; Achen, M.G.; *et al.* CCBE1 regulates VEGFC-mediated induction of VEGFR3 signaling during embryonic lymphangiogenesis. *Development* **2014**, *141*, 1239–1249.

39. Lin, F.J.; Chen, X.; Qin, J.; Hong, Y.K.; Tsai, M.J.; Tsai, S.Y. Direct transcriptional regulation of neuropilin-2 by COUP-TFII modulates multiple steps in murine lymphatic vessel development. *J. Clin. Investig.* **2010**, *120*, 1694–1707.

40. D'Amico, G.; Jones, D.T.; Nye, E.; Sapienza, K.; Ramjuan, A.R.; Reynolds, L.E.; Robinson, S.D.; Kostourou, V.; Martinez, D.; Aubyn, D.; *et al.* Regulation of lymphatic-blood vessel separation by endothelial RAC1. *Development* **2009**, *136*, 4043–4053.

41. Banerji, S.; Ni, J.; Wang, S.-X.; Clasper, S.; Su, J.; Tammi, R.; Jones, M.; Jackson, D.G. LYVE-1, a new homologue of the CD44 glycoprotein, is a lymph-specific receptor for hyaluronan. *J. Cell Biol.* **1999**, *144*, 789–801.

42. Kulkarni, R.M.; Greenberg, J.M.; Akeson, A.L. NFATC1 regulates lymphatic endothelial development. *Mech. Dev.* **2009**, *126*, 350–365.

43. Lim, K.-C.; Hosoya, T.; Brandt, W.; Ku, C.-J.; Hosoya-Ohmura, S.; Camper, S.A.; Yamamoto, M.; Engel, J.D. Conditional GATA2 inactivation results in HSC loss and lymphatic mispatterning. *J. Clin. Investig.* **2012**, *122*, 3705–3717.

44. Kazenwadel, J.; Secker, G.A.; Liu, Y.J.; Rosenfeld, J.A.; Wildin, R.S.; Cuellar-Rodriguez, J.; Hsu, A.P.; Dyack, S.; Fernandez, C.V.; Chong, C.-E.; *et al.* Loss-of-function germline GATA2 mutations in patients with MDS/AML or monomac syndrome and primary lymphedema reveal a key role for GATA2 in the lymphatic vasculature. *Blood* **2012**, *119*, 1283–1291.

45. Fritz-Six, K.L.; Dunworth, W.P.; Li, M.; Caron, K.M. Adrenomedullin signaling is necessary for murine lymphatic vascular development. *J. Clin. Investig.* **2008**, *118*, 40–50.

46. Qu, X.; Tompkins, K.; Batts, L.E.; Puri, M.; Baldwin, S. Abnormal embryonic lymphatic vessel development in TIE1 hypomorphic mice. *Development* **2010**, *137*, 1285–1295.

47. Abtahian, F.; Guerriero, A.; Sebzda, E.; Lu, M.M.; Zhou, R.; Mocsai, A.; Myers, E.E.; Huang, B.; Jackson, D.G.; Ferrari, V.A.; *et al.* Regulation of blood and lymphatic vascular separation by signaling proteins SLP-76 and SYK. *Science* **2003**, *299*, 247–251.

48. Srinivasan, R.S.; Dillard, M.E.; Lagutin, O.V.; Lin, F.-J.; Tsai, S.; Tsai, M.-J.; Samokhvalov, I.M.; Oliver, G. Lineage tracing demonstrates the venous origin of the mammalian lymphatic vasculature. *Genes Dev.* **2007**, *21*, 2422–2432.

49. Carramolino, L.; Fuentes, J.; Garcia-Andres, C.; Azcoitia, V.; Riethmacher, D.; Torres, M. Platelets play an essential role in separating the blood and lymphatic vasculatures during embryonic angiogenesis. *Circ. Res.* **2010**, *106*, 1197–1201.

50. Uhrin, P.; Zaujec, J.; Breuss, J.M.; Olcaydu, D.; Chrenek, P.; Stockinger, H.; Fuertbauer, E.; Moser, M.; Haiko, P.; Fassler, R.; *et al.* Novel function for blood platelets and podoplanin in developmental separation of blood and lymphatic circulation. *Blood* **2010**, *115*, 3997–4005.

51. Suzuki-Inoue, K.; Inoue, O.; Ding, G.; Nishimura, S.; Hokamura, K.; Eto, K.; Kashiwagi, H.; Tomiyama, Y.; Yatomi, Y.; Umemura, K.; *et al.* Essential *in vivo* roles of the c-type lectin receptor clec-2: Embryonic/neonatal lethality of CLEC-2-deficient mice by blood/lymphatic misconnections and impaired thrombus formation of CLEC-2-deficient platelets. *J. Biol. Chem.* **2010**, *285*, 24494–24507.

52. Mirza, M.; Pang, M.-F.; Zaini, M.A.; Haiko, P.; Tammela, T.; Alitalo, K.; Philipson, L.; Fuxe, J.; Sollerbrant, K. Essential role of the coxsackie- and adenovirus receptor (CAR) in development of the lymphatic system in mice. *PLoS ONE* **2012**, *7*.

53. Kizhatil, K.; Ryan, M.; Marchant, J.K.; Henrich, S.; John, S.W.M. Schlemm's canal is a unique vessel with a combination of blood vascular and lymphatic phenotypes that forms by a novel developmental process. *PLoS Biol.* **2014**, *12*.

54. Truong, T.N.; Li, H.; Hong, Y.-K.; Chen, L. Novel characterization and live imaging of Schlemm's canal expressing PROX-1. *PLoS ONE* **2014**, *9*.

55. Karpinich, N.O.; Caron, K.M. Schlemm's canal: More than meets the eye, lymphatics in disguise. *J. Clin. Investig.* **2014**, *124*, 3701–3703.

56. Grant, W.M. Further studies on facility of flow through the trabecular meshwork. *AMA Arch. Ophthalmol.* **1958**, *60*, 523–533.

57. Ramos, R.F.; Hoying, J.B.; Witte, M.H.; Daniel Stamer, W. Schlemm's canal endothelia, lymphatic, or blood vasculature? *J. Glaucoma* **2007**, *16*, 391–405.

58. Allingham, R.R.; de Kater, A.W.; Ethier, C.R. Schlemm's canal and primary open angle glaucoma: Correlation between Schlemm's canal dimensions and outflow facility. *Exp. Eye Res.* **1996**, *62*, 101–109.

59. Kagemann, L.; Wang, B.; Wollstein, G.; Ishikawa, H.; Nevins, J.E.; Nadler, Z.; Sigal, I.A.; Bilonick, R.A.; Schuman, J.S. IOP elevation reduces Schlemm's canal cross-sectional area. *Investig. Ophthalmol. Vis. Sci.* **2014**, *55*, 1805–1809.

60. Kagemann, L.; Wollstein, G.; Ishikawa, H.; Bilonick, R.A.; Brennen, P.M.; Folio, L.S.; Gabriele, M.L.; Schuman, J.S. Identification and assessment of Schlemm's canal by spectral-domain optical coherence tomography. *Investig. Ophthalmol. Vis. Sci.* **2010**, *51*, 4054–4059.

61. Kagemann, L.; Nevins, J.E.; Jan, N.-J.; Wollstein, G.; Ishikawa, H.; Kagemann, J.; Sigal, I.A.; Nadler, Z.; Ling, Y.; Schuman, J.S. Characterisation of Schlemm's canal cross-sectional area. *Br. J. Ophthalmol.* **2014**, *98*, ii10–ii14.

62. Kagemann, L.; Wollstein, G.; Ishikawa, H.; Sigal, I.A.; Folio, L.S.; Xu, J.; Gong, H.; Schuman, J.S. 3D visualization of aqueous humor outflow structures in-situ in humans. *Exp. Eye Res.* **2011**, *93*, 308–315.

63. Kagemann, L.; Wollstein, G.; Ishikawa, H.; Nadler, Z.; Sigal, I.A.; Folio, L.S.; Schuman, J.S. Visualization of the conventional outflow pathway in the living human eye. *Ophthalmology* **2012**, *119*, 1563–1568.

64. Dvorak-Theobald, G. Schlemm's canal: Its anastomoses and anatomic relations. *Trans. Am. Ophthalmol. Soc.* **1934**, *32*, 574–595.

65. Rosenquist, R.; Epstein, D.; Melamed, S.; Johnson, M.; Grant, W.M. Outflow resistance of enucleated human eyes at two different perfusion pressures and different extents of trabeculotomy. *Curr. Eye Res.* **1989**, *8*, 1233–1240.

66. Grierson, I.; Lee, W.R.; Abraham, S.; Howes, R.C. Associations between the cells of the walls of Schlemm's canal. *Albrecht Von Graefes Arch. Klin. Exp. Ophthalmol.* **1978**, *208*, 33–47.

67. Hamanaka, T.; Thornell, L.E.; Bill, A. Cytoskeleton and tissue origin in the anterior cynomolgus monkey eye. *Jpn. J. Ophthalmol.* **1997**, *41*, 138–149.

68. Ethier, C.R.; Read, A.T.; Chan, D. Biomechanics of Schlemm's canal endothelial cells: Influence on f-actin architecture. *Biophys. J.* **2004**, *87*, 2828–2837.

69. Grierson, I.; Lee, W.R. Pressure-induced changes in the ultrastructure of the endothelium lining Schlemm's canal. *Am. J. Ophthalmol.* **1975**, *80*, 863–884.

70. Bill, A.; Svedbergh, B. Scanning electron microscopic studies of the trabecular meshwork and the canal of Schlemm—An attempt to localize the main resistance to outflow of aqueous humor in man. *Acta Ophthalmol.* **1972**, *50*, 295–320.

71. Tripathi, R.C. Ultrastructure of Schlemm's canal in relation to aqueous outflow. *Exp. Eye Res.* **1968**, *7*, 335–341.

72. Holmberg, A. The fine structure of the inner wall of Schlemms canal. *Arch. Ophthalmol.* **1959**, *62*, 956–958.

73. Epstein, D.L.; Rohen, J.W. Morphology of the trabecular meshwork and inner-wall endothelium after cationized ferritin perfusion in the monkey eye. *Investig. Ophthalmol. Vis. Sci.* **1991**, *32*, 160–171.

74. Herrnberger, L.; Ebner, K.; Junglas, B.; Tamm, E.R. The role of plasmalemma vesicle-associated protein (PLVAP) in endothelial cells of Schlemm's canal and ocular capillaries. *Exp. Eye Res.* **2012**, *105*, 27–33.

75. Inomata, H.; Bill, A.; Smelser, G.K. Aqueous humor pathways through the trabecular meshwork and into Schlemm's canal in the cynomolgus monkey (*Macaca Irus*). An electron microscopic study. *Am. J. Ophthalmol.* **1972**, *73*, 760–789.

76. Overby, D.R.; Stamer, W.D.; Johnson, M. The changing paradigm of outflow resistance generation: Towards synergistic models of the JCT and inner wall endothelium. *Exp. Eye Res.* **2009**, *88*, 656–670.

77. Chang, J.Y.; Folz, S.J.; Laryea, S.N.; Overby, D.R. Multi-scale analysis of segmental outflow patterns in human trabecular meshwork with changing intraocular pressure. *J. Ocul. Pharmacol. Ther.* **2014**, *30*, 213–223.

78. Hann, C.R.; Fautsch, M.P. Preferential fluid flow in the human trabecular meshwork near collector channels. *Investig. Ophthalmol. Vis. Sci.* **2009**, *50*, 1692–1697.

79. Stamer, W.D.; Roberts, B.C.; Howell, D.N.; Epstein, D.L. Isolation, culture, and characterization of endothelial cells from Schlemm's canal. *Investig. Ophthalmol. Vis. Sci.* **1998**, *39*, 1804–1812.

80. Perkumas, K.M.; Stamer, W.D. Protein markers and differentiation in culture for Schlemm's canal endothelial cells. *Exp. Eye Res.* **2012**, *96*, 82–87.

81. Tian, B.; Geiger, B.; Epstein, D.L.; Kaufman, P.L. Cytoskeletal involvement in the regulation of aqueous humor outflow. *Investig. Ophthalmol. Vis. Sci.* **2000**, *41*, 619–623.

82. Hristov, M.; Weber, C. Endothelial progenitor cells: Characterization, pathophysiology, and possible clinical relevance. *J. Cell Mol. Med.* **2004**, *8*, 498–508.

83. Welti, J.; Loges, S.; Dimmeler, S.; Carmeliet, P. Recent molecular discoveries in angiogenesis and antiangiogenic therapies in cancer. *J. Clin. Investig.* **2013**, *123*, 3190–3200.

84. Garlanda, C.; Dejana, E. Heterogeneity of endothelial cells. Specific markers. *Arterioscler. Thromb. Vasc. Biol.* **1997**, *17*, 1193–1202.

85. Gavard, J. Endothelial permeability and VE-cadherin: A wacky comradeship. *Cell Adh. Migr.* **2014**, *8*, 158–164.

86. Lenting, P.J.; Christophe, O.D.; Denis, C.V. von Willebrand factor biosynthesis, secretion, and clearance: connecting the far ends. *Blood* **2015**, *125*, 2019–2028.

87. Ikomi, F.; Kawai, Y.; Ohhashi, T. Recent advance in lymph dynamic analysis in lymphatics and lymph nodes. *Ann. Vasc. Dis.* **2012**, *5*, 258–268.

88. Stamer, W.D.; Braakman, S.T.; Zhou, E.H.; Ethier, C.R.; Fredberg, J.J.; Overby, D.R.; Johnson, M. Biomechanics of Schlemm's canal endothelium and intraocular pressure reduction. *Prog. Retin. Eye Res.* **2015**, *44*, 86–98.

89. Miteva, D.O.; Rutkowski, J.M.; Dixon, J.B.; Kilarski, W.; Shields, J.D.; Swartz, M.A. Transmural flow modulates cell and fluid transport functions of lymphatic endothelium. *Circ. Res.* **2010**, *106*, 920–931.

90. Dautriche, C.N.; Szymanski, D.; Kerr, M.; Torrejon, K.Y.; Bergkvist, M.; Xie, Y.; Danias, J.; Stamer, W.D.; Sharfstein, S.T. A biomimetic Schlemm's canal inner wall: A model to study outflow physiology, glaucoma pathology and high-throughput drug screening. *Biomaterials* **2015**, *65*, 86–92.

91. Raviola, G.; Raviola, E. Paracellular route of aqueous outflow in the trabecular meshwork and canal of Schlemm. A freeze-fracture study of the endothelial junctions in the sclerocorneal angle of the macaque monkey eye. *Investig. Ophthalmol. Vis. Sci.* **1981**, *21*, 52–72.

92. Swartz, M.A. The physiology of the lymphatic system. *Adv. Drug Deliv. Rev.* **2001**, *50*, 3–20.

93. Overby, D.R.; Zhou, E.H.; Vargas-Pinto, R.; Pedrigi, R.M.; Fuchshofer, R.; Braakman, S.T.; Gupta, R.; Perkumas, K.M.; Sherwood, J.M.; Vahabikashi, A.; *et al.* Altered mechanobiology of Schlemm's canal endothelial cells in glaucoma. *Proc. Natl. Acad. Sci. USA* **2014**, *111*, 13876–13881.

94. Vargas-Pinto, R.; Lai, J.; Gong, H.; Ethier, C.R.; Johnson, M. Finite element analysis of the pressure-induced deformation of Schlemm's canal endothelial cells. *Biomech. Model. Mechanobiol.* **2014**, *14*, 851–863.

95. Zhou, E.H.; Krishnan, R.; Stamer, W.D.; Perkumas, K.M.; Rajendran, K.; Nabhan, J.F.; Lu, Q.; Fredberg, J.J.; Johnson, M. Mechanical responsiveness of the endothelial cell of Schlemm's canal: Scope, variability and its potential role in controlling aqueous humour outflow. *J. R. Soc. Interf.* **2012**, *9*, 1144–1155.

96. Zeng, D.; Juzkiw, T.; Read, A.T.; Chan, D.W.H.; Glucksberg, M.R.; Ethier, C.R.; Johnson, M. Young's modulus of elasticity of Schlemm's canal endothelial cells. *Biomech. Model. Mechanobiol.* **2010**, *9*, 19–33.

97. Johnson, M.; Shapiro, A.; Ethier, C.R.; Kamm, R.D. Modulation of outflow resistance by the pores of the inner wall endothelium. *Investig. Ophthalmol. Vis. Sci.* **1992**, *33*, 1670–1675.

98. Johnson, M.; Johnson, D.H.; Kamm, R.D.; DeKater, A.W.; Epstein, D.L. The filtration characteristics of the aqueous outflow system. *Exp. Eye Res.* **1990**, *50*, 407–418.

99. Allingham, R.R.; de Kater, A.W.; Ethier, C.R.; Anderson, P.J.; Hertzmark, E.; Epstein, D.L. The relationship between pore density and outflow facility in human eyes. *Investig. Ophthalmol. Vis. Sci.* **1992**, *33*, 1661–1669.

100. Alvarado, J.A.; Yeh, R.-F.; Franse-Carman, L.; Marcellino, G.; Brownstein, M.J. Interactions between endothelia of the trabecular meshwork and of Schlemm's canal: A new insight into the regulation of aqueous outflow in the eye. *Trans. Am. Ophthalmol. Soc.* **2005**, *103*, 148–163.

101. Curcio, C.A. Declining availability of human eye tissues for research. *Investig. Ophthalmol. Vis. Sci.* **2006**, *47*, 2747–2749.

102. Lei, Y.; Overby, D.R.; Read, A.T.; Stamer, W.D.; Ethier, C.R. A new method for selection of angular aqueous plexus cells from porcine eyes: A model for Schlemm's canal endothelium. *Investig. Ophthalmol. Vis. Sci.* **2010**, *51*, 5744–5750.

103. Ellis, D.Z.; Sharif, N.A.; Dismuke, W.M. Endogenous regulation of human Schlemm's canal cell volume by nitric oxide signaling. *Investig. Ophthalmol. Vis. Sci.* **2010**, *51*, 5817–5824.

104. Nishikawa, S.-I.; Goldstein, R.A.; Nierras, C.R. The promise of human induced pluripotent stem cells for research and therapy. *Nat. Rev. Mol. Cell Biol.* **2008**, *9*, 725–729.

105. Jones, D.L.; Wagers, A.J. No place like home: Anatomy and function of the stem cell niche. *Nat. Rev. Mol. Cell Biol.* **2008**, *9*, 11–21.

106. MacArthur, B.D.; Ma'ayan, A.; Lemischka, I.R. Systems biology of stem cell fate and cellular reprogramming. *Nat. Rev. Mol. Cell Biol.* **2009**, *10*, 672–681.

107. Tay, C.Y.; Sathiyanathan, P.; Chu, S.W.; Stanton, L.W.; Wong, T.T. Identification and characterization of mesenchymal stem cells derived from the trabecular meshwork of the human eye. *Stem Cells Dev.* **2012**, *21*, 1381–1390.

108. Du, Y.; Roh, D.S.; Mann, M.M.; Funderburgh, M.L.; Funderburgh, J.L.; Schuman, J.S. Multipotent stem cells from trabecular meshwork become phagocytic tm cells. *Investig. Ophthalmol. Vis. Sci.* **2012**, *53*, 1566–1575.

109. Ding, Q.J.; Zhu, W.; Cook, A.C.; Anfinson, K.R.; Tucker, B.A.; Kuehn, M.H. Induction of trabecular meshwork cells from induced pluripotent stem cells. *Investig. Ophthalmol. Vis. Sci.* **2014**, *55*, 7065–7072.

110. Du, Y.; Yun, H.; Yang, E.; Schuman, J.S. Stem cells from trabecular meshwork home to TM tissue *in vivo. Investig. Ophthalmol. Vis. Sci.* **2013**, *54*, 1450–1459.

111. Carmeliet, P. Angiogenesis in life, disease and medicine. *Nature* **2005**, *438*, 932–936.

112. Kume, T. Specification of arterial, venous, and lymphatic endothelial cells during embryonic development. *Histol. Histopathol.* **2010**, *25*, 637–646.

113. Swift, M.R.; Weinstein, B.M. Arterial-venous specification during development. *Circ. Res.* **2009**, *104*, 576–588.

114. You, L.R.; Lin, F.J.; Lee, C.T.; DeMayo, F.J.; Tsai, M.J.; Tsai, S.Y. Suppression of notch signalling by the COUP-TFII transcription factor regulates vein identity. *Nature* **2005**, *435*, 98–104.

115. Keller, K.E.; Acott, T.S. The juxtacanalicular region of ocular trabecular meshwork: A tissue with a unique extracellular matrix and specialized function. *J. Ocul. Biol.* **2013**, *1*, 10:1–10:7.

116. Adachi, T.; Osako, Y.; Tanaka, M.; Hojo, M.; Hollister, S.J. Framework for optimal design of porous scaffold microstructure by computational simulation of bone regeneration. *Biomaterials* **2006**, *27*, 3964–3972.

117. Hollister, S.J. Porous scaffold design for tissue engineering. *Nat. Mater.* **2005**, *4*, 518–524.

118. Hutmacher, D.W.; Sittinger, M.; Risbud, M.V. Scaffold-based tissue engineering: Rationale for computer-aided design and solid free-form fabrication systems. *Trends Biotechnol.* **2004**, *22*, 354–362.

119. Ravi, S.; Chaikof, E.L. Biomaterials for vascular tissue engineering. *Regen. Med.* **2010**, *5*, 107–120.

120. Ravi, S.; Qu, Z.; Chaikof, E.L. Polymeric materials for tissue engineering of arterial substitutes. *Vascular* **2009**, *17*, S45–S54.

121. Hasan, A.; Memic, A.; Annabi, N.; Hossain, M.; Paul, A.; Dokmeci, M.R.; Dehghani, F.; Khademhosseini, A. Electrospun scaffolds for tissue engineering of vascular grafts. *Acta Biomater.* **2014**, *10*, 11–25.

122. Woods, I.; Flanagan, T.C. Electrospinning of biomimetic scaffolds for tissue-engineered vascular grafts: Threading the path. *Expert Rev. Cardiovasc. Ther.* **2014**, *12*, 815–832.

123. Cheung, H.-Y.; Lau, K.-T.; Lu, T.-P.; Hui, D. A critical review on polymer-based bio-engineered materials for scaffold development. *Compos. B Eng.* **2007**, *38*, 291–300.

124. Rhee, D.J.; Haddadin, R.I.; Kang, M.H.; Oh, D.J. Matricellular proteins in the trabecular meshwork. *Exp. Eye Res.* **2009**, *88*, 694–703.

125. Genasetti, A.; Vigetti, D.; Viola, M.; Karousou, E.; Moretto, P.; Rizzi, M.; Bartolini, B.; Clerici, M.; Pallotti, F.; De Luca, G.; *et al.* Hyaluronan and human endothelial cell behavior. *Connect. Tissue Res.* **2008**, *49*, 120–123.

126. Turner, N.J.; Kielty, C.M.; Walker, M.G.; Canfield, A.E. A novel hyaluronan-based biomaterial (Hyaff-11®) as a scaffold for endothelial cells in tissue engineered vascular grafts. *Biomaterials* **2004**, *25*, 5955–5964.

127. Ibrahim, S.; Ramamurthi, A. Hyaluronic acid cues for functional endothelialization of vascular constructs. *J. Tissue Eng. Regen. Med.* **2008**, *2*, 22–32.

128. Boccafoschi, F.; Habermehl, J.; Vesentini, S.; Mantovani, D. Biological performances of collagen-based scaffolds for vascular tissue engineering. *Biomaterials* **2005**, *26*, 7410–7417.

129. Boland, E.D.; Matthews, J.A.; Pawlowski, K.J.; Simpson, D.G.; Wnek, G.E.; Bowlin, G.L. Electrospinning collagen and elastin: Preliminary vascular tissue engineering. *Front. Biosci.* **2004**, *9*, 1422–1432.

130. Berglund, J.D.; Nerem, R.M.; Sambanis, A. Incorporation of intact elastin scaffolds in tissue-engineered collagen-based vascular grafts. *Tissue Eng.* **2004**, *10*, 1526–1535.

131. Jopling, C.; Boue, S.; Izpisua Belmonte, J.C. Dedifferentiation, transdifferentiation and reprogramming: Three routes to regeneration. *Nat. Rev. Mol. Cell Biol.* **2011**, *12*, 79–89.

132. Hannum, C.; Culpepper, J.; Campbell, D.; McClanahan, T.; Zurawski, S.; Bazan, J.F.; Kastelein, R.; Hudak, S.; Wagner, J.; Mattson, J.; *et al.* Ligand for FLT3/FLK2 receptor tyrosine kinase regulates growth of haematopoietic stem cells and is encoded by variant rnas. *Nature* **1994**, *368*, 643–648.

133. Meng, X.; Leslie, P.; Zhang, Y.; Dong, J. Stem cells in a three-dimensional scaffold environment. *Springer Plus* **2014**, *3*.

134. Taylor-Weiner, H.; Schwarzbauer, J.E.; Engler, A.J. Defined extracellular matrix components are necessary for definitive endoderm induction. *Stem Cells* **2013**, *31*, 2084–2094.

135. McBeath, R.; Pirone, D.M.; Nelson, C.M.; Bhadriraju, K.; Chen, C.S. Cell shape, cytoskeletal tension, and RhoA regulate stem cell lineage commitment. *Dev. Cell* **2004**, *6*, 483–495.

136. Liersch, R.; Nay, F.; Lu, L.; Detmar, M. Induction of lymphatic endothelial cell differentiation in embryoid bodies. *Blood* **2006**, *107*, 1214–1216.

137. Alexander, J.P.; Acott, T.S. Involvement of the ERK-MAP kinase pathway in TNFalpha regulation of trabecular matrix metalloproteinases and TIMPS. *Investig. Ophthalmol. Vis. Sci.* **2003**, *44*, 164–169.

138. Bradley, J.M.; Anderssohn, A.M.; Colvis, C.M.; Parshley, D.E.; Zhu, X.H.; Ruddat, M.S.; Samples, J.R.; Acott, T.S. Mediation of laser trabeculoplasty-induced matrix metalloproteinase expression by IL-1beta and TNFalpha. *Investig. Ophthalmol. Vis. Sci.* **2000**, *41*, 422–430.

139. Chang, J.Y.; Stamer, W.D.; Bertrand, J.; Read, A.T.; Marando, C.M.; Ethier, C.R.; Overby, D.R. The role of nitric oxide in murine conventional outflow physiology. *Am. J. Physiol. Cell Physiol.* **2015**, *309*, C205–C214.

140. Huang, N.F.; Fleissner, F.; Sun, J.; Cooke, J.P. Role of nitric oxide signaling in endothelial differentiation of embryonic stem cells. *Stem Cells Dev.* **2010**, *19*, 1617–1625.

141. Bishop, C.; Kim, J.; Green, J. Biomolecule delivery to engineer the cellular microenvironment for regenerative medicine. *Ann. Biomed. Eng.* **2014**, *42*, 1557–1572.

142. Quake, S.R.; Scherer, A. From micro- to nanofabrication with soft materials. *Science* **2000**, *290*, 1536–1540.

143. Baker, B.M.; Chen, C.S. Deconstructing the third dimension: How 3D culture microenvironments alter cellular cues. *J. Cell Sci.* **2012**, *125*, 3015–3024.

144. Lee, S.-H.; Shin, H. Matrices and scaffolds for delivery of bioactive molecules in bone and cartilage tissue engineering. *Adv. Drug Deliv. Rev.* **2007**, *59*, 339–359.

145. Chung, H.J.; Park, T.G. Surface engineered and drug releasing pre-fabricated scaffolds for tissue engineering. *Adv. Drug Deliv. Rev.* **2007**, *59*, 249–262.

146. Pedrigi, R.M.; Simon, D.; Reed, A.; Stamer, W.D.; Overby, D.R. A model of giant vacuole dynamics in human Schlemm's canal endothelial cells. *Exp. Eye Res.* **2011**, *92*, 57–66.

147. Burke, A.G.; Zhou, W.; O'Brien, E.T.; Roberts, B.C.; Stamer, W.D. Effect of hydrostatic pressure gradients and Na_2EDTA on permeability of human Schlemm's canal cell monolayers. *Curr. Eye Res.* **2004**, *28*, 391–398.

148. Stamer, W.D.; Roberts, B.C.; Epstein, D.L. Hydraulic pressure stimulates adenosine 3',5'-cyclic monophosphate accumulation in endothelial cells from Schlemm's canal. *Investig. Ophthalmol. Vis. Sci.* **1999**, *40*, 1983–1988.

149. Schindler, M.; Nur, E.K.A.; Ahmed, I.; Kamal, J.; Liu, H.Y.; Amor, N.; Ponery, A.S.; Crockett, D.P.; Grafe, T.H.; Chung, H.Y.; *et al.* Living in three dimensions: 3D nanostructured environments for cell culture and regenerative medicine. *Cell Biochem. Biophys.* **2006**, *45*, 215–227.

150. Albelda, S.M.; Sampson, P.M.; Haselton, F.R.; McNiff, J.M.; Mueller, S.N.; Williams, S.K.; Fishman, A.P.; Levine, E.M. Permeability characteristics of cultured endothelial cell monolayers. *J. Appl. Physiol.* **1988**, *64*, 308–322.

151. Santaguida, S.; Janigro, D.; Hossain, M.; Oby, E.; Rapp, E.; Cucullo, L. Side by side comparison between dynamic *versus* static models of blood–brain barrier *in vitro*: A permeability study. *Brain Res.* **2006**, *1109*, 1–13.

152. Torrejon, K.Y.; Pu, D.; Bergkvist, M.; Danias, J.; Sharfstein, S.T.; Xie, Y. Recreating a human trabecular meshwork outflow system on microfabricated porous structures. *Biotechnol. Bioeng.* **2013**, *110*, 3205–3218.

153. Dautriche, C.N.; Xie, Y.; Sharfstein, S.T. Walking through trabecular meshwork biology: Toward engineering design of outflow physiology. *Biotechnol. Adv.* **2014**, *32*, 971–983.

154. Helm, C.L.; Zisch, A.; Swartz, M.A. Engineered blood and lymphatic capillaries in 3-D VEGF-fibrin-collagen matrices with interstitial flow. *Biotechnol. Bioeng.* **2007**, *96*, 167–176.

155. Ng, C.P.; Helm, C.L.; Swartz, M.A. Interstitial flow differentially stimulates blood and lymphatic endothelial cell morphogenesis *in vitro. Microvasc. Res.* **2004**, *68*, 258–264.

156. Boardman, K.C.; Swartz, M.A. Interstitial flow as a guide for lymphangiogenesis. *Circ. Res.* **2003**, *92*, 801–808.

157. Niebruegge, S.; Bauwens, C.L.; Peerani, R.; Thavandiran, N.; Masse, S.; Sevaptisidis, E.; Nanthakumar, K.; Woodhouse, K.; Husain, M.; Kumacheva, E.; *et al.* Generation of human embryonic stem cell-derived mesoderm and cardiac cells using size-specified aggregates in an oxygen-controlled bioreactor. *Biotechnol. Bioeng.* **2009**, *102*, 493–507.

158. Sargent, C.Y.; Berguig, G.Y.; McDevitt, T.C. Cardiomyogenic differentiation of embryoid bodies is promoted by rotary orbital suspension culture. *Tissue Eng. A* **2009**, *15*, 331–342.

159. Jing, D.; Parikh, A.; Tzanakakis, E.S. Cardiac cell generation from encapsulated embryonic stem cells in static and scalable culture systems. *Cell Transplant.* **2010**, *19*, 1397–1412.

160. Illi, B.; Scopece, A.; Nanni, S.; Farsetti, A.; Morgante, L.; Biglioli, P.; Capogrossi, M.C.; Gaetano, C. Epigenetic histone modification and cardiovascular lineage programming in mouse embryonic stem cells exposed to laminar shear stress. *Circ. Res.* **2005**, *96*, 501–508.

161. Hazeltine, L.B.; Selekman, J.A.; Palecek, S.P. Engineering the human pluripotent stem cell microenvironment to direct cell fate. *Biotechnol. Adv.* **2013**, *31*, 1002–1091.

Human Keratoconus Cell Contractility is Mediated by Transforming Growth Factor-Beta Isoforms

Desiree' Lyon, Tina B. McKay, Akhee Sarkar-Nag, Shrestha Priyadarsini and Dimitrios Karamichos

Abstract: Keratoconus (KC) is a progressive disease linked to defects in the structural components of the corneal stroma. The extracellular matrix (ECM) is secreted and assembled by corneal keratocytes and regulated by transforming growth factor-β (TGF-β). We have previously identified alterations in the TGF-β pathway in human keratoconus cells (HKCs) compared to normal corneal fibroblasts (HCFs). In our current study, we seeded HKCs and HCFs in 3D-collagen gels to identify variations in contractility, and expression of matrix metalloproteases (MMPs) by HKCs in response the TGF-β isoforms. HKCs showed delayed contractility with decreased Collagen I:Collagen V ratios. TGF-β1 significantly increased ECM contraction, Collagen I, and Collagen V expression by HKCs. We also found that HKCs have significantly decreased Collagen I:Collagen III ratios suggesting a potential link to altered collagen isoform expression in KC. Our findings show that HKCs have significant variations in collagen secretion in a 3D collagen gel and have delayed contraction of the matrix compared to HCFs. For the first time, we utilize a collagen gel model to characterize the contractility and MMP expression by HKCs that may contribute to the pathobiology of KC.

Reprinted from *J. Funct. Biomater.* Cite as: Lyon, D.; McKay, T.B.; Sarkar-Nag, A.; Priyadarsini, S.; Karamichos, D. Human Keratoconus Cell Contractility is Mediated by Transforming Growth Factor-Beta Isoforms. *J. Funct. Biomater.* **2015**, *6*, 422–438.

1. Introduction

Keratoconus (KC) is an ecstatic corneal thinning disease that is linked to severe dysfunction in the structural and refractive properties of the cornea [1]. KC affects over 1 in 2000 people worldwide [2]. Age-onset of KC is generally early puberty to middle age and can develop into a progressive disease with detrimental effects on visual acuity [3,4]. Corneal transplantation is the most common option for severe cases [5]. While recent advancements in collagen cross-linking have provided hope for strengthening the KC cornea, its long-term effectiveness and safety has yet to be established [6–8]. The molecular pathogenesis of KC is still unclear, and there is currently no animal model for KC. We have previously developed a 3D *in vitro* model of KC disease that mimics the *in vivo* condition [9]. We have shown that human keratoconus cells (HKCs) have an altered phenotype compared to normal human corneal fibroblasts (HCFs) characterized by decreased extracellular matrix (ECM) thickness, increased expression of fibrotic markers, and elevated oxidative stress in a self-assembled 3D-model [9,10]. In our current study, we sought to investigate the HKC disease phenotype in a floating 3D collagen gel matrix in order to measure the contractility of HKCs compared to HCFs, which may provide further insight into molecular defects present in HKCs that give rise to corneal thinning.

Floating 3D collagen gels have been used to study fibrosis and contractility in various cell types, including smooth muscle cells [11], retinal pigment epithelial cells [12], and fibroblasts [13–15]. This model utilizes a detached, free-floating collagen gel to mimic the surrounding ECM found in many tissues. Moreover, this 3D model is extremely useful in identifying the role of intracellular defects, such as those observed in HKCs, which may alter the ability of fibroblasts to attach and pull the surrounding collagen ECM. In this model, the cells begin to contract the surrounding ECM, an activity characterized by the formation of stress fibers, which are responsible for the puckering, stretching, and pulling observed when scar formation occurs [16]. Contraction of the ECM by resident cells is required in normal wound healing processes to promote wound closure [17,18]. However, variations in contractility or altered response to growth factors can contribute to development of fibrosis or inability to respond to external stimuli that may delay healing and cause permanent damage to the tissue [19,20]. An altered wound healing response enacted by KC stromal keratocytes in the presence of excessive eye rubbing has been posited to play a role in KC development [21–23].

Within the healthy cornea, stromal keratocytes reside natively in an ECM composed primarily of Collagen I (Col I) and Collagen V (Col V) in a ratio of 80:20 along with small glycoproteins and crystallins [24–26]. This assembled ECM is important in regulating intracellular processes and provides the structural integrity and refractive power of the cornea [27]. Various studies have identified significant variations in collagen lamellae organization within KC corneal buttons compared to normal controls [28,29]. Furthermore, significant variations in proteoglycan and Col I within KC corneas suggest the presence of deleterious defects in secretion and assembly of the ECM within the stroma that contribute to the KC pathology [30]. Collagen III (Col III) has been found to be upregulated in KC corneal buttons with scarring [31], and we have found that HKCs secrete [32] and assemble [9] higher Col III in a 3D *in vitro* model compared to normal HCFs. Furthermore, a mutation in the Col V locus has been linked to KC development suggesting a potential genetic association between defective collagen assembly and KC [33]. These studies suggest that altered distribution of Col I, III, and V may play an important role in the altered ECM assembled in KC.

TGF-β signaling has been shown to be an important regulator of ECM secretion [34,35], cell differentiation [36,37], and proliferation [38]. There are three primary ligands, TGF-β1, -2, and -3, which are known to modulate downstream genes expression. The pro-fibrotic ligands, TGF-β1 and TGF-β2, activate the canonical TGF-β pathway leading to expression of factors indicative of myofibroblast differentiation, including α-smooth muscle actin (α-SMA) and Collagen III [39,40]. Interestingly, TGF-β3 has been identified as promoting an anti-fibrotic wound healing response with reduced expression of fibrotic markers, but increased native ECM deposition [9,41]. We have previously identified [9,42] significant defects in HKC ECM assembly and the TGF-β pathway, and therefore we sought to investigate the effects of the TGF-β ligands on contractility, collagen deposition, and matrix metalloproteases (MMP) expression by HKCs compared to normal HCFs. To date, this is the first published report using 3D collagen gels to identify novel defects present in HKCs that may contribute to structural defects and corneal thinning.

2. Results and Discussion

2.1. Contraction Profiles of HCFs and HKCs

In order to define the role of the surrounding matrix on contractility, we utilized a pre-assembled 3D collagen gel with seeded HKCs and measured rate of contraction compared to normal HCFs. We measured changes in the area of the gel matrix biweekly for 4 weeks in control, TGF-β1, -2, and -3 treated samples using light microscopy, as shown in Figure 1 for representative control samples. At day 1, we identified a 57 mm^2 (16%) reduction in gel area by HCF controls compared to a 12 mm^2 reduction (7%) in matrix area in HKCs (Figure 2A, $p < 0.0001$). By day 12, HCFs had contracted the matrix at an average rate of 20 mm^2/day compared to a contraction rate of 15 mm^3/day by HKCs (Figure 2E,F). The initial delay in contractility by HKCs corresponded to an incremental delay in shrinkage of the matrix area compared to HCFs, both of which reached maximal contraction by day 26. However, the average rate of contraction from day 0 to day 26 were comparable between HCFs and HKCs (10 mm^2/day and 9 mm^2/day, respectively) showing that though HKCs have an initial delayed contractility compared to HCFs, the KC cells eventually reach similar HCF average contraction rate (Figure 2E,F).

In order to identify if HKCs have a differential response to the TGF-β isoforms, we stimulated HCFs and HKCs seeded in the 3D-collagen gels with the three TGF-β isoforms and measured changes in contraction rate. TGF-β1 treatment had an increased effect on contractility in HCFs with a decrease by 89 mm^2 (32%) in gel area observed from day 0 to day 1 compared to a 6 mm^2 (2%) reduction by HKCs (Figure 2B, $p < 0.0001$). This delay in contraction was resolved by day 12, at which time, HKCs had contracted to 43 mm^2, or 15% of the initial area, in the presence of TGF-β1, TGF-β2, or TGF-β3, which was similar to the contraction exhibited by HCFs (Figure 2B–D). Moreover, our results show that TGF-β1 and TGF-β3 stimulate more significant contraction at day 1 with a 89 mm^2 (32%) reduction in gel size in HCFs compared to a 48 mm^2 (17%) reduction stimulated by TGF-β2 (Figure 2B–D, $p < 0.0001$). The most significant rate of contraction by HCFs occurred on day 1 following TGF-β1, -2, or -3 stimulation with rates of 88 mm^2/day, 47 mm^2/day, and 119 mm^2/day, respectively (Figure 2E). HKCs exhibited negligible contraction at day 1 but had contraction rates of 22 mm^2/day, 26 mm^2/day, and 25 mm^2/day following stimulation by TGF-β1, -2, and -3, respectively (Figure 2F). This data shows that HKCs have reduced initial contraction, but reach similar contractility by day 12 (20 mm^2/day) in the presence of TGF-β suggesting increased responsiveness by HKCs to the TGF-β isoforms compared to HCFs. Moreover, TGF-β3 significantly increased the rate of contraction at day 1 (120 mm^2/day) by HCFs compared to the control (58 mm^2/day) (Figure 2E). This data suggests that the anti-fibrotic TGF-β3 [43] increases contraction or wound closure by normal stromal fibroblasts and mediates wound healing by directly modulating ECM secretion compared to the fibrotic nature of TGF-β1 and TGF-β2. Further studies are needed to identify the molecular mechanism by which TGF-β3 exhibits anti-fibrotic properties.

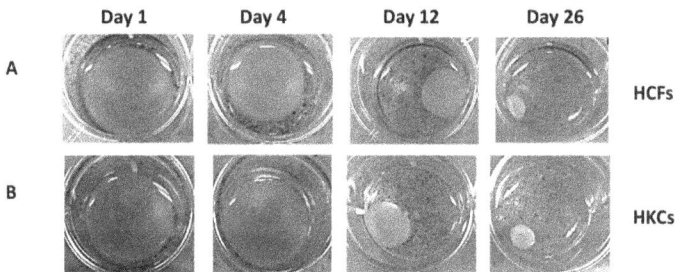

Figure 1. Floating 3D-collagen gel seeded with untreated control (**A**) human corneal fibroblasts (HCFs) and (**B**) human keratoconus cells (HKCs). Change in area of the collagen gel was measured every other day using ImageJ software. Representative images shown, $n = 3$.

Our results show that HKCs have an initial delay in contractility compared to HCFs, which suggests that HKCs are less adept to perform wound closure immediately following injury to the corneal surface. It is well-established that resident cells bind weakly to collagen fibrils directly and instead require linker-proteins, such as fibronectin, to bind to cell-surface integrins and the surrounding collagen bundles [44,45]. TGF-β1 is known to promote expression of both fibronectin [35,46] and integrin subunits important in wound healing [47–49]. Our 3D collagen model results show that HKCs are unable to establish initial binding to the collagen gel, but eventually bind and contract the matrix to similar HCF levels by day 26. This data suggests that HKCs may have altered secretion of ECM-linker proteins that delay binding to the pre-assembled ECM. TGF-β1, -2, and -3 stimulation increased the rate of contraction by HKCs and enabled similar contractility to HCFs by day 12, which suggests that TGF-β growth factors stimulate a more contractile-phenotype by HKCs, perhaps by modulating expression of fibronectin and cell-surface integrins.

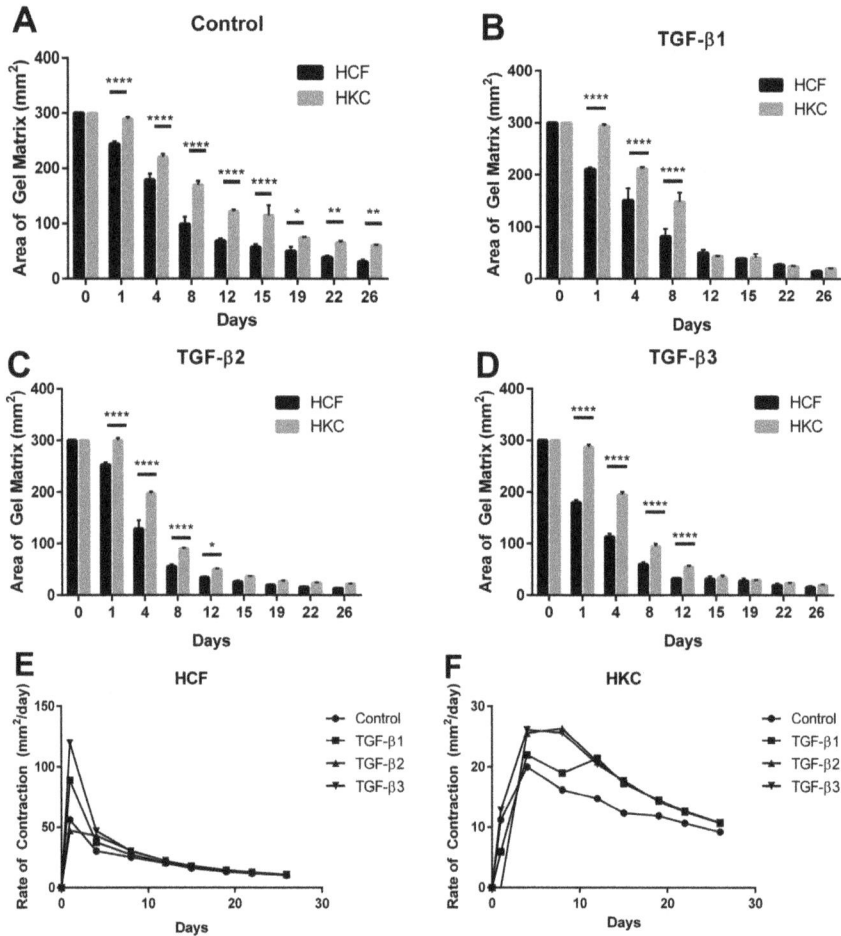

Figure 2. Quantification of the contraction of the collagen matrix in HCFs and HKCs from day 0 to 26. (**A**) control, (**B**) TGF-β1, (**C**) TGF-β2, and (**D**) TGF-β3 samples. A significant reduction in area of the collagen matrix correlates with increased contractility. Rate of contraction from day 0 to day 26 for (**E**) HCFs and (**F**) HKCs. $n = 3$, error bars represent standard error of the mean (SEM). (**** denotes $p < 0.0001$, *** denotes $p < 0.001$, ** denotes $p < 0.01$, and * denotes $p < 0.05$.)

2.2. Collagen Secretion by HCFs and HKCs

Corneal ECM organization and composition provides the structural, mechanical, and physiochemical properties that define the integrity and function of the tissue. KC is characterized by a thin corneal stroma that leads to corneal protrusion and disruption of visual acuity. The major components of the corneal stroma include collagen fibrils and the resident cell, corneal keratocytes, which secrete and assemble the surrounding matrix. The TGF-β pathway is a primary regulator of ECM production by stromal keratocytes. Several studies have identified significant defects in TGF-β signaling and ECM composition [9,43,50,51]. Col I is the dominant structural component of the corneal stroma [52]. Col V is a known regulator of collagen fibrillogenesis and is present at 20% of total collagen composition within the cornea [26,53], whereas Col III is not normally expressed in the uninjured cornea [54,55]. In order to determine the effect of the 3D-collagen gel on ECM secretion by HCFs and HKCs, we measured the amount of Col I, Col III, and Col V secreted into the media by HCFs and HKCs (Figure 3A–C). Basal secretion of Col I was reduced in HKCs by 12% compared to HCFs (Figure 3A). TGF-β1, -2, and -3 increased Col I secretion by 32%, 35%, and 52%, respectively, in HCFs, compared to an increase of 51%, 17%, and 17% by HKCs, respectively (Figure 3A, $p < 0.05$). Col III secretion did not increase significantly in HCFs following treatment with the TGF-β isoform, while HKCs showed increased Col III secretion by 49% following TGF-β2 stimulation (Figure 3B, $p < 0.05$). Col V secretion was not significantly different between the two cell types with or without TGF-β treatment suggesting a significant role for Col I and III regulation between HCFs and HKCs (Figure 3C).

Since the composition of the stromal ECM is tightly regulated and ultimately defines the structural integrity of the cornea, we measured the effect on Col I and Col V ratios. Our results show that Col I/Col V is 47% lower in control HKCs compared to HCFs (Figure 3D, $p < 0.05$). TGF-β1 treatment significantly increased this ratio by 91% in HKCs (Figure 3D, $p < 0.001$). We found that TGF-β2 and TGF-β3 treatment increased the Col I/Col V ratio in HKCs, but not HCFs, by 39% and 59%, respectively. This data shows that the TGF-β isoforms mediate increased Col I/Col V secretion by HKCs suggesting that secretion of select collagen types are regulated by TGF-β signaling, which may play an important role in the wound healing response within the KC cornea. We also measured the effect of the TGF-β isoforms on Col I/Col III secretion in both cell types. TGF-β1 and TGF-β2 treatment did not significantly increase the Col I/Col III ratio in HCFs (Figure 3E). However, TGF-β3 treatment increased this ratio by 67% (Figure 3E, $p < 0.01$) suggesting that TGF-β3 is a potent regulator of expression of specific collagen isoforms by normal HCFs. Col I/Col III increased significantly in HKCs following TGF-β1 stimulation by 70% with a lack of change in this ratio with TGF-β2 or -3 treatment (Figure 3E, $p < 0.05$).

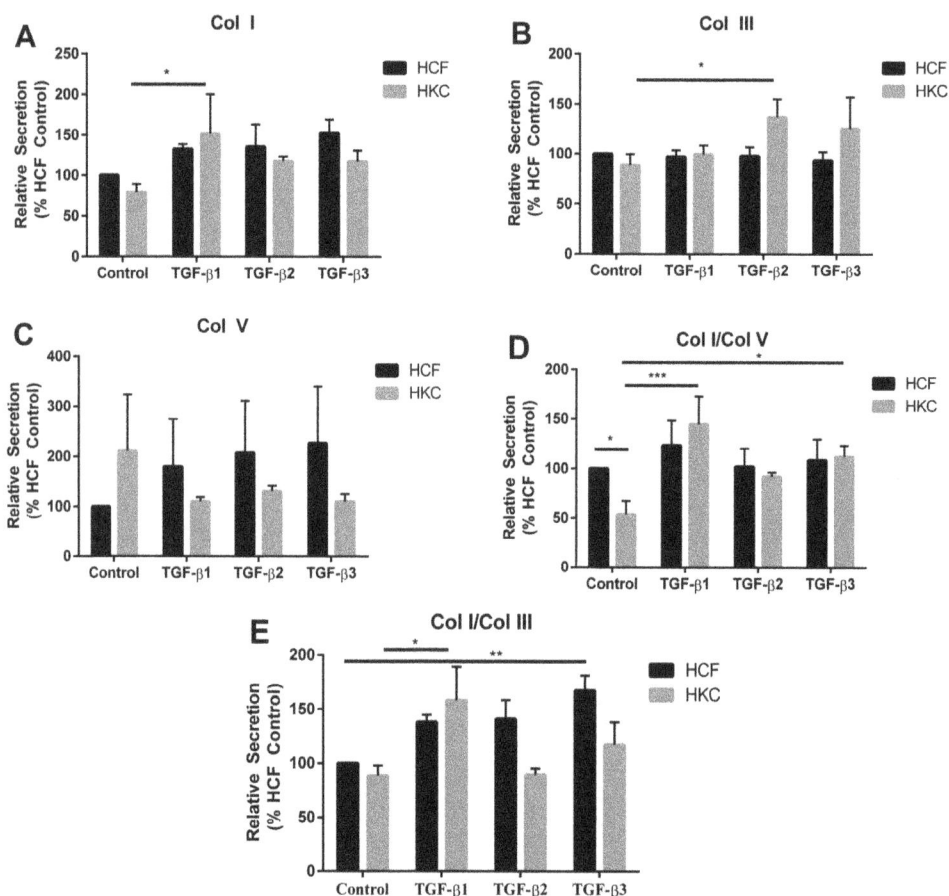

Figure 3. (**A**) Collagen I (Col I), (**B**) Collagen III (Col III), and (**C**) Collagen V (Col V) secretion measured from conditioned media by Western blot from week 1 to week 4. Data reported as ratios of (**D**) Col I/Col V and (**E**) Col I/Col III. $n = 3$. Error bars represent standard error of the mean. (*** denotes $p < 0.001$, ** denotes $p < 0.01$, and * denotes $p < 0.05$.)

Alterations in ratios of collagen isoforms from normal distributions are known to contribute to corneal dystrophies [31,56]. We have previously shown that HKCs synthesize a significantly thinner ECM compared to normal HCFs [9]. In the collagen gel, seeded HKCs secrete lower Col I/Col V, of which Col V is known to be essential for lamellae formation within the cornea [57], suggesting that collagen fibrillogenesis may be directly modulated in KC. We found that TGF-β isoform treatments increased the basal Col I/Col V ratio to HCF levels, suggesting that modulating TGF-β signaling may alter ECM secretion in KC.

2.3. mRNA Expression of Collagen I, III, and V by HCFs and HKCs

In order to determine if expression of pro-collagens correlated with collagen secretion detected in the conditioned media, we quantified the expression of Col I, Col III, and Col V, using RT-PCR in HCF and HKC at day 26 following complete contraction of the matrix. We found an increase in all three collagen types by control HKCs compared to HCFs (Figure 4A–C). We also identified a significant increase of 638% and 994% in expression of Col I and Col III, respectively, by HKCs following TGF-β1 stimulation (Figure 4A,B, $p < 0.05$). In contrast, HCFs did not significantly increase Col I, Col III, or Col V expression following stimulation with TGF-β isoforms (Figure 4A–C). Our results show that HKCs are more responsive to TGF-β isoform treatment compared to HCFs at day 26. This data suggests that fully contracted HCFs have reduced expression of ECM components compared to HKCs.

We also quantified the ratios of collagens expressed in the fully contracted ECM. At day 26, HKCs had increased Col III and Col V expression compared to Col I (Figure 4A–C), which is the dominant collagen isoform produced by normal HCFs. We measured similar Col I/Col V by HCFs and HKCs following full contraction (Figure 4D). Interestingly, we found a substantial increase in Col I/Col V ratio by HCFs following TGF-β3 stimulation. Both TGF-β1 and TGF-β3 increased the Col I/Col V ratio in HKCs by 100%, whereas TGF-β3 increased Col I/Col V secretion by 600% in HCFs (Figure 4D, $p < 0.01$). Col I/Col III ratio was significantly reduced by >50% in HKCs in the presence and absence of the TGF-β isoforms (Figure 4E, $p < 0.05$). A significant downregulation of Col I/Col III expression was noted, by 45% in control HKCs compared to HCFs (Figure 4E, $p < 0.05$). Moreover, we found significant downregulation of Col I/Col III secretion in both cell types following TGF-β1 stimulation supporting the conclusion that TGF-β1 acts as a pro-fibrotic ligand within the corneal stroma. These results show that the Col I/Col III ratio expressed by HKCs in the fully contracted matrix is significantly lower than that of the normal HCFs.

Changes in the ratios of specific collagen types can affect the structural integrity of the ECM and contribute to pathological defects in tissue structure, such as those observed in KC. We found a significant reduction in Col I/Col III ratio by HKCs compared to HCFs. This data supports earlier findings that HKCs have a myofibroblast phenotype that promotes altered ECM structure [9,43]. We have identified that HKCs have defective TGF-β signaling that contributes to expression of pro-fibrotic markers [43]. Our results in this study show that HKCs have increased responsiveness to TGF-β1, -2, and -3 stimulation with increased contractility and Col I/Col V ratios, which alters the native composition and assembly of the surrounding matrix. Since Col V is known to be important in collagen fibrillogenesis [26,53] variations in its expression would be expected to directly affect lamellae assembly. The aberrant expression of Col I/Col V and Col I/Col III in HKCs may be a source of pathogenesis and should be explored further to identify the effects on structural integrity of the KC stroma.

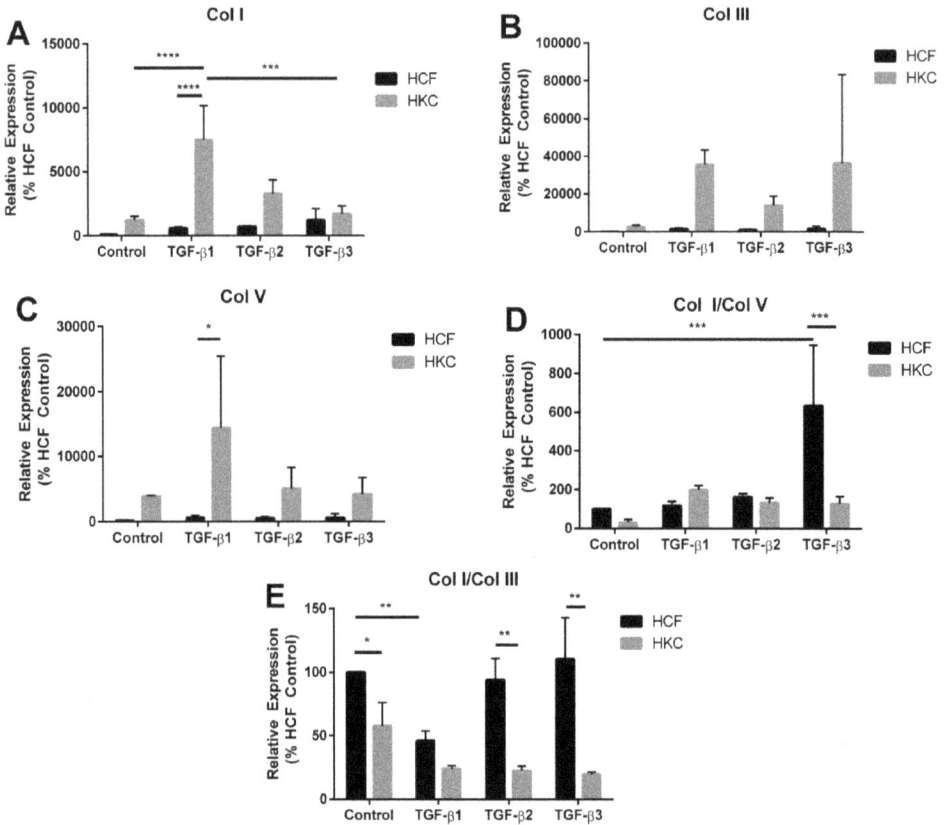

Figure 4. (**A**) Collagen I (Col I), (**B**) Collagen III (Col III), and (**C**) Collagen V (Col V) expression and (**D–E**) ratios of Col I/Col V and Col I/Col III by HCFs and HKCs at week 4 measured by RT-PCR. $n = 3$, error bars represent standard error of the mean. (**** denotes $p < 0.0001$, *** denotes $p < 0.001$, ** denotes $p < 0.01$ and * denotes $p < 0.05$.)

2.4. MMP1 and MMP3 Expression by HCFs and HKCs

MMPs are important in ECM degradation and remodeling within tissues [58]. Increased MMP activity has been posited to play a role in KC disease progression [59,60]. Previous studies have linked upregulation of MMP1 in KC corneal buttons suggesting that degradation of the resident stromal collagen may contribute to KC pathogenesis [61,62]. Furthermore, MMP1 gene expression is transcriptionally regulated with MMP3 gene expression [63], which has yet to be linked to KC. Since

KC is associated with thinning of the corneal stroma, we measured expression of MMP1 and MMP3, which are important mediators of ECM degradation in tissues [58]. Interestingly, we measured a 10-fold increase of MMP1 expression in HKCs compared to HCFs (Figure 5A, $p < 0.01$). There was no significant difference in basal expression of MMP3 between HCFs and HKCs (Figure 5B). We found a significant increase in MMP1 expression by over 10-fold with TGF-β1 treatment in both HCFs and HKCs (Figure 5A, $p < 0.01$). TGF-β2 and TGF-β3 increased MMP1 expression by 9-fold and 19-fold, respectively, in HCFs compared to a 2.4-fold and 5.3-fold increase in HKCs (Figure 5A). MMP3 expression also increased in both cell types with TGF-β1, -2, and -3 isoform treatment with a 5.5-fold, 2.4-fold, 5.3-fold, respectively, in HCFs and 8-fold, 4.4-fold, 5.9-fold, respectively, increase in HKCs (Figure 5B).

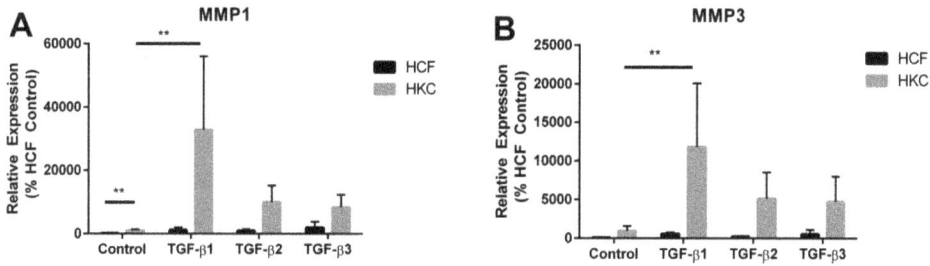

Figure 5. (**A**) MMP1 and (**B**) MMP3 expression in HCFs and HKCs measured by RT-PCR at week 4. $n = 3$, error bars represent standard error of the mean. (** denotes $p < 0.01$.)

We measured a significant increase in MMP1 expression, which agreed with earlier reports showing upregulation of MMP1 in corneal buttons [61,62]. Basal expression of MMP3 was not significantly different between HKCs and HCFs, which suggests that MMP3 does not play a prominent role in KC pathogenesis. We measured a significant increase in MMP1 and MMP3 expression in both cell types following stimulation with the TGF-β isoforms. Furthermore, the TGF-β isoforms regulate MMP1 and MMP3 expression in a similar manner between the two cell types. This data supports published reports [64,65] showing that TGF-β signaling increases MMP gene transcription. Our results suggest that an increase in basal expression of MMP1 and MMP3 may play a role in KC development; however, further work is warranted to determine if the altered ECM assembled by HKCs is primarily a result of ECM secretion, rather than degradation. Our data also suggests that activation of MMP expression via TGF-β stimulation following activation of the wound healing process may contribute to an increase in ECM degradation, which may be important in KC.

3. Experimental Section

3.1. Cell Culture

Corneas were obtained by the National Disease Research Interchange (NDRI) and processed as previously described [9,32,66]. Briefly, the endothelium and epithelium were removed by scraping briefly with a razor blade, they were then cut into ~ 2 mm × 2 mm pieces. The pieces of stroma were allowed to adhere to the bottom of a T75 flask for 30 minutes at 37 degrees Celsius before adding 10% Fetal Bovine Serum (FBS) Eagle's Minimum Essential Media (EMEM) to the flask. After 2–4 weeks the explants were passaged in 10% FBS in EMEM.

3.2. Collagen Contraction Assay

Rat-tail Collagen type I (Advanced Biomatrix) was mixed with EMEM on ice with 125 μL EMEM per 1 mL Collagen. The pH was then adjusted to pH 7–8 with 1 M NaOH. HCFs or HKCs were added at a concentration of 5×10^5 and mixed slowly to avoid air bubbles. This mixture was plated in a 12 well plate at 1 mL per well and incubated in 37 °C for 30 min to promote solidification. After congealing 1 mL of 10% FBS EMEM was added on top of the construct. The collagen matrix constructs were released after 48 h of incubation by running a sterile blade around the edges of the well. Contraction was measured every other day for 4 weeks starting at 24 h after the initial release. Treated media was supplemented with 0.1 ng/mL of TGF-β1, TGF-β2, or TGF-β3, and the area of the gel was quantified using ImageJ software following imaging by camera. Changes in contraction were measured from day 0 to day 26. The constructs were fully contracted by day 26, and we did not observe any reduction in gel area after day 26.

3.3. RT-PCR

Fully contracted constructs at day 26 were placed into 1 mL Trizol and incubated at 22 °C for 5 min. 200 μL chloroform was added before shaking vigorously and centrifuging for 15 min at 1200 rpm. The supernatant was further purified using the Ambion RNA kit (Life Technologies, Carlsbad, CA, USA), following the protocol given, with the RNA being dissolved in 30 μL RNase free water. The LVis plate (Clariostar, BMG Labtech, Ortenberg, Germany) was used to measure the concentration and purity of the extracted RNA. A 10% solution of cDNA was made with RNase free water to use for the PCR. While a ratio of 10:7 master mix to RNase free water was made along with 2 μL of a 10% cDNA sample solution and 1 μL of Taqman gene specific assay (Life Technologies) per well. This was quantified using mean cT values obtained from life technologies Real Time Thermal Cycler with standards conditions for Taqman gene expression probes (Applied Biosystems, Foster City, CA, USA) for 40 cycles. The following probes were purchased from Life Technologies: MMP1 (Hs00899658_m1), MMP3 (Hs00968305_m1), and Collagen I (Hs00164004_m1), Collagen III (Hs00943809_m1), and Collagen V (Hs00609133_m1). GAPDH (Hs99999905_m1) and 18S (Hs99999901_s1) probes were used as endogenous controls (Table 1).

Table 1. RT-PCR probes and their concentrations.

Probe	Catalogue #	Company	Final Concentration
GAPDH	Hs99999905_m1	Life Technologies	1×
18S	Hs99999901_s1	Life Technologies	1×
Col I	Hs00164004_m1	Life Technologies	1×
Col III	Hs00943809_m1	Life Technologies	1×
Col V	Hs00609133_m1	Life Technologies	1×
MMP 1	Hs00899658_m1	Life Technologies	1×
MMP 3	Hs00968305_m1	Life Technologies	1×

3.4. Western Blot

Western Blot was performed on media collected from the contracting matrix at 1 week. Total protein content within conditioned media was measured using a BCA assay (ThermoScientific, Rockford, IL, USA). Samples were then normalized to the sample containing the lowest protein content, thereby enabling equal loading onto the gel. Media samples were then run on a 4%–20% pre-cast polyacrylamide gradient gel at 130 V for 1.5 h then transferred to a nitrocellulose membrane on ice at 100 V for 1 h. The membrane was blocked in a 5% milk solution in Tris-buffered Solution with Tween20 for 1 h, then incubated overnight in a cold room with 1:1000 primary antibody. Antibodies used include: Collagen (ab34710; Abcam, Cambridge, MA, USA), Collagen III (ab7778; Abcam), Collagen V (ab94673; Abcam) (Table 2). After primary incubation, the membrane was washed for 5 min (3×) in Tris-buffered Solution with Tween20 before probing with secondary antibody Goat anti-Rb Alexafluor 568 (Life Technologies, Grand Island, NY, USA) at room temperature for 1 h with rocking. The membrane was allowed to dry before imaging using ChemiDoc-it to image. Western blots were quantified using densitometry utilizes pixels measured within each band.

Table 2. Western blot antibodies and final dilutions.

Antibody	Catalogue #	Company	Dilution
Col I	ab34710	Abcam, Cambridge, MA, USA	1/1000
Col III	ab7778	Abcam, Cambridge, MA, USA	1/1000
Col V	ab94673	Abcam, Cambridge, MA,USA	1/1000

3.5. Statistical Analysis

Statistical analyses were carried out using a two-way ANOVA test calculated by GraphPad Prism software. $p < 0.05$ were considered statistically significant. Error bars represent standard error of the mean. Data is representative of three independent experiments.

4. Conclusions

In this study, we found that HKCs have a significant reduction in initial contractility of the matrix and altered Collagen expression compared to HCFs. Contraction of the ECM is important in normal wound healing processes within the cornea [67,68]. The defect in contraction exhibited by HKCs suggests that stromal fibroblasts in KC corneas are less able to respond to external stimuli and have delayed closure of the surrounding matrix following wounding. The failure to respond properly to normal wound healing mechanisms following injury can cause significant pathologies within the cornea [69,70]. The role of eye rubbing in KC development has been posited [23], but has yet to be thoroughly explored as the causative agent of KC pathogenesis. Our study suggests that HKCs have reduced contractility and thereby are less able to perform normal wound closure within the cornea following trauma, which may occur following continual eye rubbing. Clearly, further work is warranted to identify the molecular defects present in HKCs that contribute to this phenotype. The TGF-β isoforms have been detected in the human tear film, with TGF-β1 as the dominant isoform [71]. In our study, we found that the TGF-β isoforms mediate accelerated contraction of the matrix by HKCs up to HCF levels supporting the potential role of altered TGF-β signaling in KC pathobiology. Moreover, HKCs exhibited lower Col I/Col III and Col I/Col V ratios compared to HCFs, suggesting a significant defect in collagen deposition by HKCs that may support a defected corneal stroma ECM. Our results show that MMP1, but not MMP3, was elevated in HKCs compared to HCFs suggesting that MMP1 may play a significant role in the KC pathology. Overall, our study identified novel defects in HKCs that give rise to altered ECM contractility and composition that may contribute to the pathological ECM present in KC. In future studies, we will identify the molecular mechanism supporting reduced contractility by HKCs and relate this data to the KC phenotype *in vivo*.

Acknowledgments: This work was supported by the National Institutes of Health Grants/National Eye Institute 5R01EY023568 and 5R01EY020886 (D.K.) and, in part, by an unrestricted grant (DMEI) from Research to Prevent Blindness (New York, NY, USA). We acknowledge the assistance and support of the NEI/DMEI Cellular Imaging Core Facility at OUHSC (P30EY021725). Research reported in this publication was supported by the National Eye Institute of the NIH under award number T32EY023202.

Author Contributions: Individual contributions: D.K. and D.L. conceived and designed the experiments; D.L., A.S. and S.P. performed the experiments; T.B.M., D.K., D.L. and A.S. analyzed the data; D.K. contributed reagents/materials/analysis tools; T.B.M., D.K., and D.L. wrote the paper.

Conflicts of Interest: The authors declare no conflict of interest.

References

1. Ambekar, R.; Toussaint, K.C., Jr.; Wagoner Johnson, A. The effect of keratoconus on the structural, mechanical, and optical properties of the cornea. *J. Mech. Behav. Biomed. Mater.* **2011**, *4*, 223–236.

2. Kennedy, R.H.; Bourne, W.M.; Dyer, J.A. A 48-year clinical and epidemiologic study of keratoconus. *Am. J. Ophthalmol.* **1986**, *101*, 267–273.

3. Ertan, A.; Muftuoglu, O. Keratoconus clinical findings according to different age and gender groups. *Cornea* **2008**, *27*, 1109–1113.

4. Jiménez, J.L.O.; Jurado, J.C.G.; Rodriguez, F.J.B.; Laborda, D.S. Keratoconus: Age of onset and natural history. *Optom. Vis. Sci.* **1997**, *74*, 147–151.

5. Romero-Jimenez, M.; Santodomingo-Rubido, J.; Wolffsohn, J.S. Keratoconus: A review. *Cont. Lens. Anterior. Eye.* **2010**, *33*, 157–166.

6. Ghanem, R.C.; Santhiago, M.R.; Berti, T.; Netto, M.V.; Ghanem, V.C. Topographic, corneal wavefront, and refractive outcomes 2 years after collagen crosslinking for progressive keratoconus. *Cornea* **2014**, *33*, 43–48.

7. Greenstein, S.A.; Fry, K.L.; Hersh, P.S. Corneal topography indices after corneal collagen crosslinking for keratoconus and corneal ectasia: One-year results. *J. Cataract Refract. Surg.* **2011**, *37*, 1282–1290.

8. Lesniak, S.P.; Hersh, P.S. Transepithelial corneal collagen crosslinking for keratoconus: Six-month results. *J. Cataract Refract. Surg.* **2014**, *40*, 1971–1979.

9. Karamichos, D.; Zareian, R.; Guo, X.; Hutcheon, A.E.; Ruberti, J.W.; Zieske, J.D. Novel model for keratoconus disease. *J. Funct. Biomater.* **2012**, *3*, 760–775.

10. Karamichos, D.; Hutcheon, A.E.; Rich, C.B.; Trinkaus-Randall, V.; Asara, J.M.; Zieske, J.D. *In vitro* model suggests oxidative stress involved in keratoconus disease. *Sci. Rep.* **2014**, *4*.

11. Kropp, B.P.; Zhang, Y.; Tomasek, J.J.; Cowan, R.; Furness, P.D.; Vaughan, M.B.; Parizi, M.; Cheng, E.Y. Characterization of cultured bladder smooth muscle cells: Assessment of *in vitro* contractility. *J. Urol.* **1999**, *162*, 1779–1784.

12. Kimura, K.; Orita, T.; Fujitsu, Y.; Liu, Y.; Wakuta, M.; Morishige, N.; Suzuki, K.; Sonoda, K.H. Inhibition by female sex hormones of collagen gel contraction mediated by retinal pigment epithelial cells. *Invest. Ophthalmol. Vis. Sci.* **2014**, *55*, 2621–2630.

13. Pilcher, B.K.; Kim, D.W.; Carney, D.H.; Tomasek, J.J. Thrombin stimulates fibroblast-mediated collagen lattice contraction by its proteolytically activated receptor. *Exp. Cell Res.* **1994**, *211*, 368–373.

14. Bell, E.; Ivarsson, B.; Merrill, C. Production of a tissue-like structure by contraction of collagen lattices by human fibroblasts of different proliferative potential *in vitro*. *Proc. Natl. Acad. Sci. USA* **1979**, *76*, 1274–1278.

15. Levi-Schaffer, F.; Garbuzenko, E.; Rubin, A.; Reich, R.; Pickholz, D.; Gillery, P.; Emonard, H.; Nagler, A.; Maquart, F.A. Human eosinophils regulate human lung- and skin-derived fibroblast properties *in vitro*: A role for transforming growth factor beta (TGF-beta). *Proc. Natl. Acad. Sci. USA* **1999**, *96*, 9660–9665.

16. Witte, M.B.; Barbul, A. General principles of wound healing. *Surg. Clin. North Am.* **1997**, *77*, 509–528.

17. Montesano, R.; Orci, L. Transforming growth factor beta stimulates collagen-matrix contraction by fibroblasts: Implications for wound healing. *Proc. Natl. Acad. Sci. USA* **1988**, *85*, 4894–4897.

18. Germain, L.; Jean, A.; Auger, F.A.; Garrel, D.R. Human wound healing fibroblasts have greater contractile properties than dermal fibroblasts. *J. Surg. Res.* **1994**, *57*, 268–273.

19. Diegelmann, R.F.; Evans, M.C. Wound healing: An overview of acute, fibrotic and delayed healing. *Front. Biosci.* **2004**, *9*, 283–289.

20. Werner, S.; Grose, R. Regulation of wound healing by growth factors and cytokines. *Physiol. Rev.* **2003**, *83*, 835–870.

21. Cheung, I.M.; McGhee, C.N.; Sherwin, T. A new perspective on the pathobiology of keratoconus: Interplay of stromal wound healing and reactive species-associated processes. *Clin. Exp. Optom.* **2013**, *96*, 188–196.

22. Cheung, I.M.; McGhee, C.; Sherwin, T. Deficient repair regulatory response to injury in keratoconic stromal cells. *Clin. Exp. Optom.* **2014**, *97*, 234–239.

23. McMonnies, C.W. Mechanisms of rubbing-related corneal trauma in keratoconus. *Cornea* **2009**, *28*, 607–615.

24. Segev, F.; Heon, E.; Cole, W.G.; Wenstrup, R.J.; Young, F.; Slomovic, A.R.; Rootman, D.S.; Whitaker-Menezes, D.; Chervoneva, I.; Birk, D.E. Structural abnormalities of the cornea and lid resulting from collagen v mutations. *Invest. Ophthalmol. Vis. Sci.* **2006**, *47*, 565–573.

25. Gordon, M.K.; Foley, J.W.; Birk, D.E.; Fitch, J.M.; Linsenmayer, T.F. Type v collagen and bowman's membrane. Quantitation of mrna in corneal epithelium and stroma. *J. Biol. Chem.* **1994**, *269*, 24959–24966.

26. Sun, M.; Chen, S.; Adams, S.M.; Florer, J.B.; Liu, H.; Kao, W.W.; Wenstrup, R.J.; Birk, D.E. Collagen v is a dominant regulator of collagen fibrillogenesis: Dysfunctional regulation of structure and function in a corneal-stroma-specific col5a1-null mouse model. *J. Cell Sci.* **2011**, *124*, 4096–4105.

27. Ruberti, J.W.; Roy, A.S.; Roberts, C.J. Corneal biomechanics and biomaterials. *Annu. Rev. Biomed. Eng.* **2011**, *13*, 269–295.

28. Akhtar, S.; Bron, A.J.; Salvi, S.M.; Hawksworth, N.R.; Tuft, S.J.; Meek, K.M. Ultrastructural analysis of collagen fibrils and proteoglycans in keratoconus. *Acta Ophthalmol. (Copenh.)* **2008**, *86*, 764–772.

29. Meek, K.M.; Tuft, S.J.; Huang, Y.; Gill, P.S.; Hayes, S.; Newton, R.H.; Bron, A.J. Changes in collagen orientation and distribution in keratoconus corneas. *Invest. Ophthalmol. Vis. Sci.* **2005**, *46*, 1948–1956.

30. Chaerkady, R.; Shao, H.; Scott, S.-G.; Pandey, A.; Jun, A.S.; Chakravarti, S. The keratoconus corneal proteome: Loss of epithelial integrity and stromal degeneration. *J. Proteomics* **2013**, *87*, 122–131.

31. Delaigue, O.; Arbeille, B.; Lemesle, M.; Roingeard, P.; Rossazza, C. Quantitative analysis of immunogold labellings of collagen types I, III, IV and Vi in healthy and pathological human corneas. *Graefe's Arch. Clin. Exp. Ophthalmol.* **1995**, *233*, 331–338.

32. McKay, T.B.; Lyon, D.; Sarker-Nag, A.; Priyadarsini, S.; Asara, J.M.; Karamichos, D. Quercetin attenuates lactate production and extracellular matrix secretion in keratoconus. *Sci. Rep.* **2015**, *5*.

33. Li, X.; Bykhovskaya, Y.; Canedo, A.L.; Haritunians, T.; Siscovick, D.; Aldave, A.J.; Szczotka-Flynn, L.; Iyengar, S.K.; Rotter, J.I.; Taylor, K.D.; *et al.* Genetic association of COL5A1 variants in keratoconus patients suggests a complex connection between corneal thinning and keratoconus. *Invest. Ophthalmol. Vis. Sci.* **2013**, *54*, 2696–2704.

34. Clark, R.A.; McCoy, G.A.; Folkvord, J.M.; McPherson, J.M. TGF-beta 1 stimulates cultured human fibroblasts to proliferate and produce tissue-like fibroplasia: A fibronectin matrix-dependent event. *J. Cell. Physiol.* **1997**, *170*, 69–80.

35. Roberts, C.J.; Birkenmeier, T.M.; McQuillan, J.J.; Akiyama, S.K.; Yamada, S.S.; Chen, W.T.; Yamada, K.M.; McDonald, J.A. Transforming growth factor beta stimulates the expression of fibronectin and of both subunits of the human fibronectin receptor by cultured human lung fibroblasts. *J. Biol. Chem.* **1988**, *263*, 4586–4592.

36. Zhou, L.; Lopes, J.E.; Chong, M.M.; Ivanov, I.I.; Min, R.; Victora, G.D.; Shen, Y.; Du, J.; Rubtsov, Y.P.; Rudensky, A.Y.; *et al.* TGF-beta-induced Foxp3 inhibits T(H)17 cell differentiation by antagonizing rorgammat function. *Nature* **2008**, *453*, 236–240.

37. Massague, J.; Xi, Q. TGF-beta control of stem cell differentiation genes. *FEBS Lett.* **2012**, *586*, 1953–1958.

38. Huang, S.S.; Huang, J.S. TGF-beta control of cell proliferation. *J. Cell. Biochem.* **2005**, *96*, 447–462.

39. Connor, T.B., Jr.; Roberts, A.B.; Sporn, M.B.; Danielpour, D.; Dart, L.L.; Michels, R.G.; de Bustros, S.; Enger, C.; Kato, H.; Lansing, M.; *et al.* Correlation of fibrosis and transforming growth factor-beta type 2 levels in the eye. *J. Clin. Invest.* **1989**, *83*, 1661–1666.

40. Nakatsukasa, H.; Nagy, P.; Evarts, R.P.; Hsia, C.C.; Marsden, E.; Thorgeirsson, S.S. Cellular distribution of transforming growth factor-beta 1 and procollagen types I, III, and IV transcripts in carbon tetrachloride-induced rat liver fibrosis. *J. Clin. Invest.* **1990**, *85*, 1833–1843.

41. Chang, Z.; Kishimoto, Y.; Hasan, A.; Welham, N.V. TGF-beta 3 modulates the inflammatory environment and reduces scar formation following vocal fold mucosal injury in rats. *Dis. Model. Mech.* **2014**, *7*, 83–91.

42. Priyadarsini, S.; Hjortdal, J.; Sarker-Nag, A.; Sejersen, H.; Asara, J.M.; Karamichos, D. Gross cystic disease fluid protein-15/prolactin-inducible protein as a biomarker for keratoconus disease. *PLoS One* **2014**, *9*.

43. Karamichos, D.; Hutcheon, A.E.K.; Zieske, J.D. Transforming growth factor-β3 regulates assembly of a non-fibrotic matrix in a 3D corneal model. *J. Tissue Eng. Regen. Med.* **2011**, *5*, e228–e238.

44. Bystrom, B.; Carracedo, S.; Behndig, A.; Gullberg, D.; Pedrosa-Domellof, F. Alpha11 integrin in the human cornea: Importance in development and disease. *Invest. Ophthalmol. Vis. Sci.* **2009**, *50*, 5044–5053.

45. Parapuram, S.K.; Huh, K.; Liu, S.; Leask, A. Integrin beta1 is necessary for the maintenance of corneal structural integrity. *Invest. Ophthalmol. Vis. Sci.* **2011**, *52*, 7799–7806.

46. Weston, B.S.; Wahab, N.A.; Mason, R.M. CTGF mediates TGF-beta-induced fibronectin matrix deposition by upregulating active alpha5beta1 integrin in human mesangial cells. *J. Am. Soc. Nephrol.* **2003**, *14*, 601–610.

47. Zambruno, G.; Marchisio, P.C.; Marconi, A.; Vaschieri, C.; Melchiori, A.; Giannetti, A.; De Luca, M. Transforming growth factor-beta 1 modulates beta 1 and beta 5 integrin receptors and induces the de novo expression of the alpha v beta 6 heterodimer in normal human keratinocytes: Implications for wound healing. *J. Cell Biol.* **1995**, *129*, 853–865.

48. Kagami, S.; Kuhara, T.; Yasutomo, K.; Okada, K.; Loster, K.; Reutter, W.; Kuroda, Y. Transforming growth factor-beta (TGF-beta) stimulates the expression of beta1 integrins and adhesion by rat mesangial cells. *Exp. Cell Res.* **1996**, *229*, 1–6.

49. Wang, D.; Zhou, G.H.; Birkenmeier, T.M.; Gong, J.; Sun, L.; Brattain, M.G. Autocrine transforming growth factor beta 1 modulates the expression of integrin alpha 5 beta 1 in human colon carcinoma FET cells. *J. Biol. Chem.* **1995**, *270*, 14154–14159.

50. Maier, P.; Broszinski, A.; Heizmann, U.; Bohringer, D.; Reinhardau, T. Active transforming growth factor-beta2 is increased in the aqueous humor of keratoconus patients. *Mol. Vis.* **2007**, *13*, 1198–1202.

51. Saee-Rad, S.; Raoofian, R.; Mahbod, M.; Miraftab, M.; Mojarrad, M.; Asgari, S.; Rezvan, F.; Hashemi, H. Analysis of superoxide dismutase 1, dual-specificity phosphatase 1, and transforming growth factor, beta 1 genes expression in keratoconic and non-keratoconic corneas. *Mol. Vis.* **2013**, *19*, 2501–2507.

52. Nakayasu, K.; Tanaka, M.; Konomi, H.; Hayashi, T. Distribution of types I, II, III, IV and V collagen in normal and keratoconus corneas. *Ophthalmic Res.* **1986**, *18*, 1–10.

53. Wenstrup, R.J.; Florer, J.B.; Brunskill, E.W.; Bell, S.M.; Chervoneva, I.; Birk, D.E. Type v collagen controls the initiation of collagen fibril assembly. *J. Biol. Chem.* **2004**, *279*, 53331–53337.

54. Von der Mark, K.; von der Mark, H.; Timpl, R.; Trelstad, R.L. Immunofluorescent localization of collagen types i, ii, and iii in the embryonic chick eye. *Dev. Biol.* **1977**, *59*, 75–85.

55. Malley, D.S.; Steinert, R.F.; Puliafito, C.A.; Dobi, E.T. Immunofluorescence study of corneal wound healing after excimer laser anterior keratectomy in the monkey eye. *Arch. Ophthalmol.* **1990**, *108*, 1316–1322.

56. Robert, L.; Legeais, J.M.; Robert, A.M.; Renard, G. Corneal collagens. *Path. Biol.* **2001**, *49*, 353–363.

57. Birk, D.E.; Fitch, J.M.; Babiarz, J.P.; Doane, K.J.; Linsenmayer, T.F. Collagen fibrillogenesis *in vitro*: Interaction of types I and V collagen regulates fibril diameter. *J. Cell Sci.* **1990**, *95*, 649–657.

58. Page-McCaw, A.; Ewald, A.J.; Werb, Z. Matrix metalloproteinases and the regulation of tissue remodelling. *Nat. Rev. Mol. Cell Biol.* **2007**, *8*, 221–233.

59. Collier, S.A. Is the corneal degradation in keratoconus caused by matrix-metalloproteinases? *Clin. Exp. Ophthalmol.* **2001**, *29*, 340–344.

60. Smith, V.A.; Hoh, H.B.; Littleton, M.; Easty, D.L. Over-expression of a gelatinase a activity in keratoconus. *Eye* **1995**, *9*, 429–433.

61. Seppala, H.P.; Maatta, M.; Rautia, M.; Mackiewicz, Z.; Tuisku, I.; Tervo, T.; Konttinen, Y.T. EMMPRIN and MMP-1 in keratoconus. *Cornea* **2006**, *25*, 325–330.

62. Mackiewicz, Z.; Maatta, M.; Stenman, M.; Konttinen, L.; Tervo, T.; Konttinen, Y.T. Collagenolytic proteinases in keratoconus. *Cornea* **2006**, *25*, 603–610.

63. Li, M.; Moeen Rezakhanlou, A.; Chavez-Munoz, C.; Lai, A.; Ghahary, A. Keratinocyte-releasable factors increased the expression of mmp1 and mmp3 in co-cultured fibroblasts under both 2D and 3D culture conditions. *Mol. Cell. Biochem.* **2009**, *332*, 1–8.

64. Gomes, L.R.; Terra, L.F.; Wailemann, R.A.; Labriola, L.; Sogayar, M.C. TGF-beta1 modulates the homeostasis between MMPs and MMP inhibitors through p38 MAPK and ERK1/2 in highly invasive breast cancer cells. *BMC Cancer* **2012**, *12*.

65. Zhu, G.; Kang, L.; Wei, Q.; Cui, X.; Wang, S.; Chen, Y.; Jiang, Y. Expression and regulation of MMP1, MMP3, and MMP9 in the chicken ovary in response to gonadotropins, sex hormones, and TGFB1. *Biol. Reprod.* **2014**, *90*.

66. Karamichos, D.; Guo, X.Q.; Hutcheon, A.E.; Zieske, J.D. Human corneal fibrosis: An *in vitro* model. *Invest. Ophthalmol. Vis. Sci.* **2010**, *51*, 1382–1388.

67. Midwood, K.S.; Williams, L.V.; Schwarzbauer, J.E. Tissue repair and the dynamics of the extracellular matrix. *Int. J. Biochem. Cell Biol.* **2004**, *36*, 1031–1037.

68. Netto, M.V.; Mohan, R.R.; Ambrósio, R.J.; Hutcheon, A.E.K.; Zieske, J.D.; Wilson, S.E. Wound healing in the cornea: A review of refractive surgery complications and new prospects for therapy. *Cornea* **2005**, *24*, 509–522.

69. Wilson, S.E.; Netto, M.; Ambrósio, R. Corneal cells: Chatty in development, homeostasis, wound healing, and disease. *Am. J. Ophthalmol.* **2003**, *136*, 530–536.

70. Wilson, S.E.; Kim, W.-J. Keratocyte apoptosis: Implications on corneal wound healing, tissue organization, and disease. *Investig. Ophthalmol. Vis. Sci.* **1998**, *39*, 220–226.

71. Gupta, A.; Monroy, D.; Ji, Z.; Yoshino, K.; Huang, A.; Pflugfelder, S.C. Transforming growth factor beta-1 and beta-2 in human tear fluid. *Curr. Eye Res.* **1996**, *15*, 605–614.

Incorporation of Human Recombinant Tropoelastin into Silk Fibroin Membranes with the View to Repairing Bruch's Membrane

Audra M. A. Shadforth, Shuko Suzuki, Raphaelle Alzonne, Grant A. Edwards, Neil A. Richardson, Traian V. Chirila and Damien G. Harkin

Abstract: *Bombyx mori* silk fibroin membranes provide a potential delivery vehicle for both cells and extracellular matrix (ECM) components into diseased or injured tissues. We have previously demonstrated the feasibility of growing retinal pigment epithelial cells (RPE) on fibroin membranes with the view to repairing the retina of patients afflicted with age-related macular degeneration (AMD). The goal of the present study was to investigate the feasibility of incorporating the ECM component elastin, in the form of human recombinant tropoelastin, into these same membranes. Two basic strategies were explored: (1) membranes prepared from blended solutions of fibroin and tropoelastin; and (2) layered constructs prepared from sequentially cast solutions of fibroin, tropoelastin, and fibroin. Optimal conditions for RPE attachment were achieved using a tropoelastin-fibroin blend ratio of 10 to 90 parts by weight. Retention of tropoelastin within the blend and layered constructs was confirmed by immunolabelling and Fourier-transform infrared spectroscopy (FTIR). In the layered constructs, the bulk of tropoelastin was apparently absorbed into the initially cast fibroin layer. Blend membranes displayed higher elastic modulus, percentage elongation, and tensile strength ($p < 0.01$) when compared to the layered constructs. RPE cell response to fibroin membranes was not affected by the presence of tropoelastin. These findings support the potential use of fibroin membranes for the co-delivery of RPE cells and tropoelastin.

Reprinted from *J. Funct. Biomater.* Cite as: Shadforth, A.M.A.; Suzuki, S.; Alzonne, R.; Edwards, G.A.; Richardson, N.A.; Chirila, T.V.; Harkin, D.G. Incorporation of Human Recombinant Tropoelastin into Silk Fibroin Membranes with the View to Repairing Bruch's Membrane. *J. Funct. Biomater.* **2015**, *6*, 946–962.

1. Introduction

While strategies for tissue regeneration are often based upon the replacement of lost cells, such efforts often ignore the significant contribution of extracellular matrix (ECM) components to tissue structure and function. A good example of this problem is illustrated through the attempts to treat age-related macular degeneration (AMD) of the retina. In short, although the pathology of AMD involves significant changes to both cellular and ECM components, most efforts to date have been largely focused on replacing only the cellular components and, especially, retinal pigment epithelial (RPE) cells [1–4]. In doing so, healthy RPE cells are ultimately delivered into sites containing an abnormal composition and arrangement of ECM components. In order to address this issue, a number of groups have explored the potential of a variety of biomaterials as temporary ECM substitutes to support the RPE cells during cultivation and implantation [5,6]. In our case, we have focused on the development of a substitute prepared from the silk structural protein, fibroin [7,8]. Using this strategy, we have demonstrated the feasibility of establishing functional monolayers of RPE cells grown on fibroin membranes. These RPE monolayers share several important features with those found within the healthy retina, including apical-basal polarity, patterns of growth factor secretion and phagocytic function [9]. As such, fibroin membranes have potential as a vehicle for implanting cultured RPE cells into AMD patients. Since the fibroin membranes will eventually degrade, the incorporation of ECM components, or their precursors, within the fabricated membranes may further facilitate subsequent development of a more permanent ECM. The aim of the present study, therefore, was to examine the feasibility of incorporating ECM components found naturally within the outer retina. More specifically, we have examined the incorporation of the precursor protein from which elastin fibres are produced, tropoelastin [10].

Our focus on tropoelastin arises from considering the composition of the ECM that resides immediately posterior to the RPE, a structure known as Bruch's membrane. A functional, native Bruch's membrane contains an elastin fibre-rich core that is thought to facilitate tissue compliance during cycles of tissue expansion and recoil as blood flows through the adjacent capillaries of the choriocapillaris [11]. The elastic properties of Bruch's membrane may also serve to protect the delicate connections that exist between RPE cells and the adjacent photoreceptor cells [12]. However, age-related changes, such as the accumulation of abnormal deposits referred to as drusen, disrupt the biochemical and mechanical properties of Bruch's membrane [11]. Moreover, an aged Bruch's membrane deters the survival of both endogenous, as well as implanted, RPE cells [13–17]. Importantly, RPE cells have been shown to produce microfibrils, and lysyl oxidase, the enzyme responsible for converting tropoelastin into elastin fibres [18]. Thus, by implanting RPE cells in conjunction with tropoelastin it may be possible to regenerate the core element of a functional, native Bruch's membrane following degradation of the fibroin-based delivery template. In addition, since tropoelastin shares similar elastic properties with elastin [10], it may also be possible to create fibroin-tropoelastin constructs with physical and mechanical properties that are more favourable for establishing and implanting RPE cell cultures than constructs based solely on fibroin.

Two strategies for incorporating tropoelastin into fibroin membranes were examined in this study. Membranes were produced from fibroin solutions supplemented with recombinant human tropoelastin (fibroin-tropoelastin blend) or prepared from alternating cast solutions of fibroin and tropoelastin (layered approach). In the case of the blend, we commenced by optimizing the amount of tropoelastin that can be added to fibroin solution without negatively impacting on the attachment of RPE cells to the resulting membranes. Freestanding membranes were subsequently produced from the optimal blend formulation, and by using the layered approach. The two types of biomaterial membrane were subsequently compared in parallel with standard fibroin membranes using a variety of criteria, including morphology (scanning electron microscopy), secondary structure (Fourier-transform infrared spectroscopy-attenuated total reflectance, FTIR-ATR), the distribution of tropoelastin (immunofluorescence), the cultivation of RPE cells, and mechanical properties. These studies led to some unexpected findings, especially in regard to how tropoelastin in solution interacts with cast fibroin membranes.

2. Results and Discussion

2.1. Properties of Fibroin and Tropoelastin Solutions

During their extraction from silkworm cocoons [19], a significant proportion of the native fibroin proteins (heavy chain 350 kDa and light chain 26 kDa) were cleaved into fragments of varying molecular weights (Figure 1). In contrast, human tropoelastin produced via recombinant DNA technology [20] displayed a single band by gel electrophoresis, at approximately 55 kDa (Figure 1). The aqueous solutions of fibroin and tropoelastin mixed readily with increasing ratios of up to 50% tropoelastin by weight. Phase separation was observed when combining solutions at 10% tropoelastin by weight (resulting in a cloudy solution); however, the resulting dried films were transparent and smooth when cast in plastic (polystyrene) tissue culture dishes.

2.2. Effect of Tropoelastin on RPE Cell Attachment to Fibroin

Since fibroin supports the attachment and growth of RPE cells [8] and tropoelastin has also been shown to positively influence cell attachment [21,22], we examined different blend ratios of fibroin and tropoelastin with the goal of identifying an optimal formulation for the resulting blend membrane. As demonstrated in Figure 2, a consistent trend was observed towards an optimal RPE cell attachment (as defined by DNA content), in either the presence or absence of serum (10% v/v), using a tropoelastin-fibroin ratio of 10 to 90 parts by weight. This result was consistent with prior reports [21,22] and has been explained as the optimal ratio between the two proteins.

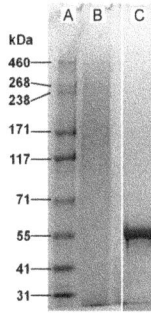

Figure 1. Relative molecular weight distribution for purified native *Bombyx mori* silk fibroin (**B**) and recombinant human tropoelastin (**C**), as displayed by gel electrophoresis. While the extracted fibroin proteins present as a broad range of peptide fragments, recombinant human tropoelastin has a defined molecular weight of approximately 55 kDa. The left lane (**A**) shows a selection of molecular weight markers.

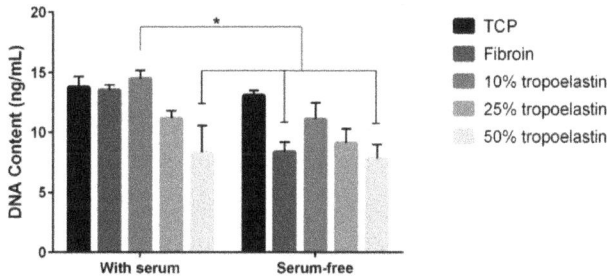

Figure 2. Comparison of cell attachment of retinal pigment epithelial cell line (ARPE-19) on tissue culture plastic (TCP) coated with either fibroin solution, or fibroin mixed with increasing concentrations of tropoelastin (proteins blended in solution before coating TCP). Evidence of cell attachment was examined after 4 h in either the presence or absence of 10% (v/v) fetal bovine serum (with washing prior to measurement). Each substrate was tested in triplicate. Bars represent mean values ± standard error of the mean from three experiments. The difference between fibroin with 10% tropoelastin used in the presence of serum and the other identified bars was statistically significant ($p < 0.05$).

2.3. Gross Morphology of the Freestanding Membranes

Having established the optimal blend ratio of fibroin to tropoelastin for RPE cells, we proceeded to test the feasibility of producing freestanding membranes from the optimal blend, as well as layered constructs produced by sequential addition and drying/stabilization of aqueous solutions containing each protein (fibroin followed by tropoelastin, then fibroin again). Both types of membrane were prepared in glass

75

Petri dishes coated with Topas® polymer as described previously [23]. In brief, the Topas® coating facilitated the subsequent removal of fibroin-based membranes from the glass Petri dishes and was itself delaminated easily, leaving behind the protein membranes. The membranes produced from the optimal blend (Figure 3B) or by layering (Figure 3C) were physically comparable to the standard fibroin membranes produced routinely in our laboratory (Figure 3A). All membranes were transparent and could be cut into the 16-mm diameter discs required for our custom-designed Teflon® cell culture chambers [8]. Nevertheless, the layered membranes (Figure 3C) were noticeably more brittle during excision, resulting in discs with uneven edges (Figure 3C). While no layers were evident within the membranes examined by scanning electron microscopy (SEM) following freeze fracture (Figure 3 D–F), a distinct band of positive immunolabelling for tropoelastin was observed within the layered construct by confocal fluorescence microscopy (Figure 3I). In contrast, an uneven distribution of staining for tropoelastin was observed within the blend membrane (Figure 3H). Unexpectedly, only a single band of fibroin autofluorescence was observed within the layered constructs (Figure 3I). This result initially suggested to us that perhaps one of the fibroin layers had detached during handling, but repeated attempts using multiple samples revealed the same result. Moreover, no evidence of a detached fibroin sheet was observed in any sample mounted for confocal microscopy. We, therefore, embarked upon an FTIR analysis of the layered composites to determine the fate of the apparently "missing" third layer.

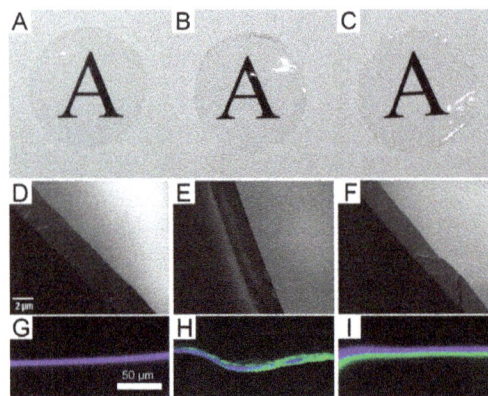

Figure 3. Physical appearance of membranes prepared from either fibroin alone (**A**, **D**, and **G**), tropoelastin-fibroin blend (10:90 ratio) (**B**, **E**, and **H**), and layered solutions of fibroin and tropoelastin (**C**, **F**, and **I**); (**A–C**): gross appearance of each membrane when placed over printed text (16-mm diameter discs); (**D–F**): internal structures revealed by scanning electron microscopy following freeze-fracture; and (**G–I**) visualization of tropoelastin (green) by immunolabelling and confocal fluorescence microscopy (the presence of fibroin revealed as blue autofluorescence).

2.4. Analysis of Membrane Structure by Fourier-Transform Infrared Spectroscopy-Attenuated Total Reflectance, "FTIR-ATR"

The FTIR-ATR spectra in the range of 1800–950 cm^{-1} were used to examine the surface structure of the different biomaterial membranes (Figure 4A). The amide I region between 1720 and 1580 cm^{-1} is traditionally used for analysis of the secondary structure in proteins, and this region has been well described for silk fibroin [24]. In the spectrum of the standard fibroin membrane (water-annealed for 6 h at 25 °C) (Figure 4A, A1), both the amide I band shape and its peak maximum at 1640 cm^{-1} indicate a significant amount of random coil component. The fibroin (Figure 4A, A2) and blend (Figure 4A, A3), membranes that were water annealed for 12 h at 60 °C, revealed a strong band at 1621 cm^{-1} and a shoulder at 1700 cm^{-1}, corresponding to β-sheet structures and their aggregates [24]. If the layered (fibroin-tropoelastin-fibroin) membrane truly had three layers as expected, both surface spectra should reveal a similar fibroin signature. One side of the layered membrane (Figure 4A, A4) did reveal a fibroin signature similar to those described above; however, the other side (Figure 4A, A5) revealed more pronounced β-sheet bands. This may be a result of the additional methanol treatment of the initial fibroin layer after tropoelastin was added. The other side (Figure 4A, A5) also revealed two weak bands at 1200 and 1135 cm^{-1} (indicative of tropoelastin) (Figure 4A, A6), suggesting that one side of the layered membrane consists of a mixture of fibroin and tropoelastin near the surface. The possibility that the methanol treatment might be removing some of the tropoelastin layer was also considered. The tropoelastin bands were used to investigate the stability of tropoelastin in two-layered (fibroin-tropoelastin) membranes before and after methanol treatment (Figure 4B). The spectrum for the tropoelastin side before methanol treatment (Figure 4B, B2) presented two bands at 1200 and 1135 cm^{-1} which correspond to the spectrum of the untreated tropoelastin membrane (Figure 4B, B5). After methanol treatment these tropoelastin bands had dramatically decreased (Figure 4B, B4). Indeed, a thin membrane (thickness of ~1 μm) of tropoelastin was readily soluble in pure methanol which was demonstrated in a separate investigation to confirm FTIR results.

In considering the differences in relative molecular weight distributions for fibroin and tropoelastin (Figure 1) and our previous studies of fibroin membrane permeability (⩽70 kDa using FITC-dextran) [25], the following explanation for the "missing layer" was devised (Figure 5). When tropoelastin solution was cast onto the first fibroin layer it is proposed that some tropoelastin penetrated through the loosely stabilized fibroin hydrogel network. These tropoelastin molecules were subsequently trapped within the fibroin network by drying and treatment with methanol. Hence, the first fibroin layer had a well-distributed content of tropoelastin, as demonstrated by immunofluorescence. A small proportion of tropoelastin remaining on top of the first fibroin layer is also likely to have been washed away by methanol treatment.

The final layer applied (second fibroin layer) would then appear as a single blue layer by autofluorescence.

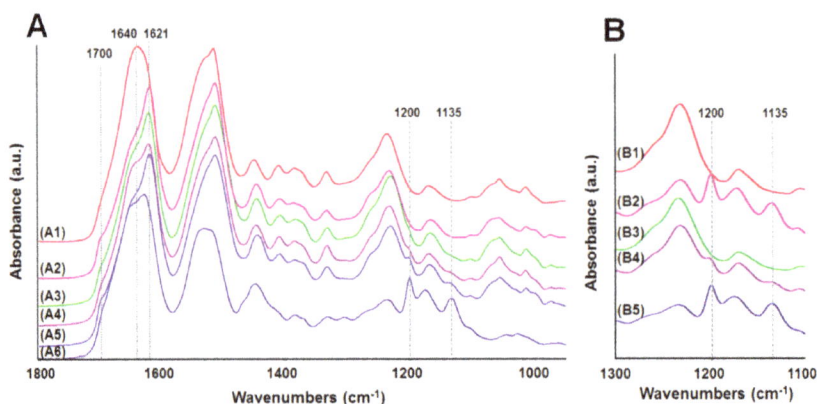

Figure 4. FTIR-ATR spectra of membranes. (**A**) (in the range of 1800–950 cm^{-1}): (A1) fibroin membrane (water annealed at 25 °C, 6 h), (A2) fibroin membrane (water annealed at 60 °C, 12 h), (A3) blend membrane (fibroin:tropoelastin = 90:10), (A4) three-layered membrane—side 1, (A5) three-layered membrane—side 2, (A6) tropoelastin membrane (untreated); and (**B**) (in the range of 1300–1100 cm^{-1}): (B1) two-layered membrane (untreated)—fibroin side, (B2) two-layered membrane (untreated)—tropoelastin side, (B3) two-layered membrane (methanol treated)—fibroin side, (B4) two-layered membrane (methanol treated)—tropoelastin side, (B5) tropoelastin membrane (untreated).

2.5. Cytocompatibility of the Membranes

The cytocompatibility of the fibroin, blend and layered membranes was examined over an extended culture period using current best practice culture conditions [26]. An assessment of cell numbers after three days' culture (Figure 6A) was quantified using the PicoGreen® assay (DNA content provides an indication of cell numbers). There was no statistically significant difference in the number of cells attached across the three biomaterial membrane types, and when compared to the TCP control substrate. The RPE cells seeded on each membrane type showed a similar appropriate morphology over the extended culture period (Figure 6B–D).

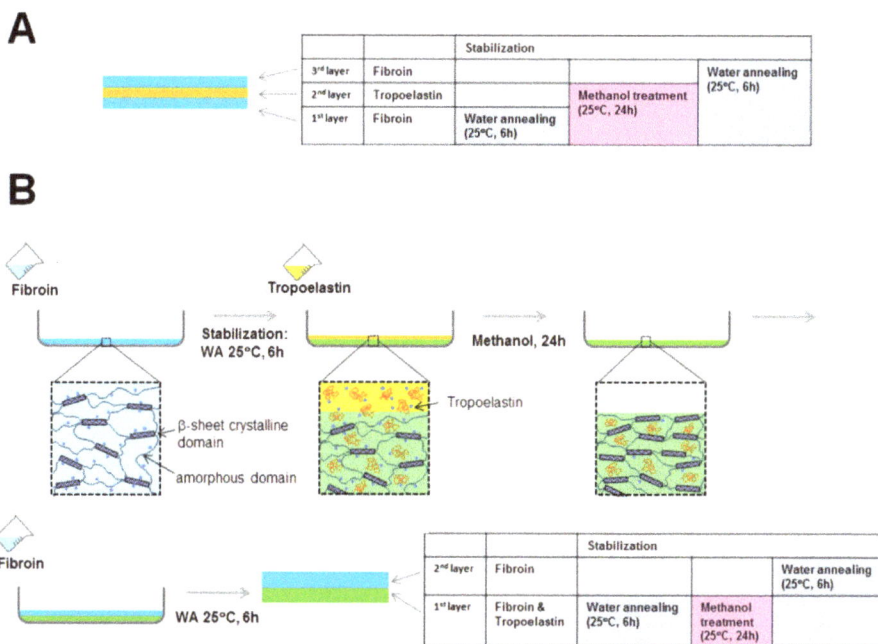

Figure 5. Schematic scenario of the predicted (**A**) and actual (**B**) outcomes achieved during the creation of a layered membrane of fibroin and tropoelastin. Based upon FTIR-ATR data, we propose that the bulk of applied tropoelastin is absorbed and subsequently trapped within the initially created fibroin membrane. Therefore, only two main layers are detected by immunofluorescence/microscopy.

2.6. Mechanical Properties of the Membranes

While our primary goal is to use fibroin as a delivery vehicle for tropoelastin, it is possible that combining the two proteins may produce changes in mechanical properties that impact upon their handling during RPE cell culture and surgical implantation. As a consequence, we compared the mechanical properties of standard fibroin membranes to those displayed by the blend and layered constructs. The results (Figure 7) revealed significant differences between the membranes. The layered membranes, while considerably thicker than the other membranes (data presented as mean values ± standard error of the mean; layered membranes 16.667 ± 0.639 µm, compared to fibroin membranes water-annealed at 25 °C 3.610 ± 0.369 µm, fibroin membranes water-annealed at 60 °C 4.612 ± 0.540 µm, and blend membranes 6.112 ± 0.362 µm), were also more brittle (Figure 7B). In contrast, membranes prepared using a 10% tropoelastin by weight blend with fibroin were the stiffest (Figure 7A), however, they were also strong (Figure 7B) and elastic (Figure 7C). The most interesting results were seen in the standard fibroin membranes that were water-annealed at 25 °C. There was no statistical difference between these membranes and the fibroin membranes water-annealed at 60 °C, however, they did show different properties. The former were the only membranes that had a Young's modulus (Figure 7A) within the range of native Bruch's membrane (7–19 MPa; [27]) and a useful combination of maximum tensile strength (Figure 7B) and elongation properties. This is especially clear when considering there was no difference in elongation at break when compared to the blend membrane (Figure 7C). There was also no statistical difference in recoil capacity of the fibroin (water-annealed at 25 °C) and blend membranes after 200 cycles (Figure 7D) of stretching.

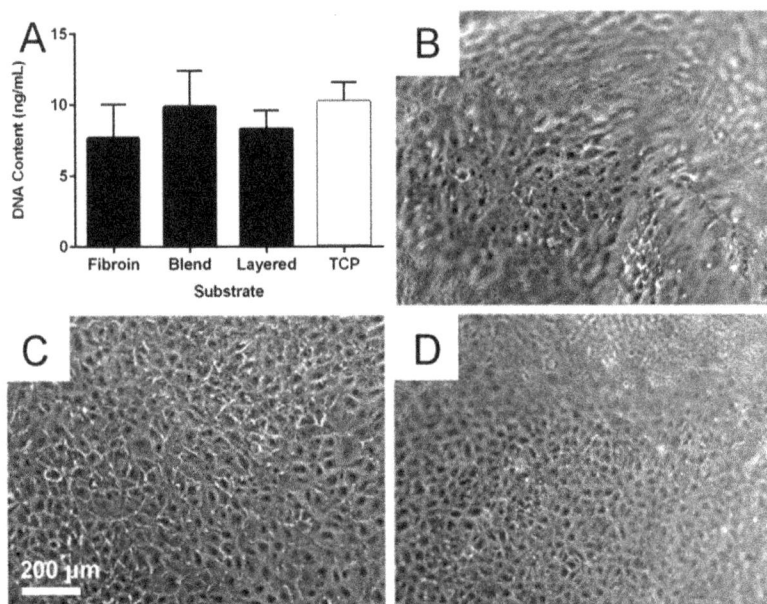

Figure 6. Retinal pigment epithelial (RPE) cell behaviour on biomaterial membranes. Quantification of RPE cell numbers (**A**) using the PicoGreen® assay after 3 days culture on fibroin, blend, and layered membranes; Tissue culture plastic (TCP) was included as control substrate. Phase contrast micrographs of RPE cells after 21 days of growth on fibroin (**B**); blend (**C**); and layered (**D**) membranes. The undulating nature of the suspended membranes is the reason some areas of panels (B) and (D) are out of focus. The scale bar represents 200 μm and applied to the micrographs.

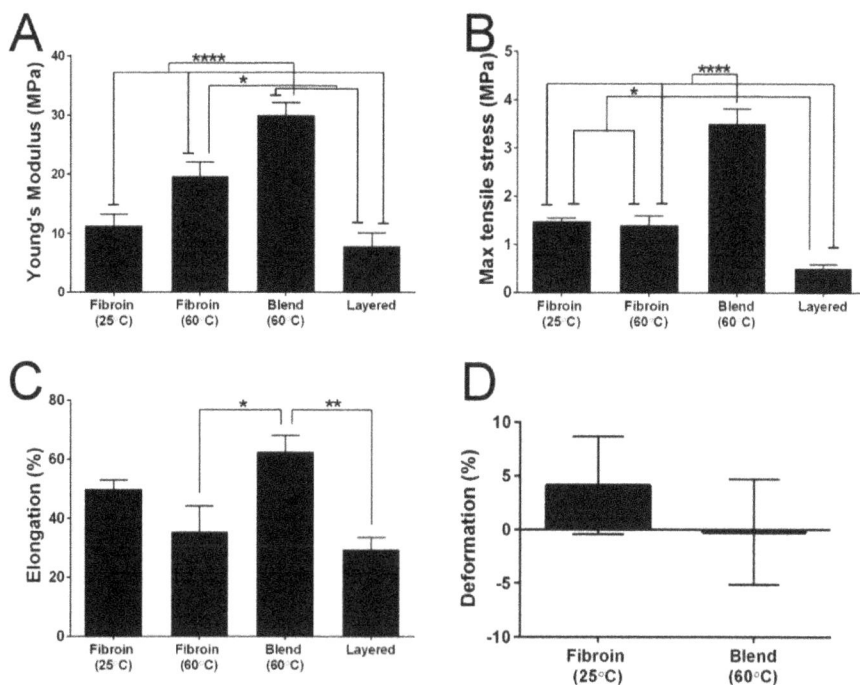

Figure 7. Quantitative comparison of the tensile properties of biomaterial membranes. (**A**) Young's modulus; (**B**) maximum tensile strength; (**C**) elongation to break; and (**D**) deformation/recoil capacity after 200 cycles. Bars represent mean values ± standard error of the mean. Asterisks indicate differences are statistically significant (* $p < 0.05$, ** $p < 0.01$, **** $p < 0.0001$).

3. Experimental Section

3.1. Production of Aqueous Solutions of Fibroin

The procedure has been previously described in detail by our group [28]. Briefly, dried *Bombyx mori* silkworm cocoons (Tajima Shoji Co. Ltd., Yokohama, Japan) were boiled in a solution of sodium carbonate containing 0.85 g of salt for each gram of cocoon material. This procedure removed the sericin outer coat from the core fibroin protein. The resulting fibrous material was washed and dried, and then dissolved (at 60 °C for 4 h) in a concentrated solution of lithium bromide (9.3 M) to obtain a silk concentration of approximately 10% wt./vol. The fibroin solution was subsequently filtered using syringe filters in succession with pore size 0.7 µm and 0.2 µm. This step is performed slowly to avoid shearing forces that could promote spontaneous gelation. The filtrate was dialyzed against water using a dialysis cassette with a molecular mass cut-off of 3.5 kDa (Slide-A-Lyzer, Pierce Biotechnology) using six

81

changes of water over three days. The resulting fibroin solution was filtered again as above and used to produce fibroin membranes.

3.2. Preparation of Films Cast in TCP Wells: Fibroin and Tropoelastin Solutions Blended in Different Ratios

Films of fibroin and tropoelastin were prepared by the method reported by [21] with some modifications. Briefly, tropoelastin (freeze-dried powder) was dissolved in cold MilliQ water (4 °C) to make the concentration of 1.78%, and kept in an ice bath for 2–3 h with occasional vortex mixing. The low temperature is required to prevent coacervation of the solution (self-aggregation of hydrophobic domains). The tropoelastin solution was slowly added to a cold fibroin solution (1.78%) by a pipet, and mixed by inverting the tube slowly. The volume ratio of the fibroin solution to the tropoelastin solution was mixed over the range 90:10, 75:25, and 50:50. The mixture solutions were cast into wells of 24-well TCP plates and dried in a fan-driven oven for 12 h at room temperature. For structural stabilization of fibroin with tropoelastin, β-sheet formation was induced by water annealing the plates in a vacuum oven at 60 °C, −80 kPa with ~100 mL water in a beaker, for 12 h, followed by drying in a fan-driven oven for 12 h at room temperature.

3.3. Cell Culture of the Human RPE Cell Line ARPE-19

ARPE-19 cells were routinely cultured using the Miller's medium formulation [29]; minimum essential medium, alpha modification (MEM-α, M-4526) supplemented with N1 supplement (N-6530), glutamine-penicillin-streptomycin (G-1146), non-essential amino acids (M-7145), taurine (T-0625), hydrocortisone (H-0396), and triiodo-thyronin (T-5516). All of these components were purchased from Sigma Aldrich. This medium formulation allows RPE cultures to be incubated at 37 °C using a standard level of 5% CO_2 air. Cultures were established in the presence of 10% fetal bovine serum, and after 24 h this serum level was decreased to 1%. Stock cultures were fed two to three times per week, and passaged routinely using Versene (15040-066, Life Technologies, Carlsbad, NM, USA) and TrypLE™ (12563-011, Life Technologies), between passages number 23 and 28. An independent STR profile analysis of our working stocks by the Garvan Institute of Medical Research revealed a 100% match with reference ARPE-19 cell line CRL-2302.

3.4. Testing the Attachment of RPE Cells on Films of Fibroin and Tropoelastin Blended in Different Ratios

The cell attachment was quantified on films prepared by blending solutions of fibroin and tropoelastin (Section 3.2). RPE cells (ARPE-19) were seeded at a density of 40,000 cells/cm^2 and incubated at 37 °C, 5% CO_2 for 4 h using Miller's medium without serum. Fibroin films and TCP (used with and without serum) were used as control substrates for ARPE-19 cell attachment. Each substrate type was tested in triplicate, with the experiment performed in triplicate and quantified using the Quant-iT PicoGreen® dsDNA kit (Molecular Probes™, Life Technologies).

3.5. Preparation of Fibroin Membranes

Fibroin membranes were cast using a custom-made casting table as described previously by our group [25]. The thickness of fibroin membranes was measured using an upright micrometer and only areas of membrane 3 μm ± 1 μm thick were used. For structural stabilisation of fibroin membranes, β-sheet formation was induced by the water-annealing of the membranes in a vacuum oven at −80 kPa with ~100 mL water (beaker) for 6 h at room temperature (25 °C). The permeability of fibroin membranes has been previously examined using a horizontal diffusion cell using three model molecules [25].

3.6. Preparation of Freestanding Membranes of Fibroin and Tropoelastin, Proteins Blended in 90:10 Solution Ratio

Freestanding membranes of fibroin and tropoelastin blend were prepared by the method outlined above (Section 3.2), except that only the 90:10 volume ratio of the fibroin solution to the tropoelastin solution was used. For casting, 45-mm Petri dishes were first coated with a Topas® (a commercial hydrophobic cyclic olefin copolymer) film (1 mL of a 7% solution) by the evaporation from a solution in cyclohexane. The Topas® solution formed a hydrophobic film on the glass, facilitating easy removal of the membranes from the dishes later. The mixture solution (1.78%, 1 mL) was poured into the dish, and dried in a fan-driven oven for 12 h at room temperature. For structural stabilisation of fibroin with tropoelastin, the blend membranes were water annealed using a vacuum oven with a container of water and kept at −80 kPa at 60 °C for 12 h, followed by drying in a fan-driven oven for 12 h at room temperature. The membranes were peeled from the Topas® film and used for cell culture and mechanical testing. The thickness of the membranes used was 2–3 μm.

3.7. Preparation of Freestanding Layered (Fibroin-Tropoelastin-Fibroin) Membranes

Layered membranes were fabricated using separate aqueous solutions of fibroin and tropoelastin, layered in sequence and followed by stabilisation after each layer. Before casting any protein solutions, 45-mm Petri dishes were first coated with a Topas® film. The layered membrane was prepared as following. Firstly, 1 mL of 0.59% fibroin solution was cast and dried in a fan-driven oven at room temperature overnight, followed by water annealing in a vacuum oven with a beaker of water at −80 kPa at room temperature, for 6 h. Then 1 mL of 0.59% tropoelastin solution was cast and dried at 4 °C for four days. The tropoelastin layer was stabilized by treatment with methanol (5 mL) for 24 h at room temperature. Finally, 1 mL of 0.59%

fibroin solution was cast on top of the tropoelastin layer, and stabilized by water annealing as above. The volumes used were calculated to generate 1 μm-thick layers of each protein, which would result in a 3 μm-thick layered membrane. A 1 μm-thick membrane of tropoelastin was cast and was not treated with methanol (untreated) as a comparison for FTIR-ATR studies.

3.8. Suspension of the Membranes in Custom-Made Teflon® Chambers

Discs (16-mm diameter) of biomaterial membrane were inserted into custom-made chambers designed by our group, which are manufactured from interlocking Telfon® rings specifically for cell culture use [8]. The combined membrane and chamber were sterilised together by immersion in 70% ethanol for 1 h at room temperature, air-dried, and washed thoroughly with phosphate-buffered solution (PBS). The custom-made chamber suspends the biomaterial membrane (reminiscent of the commercially available Transwell® insert system) creating an apical compartment (upper chamber) and a basal compartment (lower chamber) on either side of the membrane. This culture setup is required for the development of a polarised epithelial culture.

3.9. Visualization of Tropoelastin within the Membranes Using Immunofluorescence

Samples of fibroin, blend, and layered membranes were incubated with a primary monoclonal antibody to tropoelastin (BA4, 1:50, ab21599, Abcam, Cambridge, UK). The secondary antibody used was an Alexa 488-conjugated goat-anti-mouse IgG (Molecular Probes®, Life Technologies). Negative controls for immunostaining were incubated with the secondary antibody only. Confocal laser scanning microscopy (Nikon A1R, Nikon Corporation, Tokyo, Japan) was used to image immunofluorescence.

3.10. Testing Cell Growth of RPE Cells on the Membranes

Cell growth after 72 h on the fibroin, blend, and layered membranes was compared and quantified. RPE (ARPE-19) cells were seeded (4000 cells/cm^2) on discs (6-mm diameter) of the different biomaterial membranes and evaluated for total cell numbers, 72 h after seeding using the Quant-iT PicoGreen® dsDNA kit (Molecular Probes™, Life Technologies). This experiment was performed using discs of the freestanding biomaterial membranes held down by rubber o-rings in the wells of 96-well plates.

3.11. Extended Culture of RPE Cells on the Membranes

RPE (ARPE-19) cells were seeded (10,000 cells/cm^2) onto the apical surface of biomaterial membranes suspended in Teflon® chambers (Section 3.8). All membrane types; fibroin, blend, and layered membranes, were precoated with a commercial Collagen I solution obtained from porcine origin (0.3 mg/mL, Cellmatrix®, Nitta Gelatin Inc., Osaka, Japan) diluted in MilliQ water. Cultures were incubated at 37 °C and 5% CO_2, and culture media was changed twice weekly. Phase contrast light microscopy was used to examine the cultures over a two month culture period.

3.12. Fourier-Transform Infrared Spectroscopy of the Membranes

The FTIR-ATR spectra of the membranes (fibroin, blend, and layered) and tropoelastin were collected using a Nicolet FTIR spectrometer (Thermo Electron Corp, Waltham, MA, USA), equipped with a Nicolet Smart Endurance diamond ATR accessory. Each spectrum was obtained by co-adding 64 scans over the range of 4000 to 525 cm^{-1} at a resolution of 8 cm^{-1}. The OMNIC 7 software package (Thermo Electron Corp, Waltham, MA, USA) was used to analyse and plot the spectra.

3.13. Mechanical Testing of the Membranes

Strips (1 cm × 3 cm) were cut from each membrane type and subjected to tensile measurements in an Instron 5848 micrometer, equipped with a 5 N load cell and a set gauge distance of 14 mm. The membranes were mounted in pneumatic grips and submersed in PBS at 37 °C in a BioPuls™ unit for 5 min prior to testing. Stress-strain plots were recorded, and the Young's moduli were computed in the linear region. The mean values were calculated from results generated by 4–6 measurements for each specimen. In addition, cyclic tensile loading/unloading testing was carried out to evaluate recovery behaviour. The testing experiments were set up as above. However, the following method profile was used: the repeated cyclic loading/unloading was performed at strain of 20% in the stress-strain curve, which is the linear region, with ± 5% strain of loading/unloading and the rate of 14 mm/min. The number of cycles performed was 200 cycles. Four measurements were performed for each specimen. From stress-strain plots, the areas under the curve of cycles 10 and 200 were calculated and used to evaluate deformation using the following equation:

$$\text{Deformation}\,(\%) \;=\; ((\text{Area}_{\text{cycle } 10} - \text{Area}_{\text{cycle } 200})/\text{Area}_{\text{cycle } 10}) \times 100$$

3.14. Statistical Analyses

Results from cell attachment and growth assays were analysed for statistical significance using a two-way ANOVA followed by a Tukey's multiple comparisons test (with the two variables being either "substrate and serum", or "substrate and time"). Mechanical testing data were analysed using a one-way ANOVA with Tukey's test comparing membrane types (with the variable being "membrane type"). Recoil testing data for fibroin and blend membranes were analysed using an unpaired *t* test (since comparing only two independent samples). All statistical analyses were performed using GraphPad Prism, V 6.

4. General Discussion

AMD is a leading cause of permanent vision loss in the elderly. Significant efforts are therefore underway in countries with ageing populations to address this disease. Consideration of the underlying histopathology indicates that therapies based upon the replacement of both cellular (e.g., RPE cells), as well as extracellular tissue components, may well be required. To this end, we have previously demonstrated that freestanding membranes prepared from silk fibroin provide a potential vehicle for delivering RPE cells into the subretinal space [8,9]. The present study builds upon this research by examining the feasibility of incorporating elastin (in the form of tropoelastin) into these same fibroin membranes. In doing so, we have proposed that fibroin membranes may provide a vehicle for co-delivering RPE cells and tropoelastin to the subretinal space. Moreover, since tropoelastin displays similar elastic properties to elastin, we considered that the mechanical properties of fibroin membranes may be significantly altered when combined with tropoelastin.

With regard to our first aim, our data confirms the feasibility of incorporating human recombinant tropoelastin into fibroin membranes. Varying results, however, are achieved according to the methods used. In short, membranes prepared from blended solutions of the two proteins displayed a more heterogeneous composition than those produced using a sequential layering method. We propose that the patchy distribution most likely results from either phase separation or specific molecular interactions between the two proteins when present together in solution. By comparison, our subsequent analyses by FTIR suggest that the more homogenous distribution of tropoelastin achieved using the layering approach is due to absorption and subsequent fixation of this protein within the originally cast fibroin membrane (by treatment with methanol). This result suggests that membranes prepared via the absorption/fixation method should theoretically support a more even profile of tropoelastin delivery following implantation to the subretinal space. Nevertheless, the choice of technique is also likely to be influenced by consideration of membrane mechanical properties.

Our study of the effects of tropoelastin on the mechanical properties of fibroin membranes, when blended, led to some unexpected results. While others have reported reduced stiffness of fibroin membranes following inclusion of tropoelastin [21,22], we have presently reported the opposite result. On the surface, this conflicting data seems quite difficult to resolve. A close comparison of the methods used, however, reveals several significant variations including the source of cocoons, fibroin isolation protocol, water annealing temperature and the thickness of membranes used. In our experience, any one of the parameters alone can have significant effects on the properties of fibroin membranes. Thus, in combination, the differing processes could well have been responsible for the variations in response to the tropoelastin observed between each study.

A comparison of blended *versus* "layered" strategies for incorporating fibroin and tropoelastin is also an interesting exercise. On the basis of their superior strength and elasticity, it could be concluded that the blended membranes are superior to the more brittle "layered" constructs. Nevertheless, a revised formulation whereby the tropoelastin is simply absorbed and trapped, without an additional fibroin layer being deposited, is worthy of investigation. In any case, the key comparison to make is how closely each membrane resembles the mechanical properties of Bruch's membrane. It is, thus, significant that fibroin membranes water-annealed at 25 °C and membranes prepared using the layered approach are closest to native Bruch's membrane in terms of Young's modulus [27]. Therefore, on this basis, and in combination with the more uniform distribution of tropoelastin, we propose that the layered membranes are at present the better option to pursue in order to address both issues of ECM delivery, as well as matching the desired mechanical properties.

5. Conclusions

Reconstructing both the cellular and ECM components of diseased and injured tissues is an important area of tissue engineering and regenerative medicine. The incorporation of a tropoelastin component in fibroin membranes, while maybe not bestowing benefits to mechanical properties, offers a potential vehicle for the delivery of RPE cells and Bruch's membrane ECM components into the subretinal environment of patients with AMD. Future studies will need to investigate the suitability of these membranes in a pre-clinical animal model.

Acknowledgments: This work was supported by the Queensland Eye Institute Foundation (formerly Prevent Blindness Foundation), Australia, and by funding from the National Health and Medical Research Council (NHMRC) of Australia, and the Macular Disease Foundation Australia. A.M.A.S. was supported by a Dora Lush Biomedical Postgraduate Research Scholarship from NHMRC. We thank Anthony Weiss at The University of Sydney for his time and invaluable advice over the course of the study, and for his gift of the human recombinant tropoelastin protein. We also thank Sanjleena Singh at the Central Analytical Research Facility, Queensland University of Technology for help with microscopic analyses.

Author Contributions: Audra M. A. Shadforth contributed to the design of the study and interpretation of results, carried out the cell culture work and morphology testing of membranes, prepared graphic matter, drafted and prepared the manuscript. Shuko Suzuki contributed to the design of the study and interpretation of results, carried out the production and analysis of silk materials, drafted a section of the manuscript, and prepared graphic matter. Raphaelle Alzonne assisted in the production and analysis of silk materials. Grant A. Edwards assisted with the mechanical testing experiments and analysis of results. Neil A. Richardson assisted with supervision of Audra M. A. Shadforth, and preparation of manuscript. Traian V. Chirila contributed to the design of the study and interpretation of results, and revised the final draft of the manuscript. Damien G. Harkin proposed the study, contributed to the design of experiments and interpretation of results, performed the confocal microscopic examination, and drafted the manuscript. All authors read the manuscript and approved its final version.

Conflicts of Interest: The authors declare no conflict of interest.

References

1. Binder, S.; Stanzel, B.V.; Krebs, I.; Glittenberg, C. Transplantation of the RPE in AMD. *Prog. Retin. Eye Res.* **2007**, *26*, 516–554.
2. Da Cruz, L.; Chen, F.K.; Ahmado, A.; Greenwood, J.; Coffey, P. RPE transplantation and its role in retinal disease. *Prog. Retin. Eye Res.* **2007**, *26*, 598–635.
3. Lee, E.; Maclaren, R.E. Sources of retinal pigment epithelium (RPE) for replacement therapy. *Br. J. Ophthalmol.* **2011**, *95*, 445–449.
4. Jha, B.S.; Bharti, K. Regenerating retinal pigment epithelial cells to cure blindness: a road towards personalized artificial tissue. *Curr. Stem Cell Rep.* **2015**, *1*, 79–91.
5. Hynes, S.R.; Lavik, E.B. A tissue-engineered approach towards retinal repair: Scaffolds for cell transplantation to the subretinal space. *Graefes Arch. Clin. Exp. Ophthalmol.* **2010**, *248*, 763–78.
6. Binder, S. Scaffolds for retinal pigment epithelium (RPE) replacement therapy. *Br. J. Ophthalmol.* **2011**, *95*, 441–442.
7. Harkin, D.G.; George, K.A.; Madden, P.W.; Schwab, I.R.; Hutmacher, D.W.; Chirila, T.V. Silk fibroin in ocular tissue reconstruction. *Biomaterials* **2011**, *32*, 2445–2458.
8. Shadforth, A.M.A.; George, K.A.; Kwan, A.S.; Chirila, T.V.; Harkin, D.G. The cultivation of human retinal pigment epithelial cells on Bombyx mori silk fibroin. *Biomaterials* **2012**, *33*, 4110–4117.
9. Shadforth, A.M.A.; Suzuki, S.; Theodoropoulos, C.; Richardson, N.A.; Chirila, T.V.; Harkin, D.G. A Bruch's membrane substitute fabricated from silk fibroin supports the function of retinal pigment epithelial cells *in vitro*. *J. Tissue Eng. Regen. Med.* **2015**. in press.
10. Wise, S.G.; Mithieux, S.M.; Weiss, A.S. Engineered tropoelastin and elastin-based biomaterials. In *Advances in Protein Chemistry and Structural Biology*; McPherson, A., Ed.; Elsevier Science: Oxford, UK, 2009; Volume 78, pp. 1–24.
11. Booij, J.C.; Baas, D.C.; Beisekeeva, J.; Gorgels, T.G.; Bergen, A.A. The dynamic nature of Bruch's membrane. *Prog. Retin. Eye Res.* **2010**, *29*, 1–18.
12. Leure-duPree, A. Ultrastructure of the retinal pigment epithelium in domestic sheep. *Am. J. Ophthalmol.* **1968**, *65*, 383–398.
13. Bhutto, I.; Lutty, G. Understanding age-related macular degeneration (AMD): Relationships between the photoreceptor/retinal pigment epithelium/Bruch's membrane/choriocapillaris complex. *Mol. Aspects Med.* **2012**, *33*, 295–317.
14. Tezel, T.H.; Del Priore, L.V. Reattachment to a substrate prevents apoptosis of human retinal pigment epithelium. *Graefes Arch. Clin. Exp. Ophthalmol.* **1997**, *235*, 41–47.
15. Tezel, T.H.; Kaplan, H.J.; Del Priore, L.V. Fate of human retinal pigment epithelial cells seeded onto layers of human Bruch's membrane. *Invest. Ophthalmol. Vis. Sci.* **1999**, *40*, 467–476.

16. Tezel, T.H.; Del Priore, L.V. Repopulation of different layers of host human Bruch's membrane by retinal pigment epithelial cell grafts. *Invest. Ophthalmol. Vis. Sci.* **1999**, *40*, 767–774.

17. Sugino, I.K.; Sun, Q.; Wang, J.; Nunes, C.F.; Cheewatrakoolpong, N.; Rapista, A.; Johnson, A.C.; Malcuit, C.; Klimanskaya, I.; Lanza, R.; *et al.* Comparison of fRPE and human embryonic stem cell-derived RPE behavior on aged human Bruch's membrane. *Invest. Ophthalmol. Vis. Sci.* **2011**, *52*, 4979–4997.

18. Wachi, H.; Sato, F.; Murata, H.; Nakazawa, J.; Starcher, B.C.; Seyama, Y. Development of a new *in vitro* model of elastic fiber assembly in human pigmented epithelial cells. *Clin. Biochem.* **2005**, *38*, 643–653.

19. Wray, L.S.; Hu, X.; Gallego, J.; Georgakoudi, I.; Omenetto, F.G.; Schmidt, D.; Kaplan, D.L. Effect of processing on silk-based biomaterials: Reproducibility and biocompatibility. *J. Biomed. Mater. Res. B Appl. Biomater.* **2011**, *99*, 89–101.

20. Martin, S.L.; Vrhovski, B.; Weiss, A.S. Total synthesis and expression in Escherichia coli of a gene encoding human tropoelastin. *Gene* **1995**, *154*, 159–166.

21. Hu, X.; Park, S.-H.; Gil, E.S.; Xia, X.-X.; Weiss, A.S.; Kaplan, D.L. The influence of elasticity and surface roughness on myogenic and osteogenic-differentiation of cells on silk-elastin biomaterials. *Biomaterials* **2011**, *32*, 8979–8989.

22. Hu, X.; Wang, X.; Rnjak, J.; Weiss, A.S.; Kaplan, D.L. Biomaterials derived from silk-tropoelastin protein systems. *Biomaterials* **2010**, *31*, 8121–8131.

23. Bray, L.J.; George, K.A.; Suzuki, S.; Chirila, T.V.; Harkin, D.G. Fabrication of a corneal-limbal tissue substitute using silk fibroin. In *Corneal Regenerative Medicine: Methods and Protocols*; Wright, B., Connon, C.J., Eds.; Springer Science+Business Media: New York, NY, USA, 2013; Volume 1014, pp. 131–139.

24. Hu, X.; Kaplan, D.; Cebe, P. Determining beta-sheet crystallinity in fibrous proteins by thermal analysis and infrared spectroscopy. *Macromolecules* **2006**, *39*, 6161–6170.

25. Bray, L.J.; George, K.A.; Ainscough, S.L.; Hutmacher, D.W.; Chirila, T.V.; Harkin, D.G. Human corneal epithelial equivalents constructed on Bombyx mori silk fibroin membranes. *Biomaterials* **2011**, *32*, 5086–5091.

26. Pfeffer, B.A.; Philp, N.J. Cell culture of retinal pigment epithelium: Special Issue. *Exp. Eye Res.* **2014**, *126*, 1–4.

27. Curcio, C.A.; Johnson, M. Structure, function, and pathology of Bruch's membrane. In *Retina*; Ryan, S.J., Schachat, A.P., Wilkinson, C.P., Hinton, D.R., Sadda, S., Wiedemann, P., Eds.; Elsevier: London, UK, 2013; Volume 1, pp. 465–481.

28. Chirila, T.V.; Barnard, Z.; Harkin, D.G.; Schwab, I.R.; Hirst, L.W. Bombyx mori silk fibroin membranes as potential substrata for epithelial constructs used in the management of ocular surface disorders. *Tissue Eng. A* **2008**, *14*, 1203–1211.

29. Maminishkis, A.; Chen, S.; Jalickee, S.; Banzon, T.; Shi, G.; Wang, F.E.; Ehalt, T.; Hammer, J.A.; Miller, S.S. Confluent monolayers of cultured human fetal retinal pigment epithelium exhibit morphology and physiology of native tissue. *Invest. Ophthalmol. Vis. Sci.* **2006**, *47*, 3612–3624.

Substrates for Expansion of Corneal Endothelial Cells towards Bioengineering of Human Corneal Endothelium

Jesintha Navaratnam, Tor P. Utheim, Vinagolu K. Rajasekhar and Aboulghassem Shahdadfar

Abstract: Corneal endothelium is a single layer of specialized cells that lines the posterior surface of cornea and maintains corneal hydration and corneal transparency essential for vision. Currently, transplantation is the only therapeutic option for diseases affecting the corneal endothelium. Transplantation of corneal endothelium, called endothelial keratoplasty, is widely used for corneal endothelial diseases. However, corneal transplantation is limited by global donor shortage. Therefore, there is a need to overcome the deficiency of sufficient donor corneal tissue. New approaches are being explored to engineer corneal tissues such that sufficient amount of corneal endothelium becomes available to offset the present shortage of functional cornea. Although human corneal endothelial cells have limited proliferative capacity *in vivo*, several laboratories have been successful in *in vitro* expansion of human corneal endothelial cells. Here we provide a comprehensive analysis of different substrates employed for *in vitro* cultivation of human corneal endothelial cells. Advances and emerging challenges with *ex vivo* cultured corneal endothelial layer for the ultimate goal of therapeutic replacement of dysfunctional corneal endothelium in humans with functional corneal endothelium are also presented.

Reprinted from *J. Funct. Biomater.* Cite as: Navaratnam, J.; Utheim, T.P.; Rajasekhar, V.K.; Shahdadfar, A. Substrates for Expansion of Corneal Endothelial Cells towards Bioengineering of Human Corneal Endothelium. *J. Funct. Biomater.* **2015**, *6*, 917–945.

1. Introduction

The cornea is the transparent anterior part of the eye that transmits and focuses light onto the retina. From anterior to posterior (Figure 1), the cornea is composed of the corneal epithelium (50 μm thick), the Bowman's membrane (12 μm), the stroma (480–500 μm), the Descemet's membrane (8–10 μm), and the endothelium (5 μm) [1]. Recently, a new layer of the cornea, Dua's layer, was also described [2].

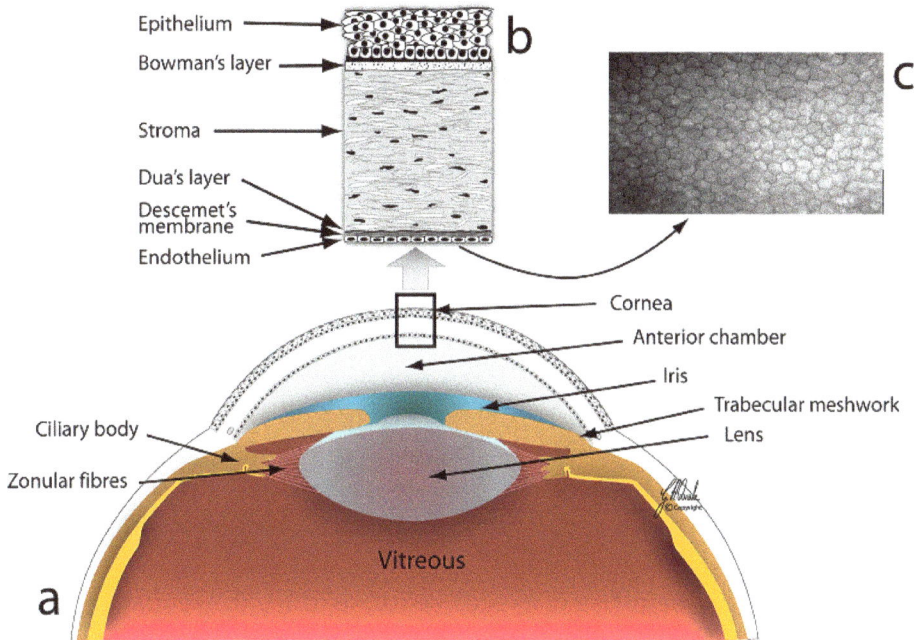

Figure 1. Anatomy of the cornea. (**a**) Section of the anterior part of the eye; (**b**) Section of the cornea illustrating six layers; (**c**) *In vivo* confocal microscopy image of the corneal endothelium. Courtesy of Geir A. Qvale.

The human cornea has a thickness of 0.5–0.6 mm centrally and 0.6–0.8 mm peripherally [3]. The horizontal diameter of an average adult human cornea is 11.7 mm and the vertical diameter is approximately 1 mm less than the horizontal diameter. The cornea is one of the few avascular tissues in the body. The cornea is also one of the most heavily innervated and sensitive tissues in the body, with a density of nerve endings about 300–400 times greater than the skin [1,4], thus diseases of the cornea may be extremely painful. It has several functions that are essential for clear vision: The integrity and functionality of the epithelium [5] and endothelium [6], corneal shape [1], and transparency [1]. The corneal endothelium maintains corneal transparency by regulating water content of corneal stroma. The cornea provides approximately two-thirds of the total refractive power of the eye (Figure 2) [6]; thus, even a small change in corneal contour may result in refractive errors.

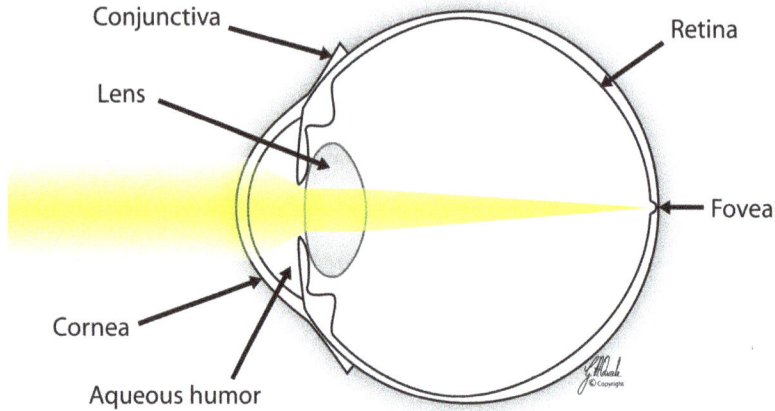

Figure 2. The refraction of light. The cornea provides more than two-thirds of the total refractive power of the eye. Courtesy of Geir A. Qvale.

According to the World Health Organization's global estimation of blindness and visual impairment in 2010, 285 million people were reported to be visually impaired [7]. Corneal diseases are the fourth leading cause of blindness worldwide [7]. Causes of corneal endothelial disease (CED) include endothelial dystrophies, iridocorneal endothelial syndrome, and endothelial dysfunction following cataract surgery and corneal transplantation. Corneal endothelial disease usually presents with a gradual onset of decreased vision. Advanced CED can cause recurrent corneal epithelial erosions, resulting in episodes of severe pain. The corneal endothelium is derived from embryonic neural crest cells [8]. Human corneal endothelial cells (HCECs) have limited proliferative capacity *in vivo* and are suggested to be arrested in the G1-phase of cell cycle [9]. In addition, age-related decrease in corneal endothelial cell density is reported. The mean corneal endothelial cell density decreases from 3600 cells per square millimeter (cells/mm^2) at age 5 years to 2700 cells/mm^2 at age 15 years [10]. Further reduction of the central corneal endothelial density in adults is reported at the rate of 0.6% yearly with gradual change in cell shape and size [11]. Corneal endothelial cell density below critical level of approximately 500 cells/mm^2 results in corneal edema and thereby decreased visual acuity. Significant HCEC loss and inadequate replacement of corneal endothelial cells *in vivo* suggest there is a lack of or inefficient cell division. In corneal endothelial wound healing in humans, the endothelial cells adjacent to the wound enlarge as they elongate and slide to the wound area [12]. At present, transplantation is the only available treatment for diseases affecting the corneal endothelium. There are two main types of corneal transplantation for CED: Penetrating keratoplasty and endothelial keratoplasty. Penetrating keratoplasty refers to the replacement of all corneal layers of the recipient's cornea with a donor cornea. Selective replacement of the diseased posterior layer of the cornea is called endothelial keratoplasty [13]. The above surgical advancements are, however, hindered by the worldwide scarcity of available healthy donor corneas.

A considerable research effort has been put into developing alternative methods for treatment of CED. The remaining HCECs may be stimulated to proliferate or enhance their function with topical eye drops [14] or cell suspension injection into the anterior chamber [15,16]. Magnetic field-guided *in vitro* cultivated HCEC delivery is thought to attract the cells towards Descemet's membrane [15,17,18]. However, the possible complications of injection of cell suspension into the anterior chamber, such as an increase in intracellular pressure due to clogging of the trabecular meshwork, should be further investigated before human trials are initiated. Although growth factors may promote corneal endothelial wound healing [19], it does not induce HCEC proliferation [20]. Thus, there is a clinical interest for engineering corneal endothelium for transplantation purposes. With increasing advances in regenerative medicine, several research groups have investigated on expansion of corneal endothelial cells and transplantation of tissue engineered corneal endothelium in experimental animal models [21–36]. The present review focuses on emerging substrates for improved culturing of HCECs. To provide a background for the current use of substrates, cell sources for tissue engineering of corneal endothelium are also described.

2. Cell Sources for Tissue Engineering of Corneal Endothelium

Various efforts have been made to increase the availability of human HCEC lines. These include immortalization of retroviral transduction by simian virus 40 (SV40) T antigen [37,38], Cdk4R24c/CyclinD1 [39], and/or human papilloma virus 16 E6/E7 [40]. Immortalized cells increase the risk of tumor formation, aneuploidy [41], and structural rearrangements [42]. Recently, the establishment of untransfected HCEC line [43] and immortalization of HCECs with human telomerase reverse transcriptase have been explored [44]. However, the limitation of methods not using transfection is immortalization of only a subpopulation of the primary culture.

Although the HCECs have limited proliferative capacity *in vivo*, these cells have the ability to proliferate under *in vitro* culture conditions [18,25,29–32,36,45–119]. Primary HCECs, human HCEC lines, and stem cells have been utilized for tissue engineering of corneal endothelium. Donor corneoscleral rims, which remain after corneal trephination for corneal transplantation, and human cadaver corneas that are unsuitable for corneal transplantation provide sources for primary HCECs. The age of donors used in tissue engineering of corneal endothelium varies substantially in the literature. Primary HCEC cultures have been established from corneas from 8-week-old human embryos [120] up to age 80 [49,62]. The proliferative response of HCECs tends to decrease in older donors [59,64,69,75,121–123]. Interestingly,

Gao *et al.*, in 2010, were not able to demonstrate a high proliferative rate in human fetal corneal endothelial cells [124]. Regardless of age, human corneal endothelial cells from peripheral areas of the cornea are reported to exhibit a higher replication competency compared to the central area [60,73,121–124]. The lower proliferative capacity of HCECs from the central area may be due to senescence-like characteristics of central HCECs, including stress-induced premature senescence [122]. In addition, it remains interesting to investigate if there are potential stem like progenitors of corneal endothelium that may have more proliferative capacity to produce more corneal endothelial cells than their progency with limited proliferative capacity in center areas of the growing colonies *in vitro*. Isolation of sphere forming HCECs has in fact been reported vividly and has been considered as a potential source of progenitor cells [72,80,108,125–127]. It is possible that such progenitor cells in the central region of the colonies in culture may acquire altered epigenetic modifications which could in turn inhibit their further proliferation or result in their terminal differentiation followed by senescence similar to that was reported with many instances of embryonic stem cell colonies in cell cultures [128].

Stem cells are a potential source for engineering of many organs including corneal endothelium. Organ specific adult stem cells, as well as directed differentiation competent embryonic stem cells, and induced pluripotent stem cells (iPS cells) form such sources. Adult stem cells are suggested to reside in the junction between the peripheral corneal endothelium and anterior part of the trabecular meshwork [129]. Embryonic stem cells [130] have the major advantages due to their characteristics of pluripotency and an unlimited proliferation capacity. However, ethical concerns, immune rejection, and risk of teratoma formation have limited the application of embryonic stem cells in clinical trials. The use of iPS cells in clinical trials is also limited because of bio-safety concerns, epigenetic memory from somatic cells, unintended genomic alterations, and related oncogenesis exacerbated by the use of retroviral or lentiviral transducing vectors. The above said sphere forming HCECs [72,80,108,125–127] and also human corneal stromal precursors may represent a potential source for corneal endothelial cells [131]. Other sources of human corneal endothelial-like cells for tissue engineering of corneal endothelium include umbilical cord blood mesenchymal stem cells [132], adipose-derived stem cells [133], and bone marrow-derived endothelial progenitor cells [134]. Functional corneal endothelium tissue engineered from corneal stromal derived stem cells of neural crest origin in humans and mice [131], and corneal endothelial like cells from neural crest origin in rats [135], are also reported. However, there are no specific bio-markers for identification of corneal endothelial cells. Although sodium-potassium adenosine triphosphatase ($Na^+K^+ATPase$) and zonula occludens-1 (ZO-1) are located on the corneal endothelial cell membrane, both are also present in other type of cells. Therefore, the isolation of HCECs from donor corneas has been widely followed.

In 1965, Mannagh *et al.* reported successful expansion of HCECs [45]. Following this report, different isolation techniques and culture media have been introduced to harvest and expand HCECs. At present time, isolation of HCECs technique consists of two steps. At first the Descemet's membrane is peeled with HCECs, thereafter the peeled membranes undergo enzymatic treatment to dissociate the HCECs. Human corneal endothelial cells have largely been a challenging task to culture and expand. So far, there is no superior culture medium for consistent expanding of HCECs.

3. Substrates for Cultivation of Human Corneal Endothelial Cells

In vitro expansion of HCECs is challenging, and the cells require native-like favorable growth conditions. The cultivated corneal endothelium is fragile and difficult to handle. Therefore, the use of substrates provides mechanical support during transplantation of *ex vivo* engineered human corneal endothelial sheets. In addition, they may create a favorable microenvironment needed for cellular activity. Ideally, the substrate should mimic Descemet's membrane in its biological, mechanical, chemical, and physiological characteristics. A spectrum of substrates is used in *in vitro* expansion of HCECs and in reconstruction of human corneal endothelial layer. These include biological, synthetic, and biosynthetic materials (Table 1).

For bioengineering of corneal endothelium the substrate materials should preferably fulfil the following criteria: (i) provide favorable microenvironment for corneal endothelial cellular activity; (ii) provide mechanical support; (iii) promote cell layer-carrier interactions, cell adhesion, and extracellular matrix deposition; (iv) be non-toxic; (v) allow transport of gases, nutrients, and molecules; (vi) be easy to handle during cell layer transport or surgery (endothelial keratoplasty); (vii) be transparent; (viii) be easily reproducible (*i.e.,* (v)–(vii) are applicable for transplantation for tissue engineered corneal endothelial grafts).

The substrate should preferably create desired microenvironment for HCEC viability, cell proliferation, and signaling pathways. The corneal endothelium displays high pump capacity and barrier function *in vivo* in order to maintain the cornea in its relatively dehydrated physiological state. The substrate materials must enable support of these principle functions of HCECs and corneal endothelium. Following transplantation of tissue engineered corneal endothelial graft; the substrate should allow sufficient transport of gases, nutrients and molecules between corneal endothelium and stroma.

The Descemet's membrane is a specialized basement membrane. After birth, the corneal endothelium secretes Descemet's membrane consisting of non-banded collagen in physiological conditions [136]. In tissue engineering, it is difficult to reconstruct a substrate that totally mimics complex composition, dynamic nature and multiple function of a native basement membrane. Therefore, it might be beneficial if the substrates are able to stimulate collagen secretion.

The substrates should be easy to reproduce, and either degradable or non-degradable substrates may be used in transplantation. If biodegradable, the substrate dissolution rate must be at a preset value that does not give adverse effect on rest of the eye. As microsurgery and minimal incision operations are increasingly used, the tissue engineered corneal endothelium on substrate should be easy to handle and fold/unfold under the surgery.

There are various substrates applied for cultivated HCECs in experimental models (Table 1). In this review, the substrates are classified for convenience into biological, synthetic, and biosynthetic groups of substrates.

Table 1. Cultivation of primary human corneal endothelial cells on different types of substrate.

Main Groups of Substrate	Specific Substrates	Author(s)/Year	Cell Suspension/Sheet on Substrate	Cell Density */Suspension at Time of Seeding on Substrate	Final Cell Density on the Substrate *	Morphology	Phenotype
Biological Substrates							
Amniotic membrane	Denuded human AM	Ishino et al., 2004 [68]	Cell suspension (trypsinized)	3285 ± 62	2410 ± 31	Polygonal, uniformly sized cells with cell-cell and cell-AM contact	ZO-1
	Culture flasks + ** human cornea denuded of endothelium	Insler and Lopez, 1986 [29]	Cell suspension (trypsinized)	$100\ \mu L$ of 7.5×10^5 cells	560–1650	–	–
	Culture flasks + human cornea denuded of endothelium	Insler and Lopez, 1991 [30]	Cell suspension (trypsinized)	–	–	–	–
	Culture flasks + human cornea denuded of endothelium	Insler and Lopez, 1991 [31]	Cell suspension (trypsinized)	2000–2200	1000–1600	–	–
Decellularized/ devitalized corneal materials	Culture plates + human cornea denuded of endothelium	Chen et al., 2001 [62]	Cell suspension (trypsinized)	1503–2159	1895 ± 178	Polygonal with cell-cell adhesion complexes and gap junction	ZO-1
	Bovine ECM coated culture dishes + human cornea denuded of endothelium	Amano, 2003 [67]	Cell suspension (trypsinized)	Cell suspension 2×10^5 in 2 mL	2380 ± 264	Uniform in size and shape	–
	Bovine ECM coated culture dishes + rat cornea denuded of endothelium and coated with fibronectin	Mimura et al., 2004 [27]	Cell suspension (trypsinized)	$300\ \mu L$ of 1×10^6 cells	2744 ± 337	Polygonal	–
	Bovine ECM coated culture dishes + human cornea denuded of endothelium	Amano et al., 2005 [71]	Cell suspension (trypsinized)	2 mL of 2×10^5 cells	2380 ± 264	*In vivo* morphology with cell-cell contact	–
	Decellularized human corneal stroma	Choi et al., 2010 [90]	Cell suspension (trypsinized)	130–3000	–	Compact cells	ZO-1, Na+ K+ ATPase, connexin 43

Table 1. *Cont.*

Main Groups of Substrate	Specific Substrates	Author(s)/Year	Cell Suspension/Sheet on Substrate	Cell Density */Suspension at Time of Seeding on Substrate	Final Cell Density on the Substrate *	Morphology	Phenotype
Decellularized/devitalized corneal materials	Culture plates + decellularized posterior lamellae of bovine cornea	Bayyoud et al., 2012 [137]	Cell suspension (trypsinized)	5×10^4 cells/well	2380 ± 179	Polygonal	ZO-1, Na^+K^+ATPase, $Na^+HCO_3^-$, connexin 43
	Culture plates + decellularized porcine cornea	Yoeruek et al., 2012 [138]	Cell suspension (trypsinized)	–	–	–	–
Lens capsule	Deepithelialized human anterior lens capsule	Yoeruek et al., 2009 [88]	Cell suspension (trypsinized)	5×10^4 cells/well	3012 ± 109	Polygonal	ZO-1, Na^+K^+ATPase, connexin 43
	Culture plates + deepithelialized human anterior lens capsule	Kopsachilis et al., 2012 [99]	Cell suspension (trypsinized)	5×10^4 cells/well	2455 ± 284	Hexagonal	ZO-1, Na^+K^+ATPase
	Collagen-coated, dextran-based microcarrier beads	Insler and Lopez, 1990 [56]	Cell suspension (isolated cells)	–	–	Cobbelstone	–
	Collagen membranes	Kopsachilis et al., 2012 [99]	Cell suspension (trypsinized)	5×10^4 cells/well	2072 ± 325	Hexagonal	–
	Atelocollagen coated culture dishes + collagen vitrigel	Yoshida et al., 2014 [139]	Cell suspension (trypsinized)	1.3×10^6 cells/well	2650 ± 100	–	–
Natural polymers	Type I collagen sponges	Orwin and Hubel, 2000 [65]	Cell suspension (trypsinized)	–	–	Cobbelstone	–
	Bovine ECM coated culture dishes + type I collagen sheet	Mimura et al., 2004 [28]	Cell suspension (trypsinized)	300 μL of 1×10^6 cells	–	Also fibroblast like cells	–
	Type I collagen coated culture dishes	Choi et al., 2013 [97]	Cell suspension (trypsinized)	–	–	–	ZO-1, Na^+K^+ATPase
	Type I collagen coated culture plates	Numata et al., 2014 [110]	Cell suspension (trypsinized)	–	–	Hexagonal	ZO-1, Na^+K^+ATPase

Table 1. *Cont.*

Main Groups of Substrate	Specific Substrates	Author(s)/Year	Cell Suspension/Sheet on Substrate	Cell Density */Suspension at Time of Seeding on Substrate	Final Cell Density on the Substrate *	Morphology	Phenotype
	Type IV collagen coated culture dishes	Choi et al., 2010 [90]	Cell suspension (trypsinized)	–	–	Compact	–
	Type IV collagen coated culture dishes	Yamaguchi et al., 2011 [93]	Cell suspension (trypsinized)	6000	–	–	–
	Type IV collagen coated culture dishes	Choi et al., 2013 [97]	Cell suspension (trypsinized)	–	–	–	ZO-1, Na^+K^+ATPase
	Type IV collagen coated culture plates	Numata et al., 2014 [110]	Cell suspension (trypsinized)	–	–	Hexagonal	ZO-1, Na^+K^+ATPase
	Bovine ECM coated culture plates	Blake et al., 1997 [59]	Cell suspension (trypsinized)	–	–	Hexagonal	–
	Bovine ECM coated culture plates	Yamaguchi et al., 2011 [93]	Cell suspension (trypsinized)	6000	–	–	–
Natural polymers	Fibronectin coated culture plates	Blake et al., 1997 [59]	Cell suspension (trypsinized)	–	–	Hexagonal	–
	Fibronectin coated culture plates	Choi et al., 2010 [90]	Cell suspension (trypsinized)	–	–	Compact	–
	Fibronectin coated culture dishes	Yamaguchi et al., 2011 [93]	Cell suspension (trypsinized)	6000	–	–	–
	Fibronectin coated culture plates	Choi et al., 2013 [97]	Cell suspension (trypsinized)	–	–	–	ZO-1, Na^+K^+ATPase
	Fibronectin coated culture plates	Numata et al., 2014 [110]	Cell suspension (trypsinized)	–	–	Hexagonal	ZO-1, Na^+K^+ATPase
	FNC coating mix coated culture plates	Choi et al., 2013 [97]	Cell suspension (trypsinized)	–	–	–	ZO-1, Na^+K^+ATPase
	Gelatin coated culture flasks	Nayak and Binder, 1984 [50]	Cell suspension (trypsinized)	–	–	Flattened and polygonal	–
	A mixture of laminin and chondroitin sulfate coated culture plates	Engelmann et al., 1988 [51]	Cell suspension (trypsinized)	–	–	Mosaic pattern	–

Table 1. *Cont.*

Main Groups of Substrate	Specific Substrates	Author(s)/Year	Cell Suspension/Sheet on Substrate	Cell Density */Suspension at Time of Seeding on Substrate	Final Cell Density on the Substrate *	Morphology	Phenotype
Natural polymers	Thermoresponsive PIPAAm–grafted surfaces + gelatin discs	Hsiue et al., 2006 [76]	Cell sheet	–	–	Polygonal	ZO-1
	Thermoresponsive PIPAAm–grafted surfaces + gelatin discs	Lai et al., 2007 [32]	Cell sheet	4×10^4 cells	2587 ± 272	Polygonal with cell-cell contact	ZO-1, Na$^+$K$^+$ATPase
	Type IV collagen coated culture dishes + gelatin hydrogel sheets	Watanabe et al., 2011 [94]	Cell suspension (trypsinized)	3-5×10^3	–	Mosaic pattern with ruffled borders	ZO-1, Na$^+$K$^+$ATPase, N-cadherin
	Laminin-5 coated culture dishes	Yamaguchi et al., 2011 [93]	Cell suspension (trypsinized)	6000	–	–	–
	Laminin coated culture plates	Choi et al., 2013 [97]	Cell suspension (trypsinized)	–	–	–	ZO-1, Na$^+$K$^+$ATPase
Synthetic Substrates	Rose chamber	Mannagh and Irving, 1965 [45]	Cell suspension (isolated cells)	–	–	Elongated with cell-cell contact	–
	Tissue culture dishes or flasks	Newsome et al., 1974 [46]	Endothelium-Descemet's membrane explant	–	–	Flat and polygonal	–
	Culture flasks or Petri culture dishes	Baum et al., 1979 [47]	Endothelium-Descemet's membrane explant	–	–	Small and uniform in young donors. Large and pleomorphic in older donors	–
	Coverglass of disposable tissue culture chamber	Tripathi and Tripathi, 1982 [49]	Isolation of cells by scraping and Descemet's membrane explant	–	–	Flattened and hexagonal or polygonal	–
	Culture plates	Blake et al., 1997 [59]	Cell suspension (trypsinized)	–	–	Hexagonal	–

Table 1. *Cont.*

Main Groups of Substrate	Specific Substrates	Author(s)/Year	Cell Suspension/Sheet on Substrate	Cell Density */Suspension at Time of Seeding on Substrate	Final Cell Density on the Substrate *	Morphology	Phenotype
	Collagen type IV coated culture dishes + thermoresponsive PIPAAm–grafted surfaces	Sumide et al., 2006 [74]	Cell sheet	3×10^6 cells/dish	3000	Hexagonal with cilia and microvilli	–
	Thermoresponsive PIPAAm-grafted culture dishes	Ide et al., 2006 [140]	Cell sheet	–	–	Polygonal with cilia and microvilli	–
	Thermoresponsive PIPAAm-grafted culture dishes	Lai et al., 2006 [141]	Cell sheet	4×10^4 cells	2500	Hexagonal	ZO-1, Na+K+ATPase
	Bovine ECM coated culture dishes + culture plates and culture inserts	Hitani et al., 2008 [25]	Cell sheet	600 μL of 4×10^6 cells	2425 ± 83	Uniformly sized cells	ZO-1, Na+K+ATPase
	Culture plates	Choi et al., 2010 [90]	Cell suspension (trypsinized)	–	–	Compact	–
	Culture plates	Yamaguchi et al., 2011 [93]	Cell suspension (trypsinized)	6000	–	–	–
	Culture plates	Kopsachilis et al., 2012 [99]	Cell suspension (trypsinized)	5×10^4 cells/well	2507 ± 303	Hexagonal	–
	Culture plates	Choi et al., 2013 [97]	Cell suspension (trypsinized)	–	–	–	ZO-1, Na+K+ATPase
Biosynthetic Substrate							
	FNC coating mix coated culture dishes + FNC coated RAFT + collagen gel (compressed plastic and type I collagen)	Levis et al., 2012 [101]	Cell suspension (trypsinized)	2000	1941	Polygonal	ZO-1, Na+K+ATPase

Notes: AM: Amniotic membrane; DM: Descemet's membrane; ECM: Extracellular matrix; FNC Coating Mix®: A commercial available coating mixture consisting of fibronectin, collagen and albumin; Na+HCO₃⁻: Sodium bicarbonate; Na+K+ATPase: Sodium-potassium adenosine triphosphatase; PIPAAm: Poly (*N*-isopropylacrylamide); Pos: Positive; RAFT: Real Architecture For 3D Tissues; ZO-1: Zona occludens. * cell density in cells/mm² if otherwise not stated; ** change of substrate for cultivation of cells.

3.1. Biological Substrates

3.1.1. Amniotic Membrane

Amniotic membrane (AM) is a membrane composed of collagen type IV similar to basement membrane of conjunctiva but not cornea [142]. The anti-inflammatory [143] and non-immunogenic [144] properties of AM are believed to be important factors that make it a suitable substrate. The AM is used in treatment of different ocular surface diseases, and it is applied as substrate for limbal transplantation in patients with limbal stem cell deficiency [5]. Ishino *et al.* used denuded AM as a substrate for cultivated HCECs and transplanted onto rabbit corneas denuded of corneal endothelium and Descemet's membrane [68]. The authors demonstrated that the corneal endothelial cell density and function of reconstructed corneal endothelial graft were similar to normal corneas. However, the tissue-engineered grafts consisting of HCECs sheet on AM had some edema. In another study, the basement membrane of AM was used as a carrier for transplantation of cultivated cat corneal endothelial cells on cat cornea denuded of Descemet's membrane and endothelium [145]. The cultivated cells predominantly displayed hexagonal shape, and the reconstructed corneal endothelial layer maintained corneal graft thickness and remained transparent for six weeks.

Although AM provides good biocompatibility, dependency on donor tissue is a limitation. However, AM displays several challenges for clinical use, and thus efforts to identify alternative culture substrates should be encouraged. First, it is semi-opaque; second, preparation is time-consuming; third, there is possible transfer of pathogens from AM; and fourth, inter-donor and intra-donor variations and rate of biodegradability may influence the outcome of its clinical use [146].

3.1.2. Decellularized/Devitalized Corneal Materials

The feasibility of using devitalized corneas or corneas denuded of endothelium as substrate for HCECs is studied extensively [27,29–31,62,67,71,97,137,138]. They are applicable without substantial redesign as they provide the desired shape, mechanical support, and transparency. Reconstructed human corneal endothelial graft with *in vitro* cultivated HCECs seeded on decellularized human corneal stroma expressed ZO-1, $Na^+K^+ATPase$ and connexin 43. Proulx *et al.* studied the function of tissue engineered corneal endothelium [33]. In experimental animal models they transplanted tissue engineered corneal endothelial grafts consisting of cultivated feline corneal endothelial cells on devitalized human cornea denuded of endothelial cells. The follow-up time after transplantation was only 7 days. In this study, 9 of 11 reconstructed corneal endothelial grafts were clear at the end of the follow-up time. The pump function of the reconstructed corneal endothelial graft must have remained functional in order to maintain the cornea transparent. In addition, the reconstructed corneal endothelial layers expressed proteins related to function such as ZO-1 and $Na^+K^+ATPase$ and sodium bicarbonate $(Na^+HCO_3^-)$ transporter [33]. The same research group performed ultrastructural and immunohistochemical studies of cultivated feline corneal endothelial layer on devitalized cornea [147]. Scanning and transmission electron microscopy demonstrated a monolayer of corneal endothelium, and the tissue engineered endothelium expressed function related proteins including ZO-1 and $Na^+K^+ATPase$ and $Na^+HCO_3^-$ transporter.

Bayyoud *et al.* seeded *in vitro* expanded HCECs on devitalized posterior corneal stromal lamellae. The reconstructed corneal endothelial graft had intact barrier and expressed positive staining for sodium-potassium pump $(Na^+K^+ATPase)$, membrane transporter $(Na^+HCO_3^-)$, tight junction (ZO-1), gap junction (connexin 43), and extracellular matrix protein (collagen VIII) [137].

Current methods to decellularize or devitalize cornea include scraping off corneal endothelium mechanically [67,71], use of chemicals [62,137], or freeze/thaw method [33,147]. High-hydrostatic pressurization is an alternative technique to decellularize cornea [148]. However, the following are some inherent technicalities to be aware of using this approach. First, resident viable keratocytes may potentially give raise to fibroblastic contamination. Second, biological tissues may transfer infections. Third, stroma from donor corneas does not reduce the dependency of donor tissues.

3.1.3. Lens Capsule

The human anterior lens capsule has been evaluated as potential substrate for tissue engineered corneal endothelium. Yoeruek *et al.* received human anterior lens capsule from patients who had undergone cataract surgery [88]. They seeded HCECs on de-epithelialized anterior lens capsule and demonstrated that the HCECs grew to confluency. The *in vitro* bioengineered corneal endothelium strongly expressed staining for ZO-1, $Na^+K^+ATPase$, and connexin 43. Kopsachilis *et al.* compared three different substrates; these included de-epithelialized human anterior lens capsule, collagen membrane, and polystyrene culture plates [99]. They obtained human anterior lens capsule of a mean diameter of 10 mm from cornea donors. The cultivated cells displayed hexagonal morphology in all groups, and the cells formed a monolayer of corneal endothelium at two weeks. They reported higher cell density on anterior lens capsule and culture plates in comparison to collagen membrane (Table 1). However, no statistically significant difference in cell density was shown among all three groups. Although the de-epithelialized human anterior lens capsule is a biocompatible substrate, it does not reduce donor dependency. The diameter of anterior lens capsule following capsulorhexis in cataract surgery is approximately half the size needed for a carrier for cultivated corneal endothelium for endothelial keratoplasty.

3.1.4. Natural Polymers

Extracellular protein coatings are composed of single proteins (e.g., collagen, gelatin) or combination of different proteins (e.g., FNC coating mix®). Although the exact components and composition of the coatings are known, the biological activity of HCECs on these coatings varies. The coating proteins influence HCEC adhesion, proliferation, morphology, and function of HCECs. There are many different types of coating materials available for expansion of HCECs. These include collagen [28,56,65,74,90,93,97,99,110,139], fibronectin [59,90,93,97,110], gelatin [32,50,76,94], laminin [93,97], extracellular matrix (ECM) from cultured bovine corneal endothelial cells [25,27,59,67,71,93], a mixture of laminin and chondroitin sulfate [51], and a mixture of fibronectin, collagen, and albumin (FNC Coating Mix®) [97,101].

Choi *et al.* evaluated adhesion, proliferation, and phenotypic maintenance of HCECs on ECM coated culture plates [97]. They studied collagen type I, collagen type IV, fibronectin, laminin, and FNC coating mix. The HCECs expressed a number of integrin genes (integrin $\alpha 1$, $\alpha 2$, $\alpha 3$, αv, $\beta 1$ and $\beta 5$), but not integrin gene $\beta 3$. High expression of integrin genes supports HCEC binding to ECM. Although cells on collagen type IV and fibronectin showed the highest expression and cells on collagen type I exhibited the least expression, there were no statistically significant differences. Compared to uncoated control plates, HCECs adhered more tightly to culture plates coated with coating proteins such as collagen I, collagen IV, fibronectin, and FNC coating mix. The authors also investigated the cell adhesion strength, and showed that all the coating proteins increased the adhesion strength compared to uncoated controls, except for laminin. They were able to demonstrate that HCECs could grow into a confluent layer in a week on all ECM tested, including uncoated culture plates. Gene expression of ZO-1 and $Na^+K^+ATPase$ was found in all conditions, but $Na^+K^+ATPase$ expression was significantly higher in collagen type I, fibronectin and laminin coated culture plates [97]. In a previous study of Choi *et al.*, it was demonstrated that proliferation of HCECs on fibronectin coated culture plates was significantly higher on day 2 after seeding compared to collagen type IV coated culture plates and uncoated culture plates [90]. On day 4 after seeding, however, there was no significant difference in the growth rate in any of the experimental groups.

Yamaguchi *et al.* studied HCEC adhesion and proliferation in the presence of recombinant laminin-5 [93]. Their results showed significantly higher adhesion of HCECs on recombinant laminin-5 coated dishes compared to uncoated control culture dishes. Furthermore, HCECs did not proliferate on collagen type IV coated culture dishes, and the number of adherent HCECs on laminin-5 coated culture dishes increased 1.5 times after 7 days of cell culture.

In few studies gelatin as substrate for HCECs was evaluated [50,76,94]. Hsiue *et al.* were able to demonstrate that gelatin discs dissolved and the HCEC sheet was adherent to posterior part of corneal stroma two weeks after transplantation of HCEC sheet [76]. Silkworm fibroin can be prepared as a transparent membrane and used as carrier for cultivated corneal endothelial cells [149]. However, higher cell density of B4G12 cell line was achieved on uncoated tissue culture compared to on fibroin. Human corneal endothelial cells grew to confluency with polygonal morphology only on collagen type IV coated fibroin [149].

Extracellular matrices from cultured bovine corneal endothelial cells are used as coating material for *in vitro* cultivation HCECs [25,27,59,67,71,93]. In a study the HCECs were cultured initially on bovine ECM coated culture dishes following seeding of the cells on type I collagen sheet [28]. The HCEC sheet was reported to also have cells with fibroblastic-like morphology. Extracellular matrices produced by bovine corneal endothelial cells may be reservoir for progelatinase A, a matrix metalloproteinase, which is important for turnover of ECM and is involved in inflammation, wound healing, angiogenesis, and metastasis [150].

Studies were carried out by using collagen type I and IV as a substrate for HCECs [74,90,93,97,110]. Cultivated monkey corneal endothelial cells were further cultured on collagen type I carrier for 4 weeks and transplanted into monkeys. The cultivated corneal endothelial layer produced confluent monolayer expressing ZO-1 and Na^+K^+ATPase. The transplanted tissue-engineered corneal endothelial graft remained clear and had an endothelial cell density of 1992 to 2475 cells/mm^2 on examination using *in vivo* specular microscopy six months after transplantation [35].

Cultured HCECs on collagen sheets composed of cross-linked collagen type I were transplanted into rabbits. Pump function was evaluated using Ussing chamber and ouabain, a Na^+K^+ATPase inhibitor. The results showed that the cultured HCECs on collagen sheets maintained 76%–95% of pump function of human donor corneas [28].

The difference in adhesion, proliferation, and phenotype displayed by HCECs on the same type of coating in different studies can be related to different culture techniques and media used. However, further studies should be conducted to assess the consistency of the different types of coatings. The use of these coatings in clinical setting remains to be rigorously verified as the coatings are derived from animals.

3.2. Synthetic Substrates

Synthetic polymers have the advantage of high purity with known chemical composition, structure and properties. They can be reproduced at controlled conditions with known mechanical and physical properties. Coated hydrogel lens was used as carrier for cultivated kitten and rabbit corneal endothelial cells, and these constructs were transplanted into adult cats and rabbits with induced corneal edema, respectively. The transplanted corneas became clear within three days after transplantation in both cats and rabbits, and the cornea remained clear for 50 days in cats and 40 days in rabbits [23].

In few studies the HCECs were cultured on plastic culture plates without coating [45–47,49]. These studies do not reveal details of adhesion and proliferation profiles of HCECs. The cyclic dimers of glycolic and lactic acids are monomers used in production of biomedical devices. Glycolic and lactic acids are by-products of metabolic pathway in normal physiological conditions. Therefore, they are regarded as highly biocompatible with minimal systemic toxicity. Poly(lactic acid) (PLLA) and poly(lactic-co-glycolic acid) (PLGA) are synthetic polymers extensively studied owing to their biocompatibility and biodegradability [151]. Hadlock *et al.* seeded *in vitro* expanded rabbit corneal endothelial cells on PLLA and PLGA [152]. In tissue culture conditions the cells grew into confluency on the synthetic materials and stained for ZO-1 along the lateral cell borders.

Synthetic polymers are used commonly as drug delivery devices. In few ocular diseases dexamethasone can be delivered into the vitreous cavity as an implant. Ozurdex, consisting of dexamethasone and PLGA with hydroxypropyl methylcellulose, is injected intravitreally in patients with e.g., macular edema secondary to retinal vein occlusion. Poly(lactic-co-glycolic acid) polymer matrix degrades slowly to lactic and glycolic acids meaning the final degradation products are water and carbon dioxide [153]. Another dexamethasone delivery device, Surodex, consisting of PLGA with hydroxypropyl methylcellulose is inserted into the anterior chamber following cataract surgery to treat postoperative inflammation. In a comparative single-masked parallel-group study, Wadood and coauthors compared the safety and efficacy of dexamethasone eye drops and Surodex inserted into the anterior chamber in patients following phacoemulsification cataract extraction and posterior chamber intraocular lens implantation [154]. Out of 19 patients in this study, 11 patients received Surodex. Surodex remnants were present in all eyes at 60-day post-operative control, and in 3 patients the traces of remnants were present at 32–36 months. However, no significant complications were reported during the follow-up time of 3 years. The authors reported peripheral anterior synechias of less than 1 clock hour at the site of Surodex implantat in 1 patient, and they regarded this as an adverse event. One patient developed high intraocular pressure after Surodex implantation. The authors considered the patient to be a steroid responder. The intraocular pressure normalized without treatment during the follow-up time of 36 months. Although PLGA is considered to be well-tolerated by patients when inserted into the anterior chamber or vitreous cavity, the removal of device in e.g., cases of endophthalmitis remains a major concern due to residing remnants.

In the early phase of cultivation of HCECs, adherence of the cells to the substrate is of great importance to initiate cell growth, while detachment of an intact and confluent cell layer in a later phase is necessary for transplantation purposes. Stimuli-responsive polymers have the ability to change their molecular structures or physicochemical properties according to the variation in the environment they are in. The design of these polymers with associated processes is highly specialized. Major changes, such as alteration in the shape, transparency and permeability to water, can be achieved by a small stimulus, such as change in temperature, pH or wavelength of light.

Research groups have cultivated HCECs on culture dishes grafted with temperature-responsive polymer poly(N-isopropylacrylamide) (PIPAAm) which reversibly alter its hydrophobicity/hydrophilicity dependent on incubation temperature [32,74,76,140,141]. They have the advantage of providing both initial cell adhesion and later cell layer detachment. At 37 °C the seeded cells adhere and proliferate on hydrophobic PIPAAm-grafted surfaces. The HCEC sheet detaches from PIPAAm-grafted culture dishes as surfaces become hydrophilic when temperature is reduced below the lower critical solution temperature of 32 °C. A circular portion of 18 mm in diameter in the center of 35 mm of culture dishes was grafted with temperature responsive polymer, PIPAAm [74,76]. The *in vitro* cultivated HCECs were seeded on PIPAAm-grafted culture dishes, and the cells reached confluency in 1–3 weeks [32,74,76,140,141]. The gross appearance of confluent HCEC layer on hydrophobic PIPAAm-grafted surfaces was whitish gray, and the authors related this to accumulation of ECM [32]. Upon reduction of incubation temperature from 37 to 20 °C, the HCEC sheets detached from culture dish surfaces within 45–60 min [32,74,140,141]. Although the HCEC sheets detached as single contiguous layers, their surfaces were reported as wrinkled by Lai *et al.* [140] and as having a white paper-like texture by Hsiue *et al.* [76]. The monolayered cell sheets expressed ZO-1 [32,76,140,141] and $Na^+K^+ATPase$ [32,74,141] proteins. Deposit of ECM on basal surface of HCEC sheets were observed [32,74,76,140,141], and the ECM components, collagen type IV and fibronectin, were detected by immunostaining [140]. Scanning electron microscopy micrographs showed polygonal cells with cellular interconnections [74,76,140,141] and microvilli and cilia [74,140], and transmission electron microscopy micrographs revealed abundant cytoplasmic organelles, rough endoplasmic reticulum and mitochondria [32,140]. However, Hsiue *et al.* demonstrated the absence of clear cell boundaries [76].

In two studies, the harvested HCEC sheets from PIPAAm-grafted surfaces were immediately transferred to gelatin disc carriers (7 mm in diameter and 700–800 μm in thickness) [32,76]. The reconstructed corneal endothelium was transplanted into experimental rabbit models denuded of corneal endothelium. The gelatin discs dissolved in two weeks, and the corneas transplanted with reconstructed corneal endothelium were clear with near normal corneal thickness at four weeks [76]. In rabbits transplanted with tissue engineered corneal endothelium, the corneal thickness increased to 892 µm at post-operative day 1, and then decreased to near normal corneal thickness of approximately 500 µm at post-operative day 168 [32]. Sumide *et al.* transplanted HCEC sheet attached to cornea denuded of corneal endothelium and Descemet's membrane into rabbit models [74]. Control rabbits underwent all procedures except for having HCEC sheet on corneal button. Minimal corneal edema was reported in rabbits in HCEC sheet transplant group at day 7. In contrast, the corneas were opaque in control group. The average corneal thickness in HCEC sheet transplant group was significantly lower compared to control group at day 7. Even though the stimuli-responsive polymers are investigated extensively, their role in corneal endothelial layer transplantations and the effect of the temperature change on the HCEC bioactivity remains to be investigated.

3.3. Biosynthetic Substrates

Substrates made from a mixture of natural and synthetic polymers are referred to as biosynthetic substrates in this review. Gao *et al.* evaluated biocompatibility and biodegradability of substrate composed of hydroxypropyl chitosan, gelatin, and chondroitin sulfate [155]. Scanning electron microscopy images revealed a porous structure without fibrils, and the light transmission (wavelength ranging 400–800 nm) measurements through the substrate showed transmittance of more than 90%; both indicating the membrane transparency. They demonstrated comparable or better glucose permeability through the substrate in comparison to native corneas. Cultivated rabbit corneal endothelial cells on this substrate reached confluency on day 4, and displayed characteristic cobblestone appearance. Histocompability and biodegradability were assessed by implanting the substrates into skeletal muscle of rats. Sign of inflammation was seen during post-mortem examination at the interface between the host tissue and substrate even at the end of observation period of 2 months. Degradation of substrate was observed from day 30.

Plastic compressed collagen gels [101] and a blending of chitosan and polycaprolactone [156] may give the necessary mechanical strength as a carrier. Synthetic polymers have the advantage of being reproduced under controlled conditions with known mechanical and physical properties. Different ratio of chitosan and polycaprolactone in a substrate were examined. A composition of 75% chitosan and 25% polycaprolactone supported cultivation of bovine corneal endothelial cells and gave the necessary mechanical strength of a substrate. The cells reached confluency on day 7 and expressed ZO-1 protein on substrate composed of chitosan and polycaprolactone at raio of 75:25 [156].

Plastic compressed collagen type I, termed Real Architecture For 3D Tissues (RAFT), can be easily reproduced and trephined into the size required [101]. Scanning and transmission electron microscopy imaging revealed a confluent monolayer of corneal endothelial cells on RAFT. Human corneal endothelial cells cultivated on RAFT stained for ZO-1 and Na$^+$K$^+$ATPase proteins [101].

The synthetic polymers degrade slowly, and hence potential adverse effects on the eyes over long time course remains to be investigated. Biosynthetic substrate is reported to give raise to inflammation in experimental animal models [155]. Therefore, histocompability studies should be performed before use of biosynthetic substrates in humans.

4. Conclusions and Future Perspective

It is obvious to date that the development and utility of different substrates in tissue engineering of corneal endothelium is slowly evolving. Functional realization of the bioengineered corneal endothelium has not yet been optimal due to the current limited knowledge of molecular mechanism of proliferation of HCECs and their associated inter- and intra-signaling pathways that maintain the corneal endothelial tissue homeostasis. Ideal substrates for cultivation of HCECs should mimic Descemet's membrane in molecular, physiological and mechanical terms. Therefore, it is essential to have thorough molecular and functional insights into the microenvironment of human corneal endothelium *in vivo* and engineer such characteristics into the deriving HCEC grafts. Identification of specific marker(s) of HCEC will be extremely advantageous in optimizing differentiation of large numbers of HCECs from a variety of available cell sources. In addition, even perhaps the patient specific iPS cells with the eventual goal of prospectively circumventing the need for increasingly limiting donor corneas. Finally, though the preliminary xenotransplantation studies appear promising, focused research on the discovery and derivation of suitable substrates, optimization of HCEC culture techniques and identification of specific marker(s) of HCESs appears very valuable before any bioengineered human corneal endothelial graft is used clinically.

Acknowledgments: The authors would like to thank Geir A. Qvale at the Center for Eye Research, Department of Ophthalmology, Oslo University Hospital, for contributing with the figures. We thank many investigators that have contributed to this subject, while regretting to be unable to include all the works of others due to space constraint.

Conflicts of Interest: The authors declare no conflict of interest.

References

1. Nishida, T. Neurotrophic mediators and corneal wound healing. *Ocul. Surf.* **2005**, *3*, 194–202.
2. Dua, H.S.; Faraj, L.A.; Said, D.G.; Gray, T.; Lowe, J. Human corneal anatomy redefined: A novel pre-Descemet's layer (dua's layer). *Ophthalmology* **2013**, *120*, 1778–1785.
3. Rüfer, F.; Schröder, A.; Erb, C. White-to-white corneal diameter: Normal values in healthy humans obtained with the orbscan II topography system. *Cornea* **2005**, *24*, 259–261.
4. Rozsa, A.J.; Beuerman, R.W. Density and organization of free nerve endings in the corneal epithelium of the rabbit. *Pain* **1982**, *14*, 105–120.
5. Dua, H.S.; Azuara-Blanco, A. Limbal stem cells of the corneal epithelium. *Surv. Ophthalmol.* **2000**, *44*, 415–425.
6. Meek, K.M.; Dennis, S.; Khan, S. Changes in the refractive index of the stroma and its extrafibrillar matrix when the cornea swells. *Biophys. J.* **2003**, *85*, 2205–2212.
7. Pascolini, D.; Mariotti, S.P. Global estimates of visual impairment: 2010. *Br. J. Ophthalmol.* **2012**, *96*, 614–618.
8. Tuft, S.J.; Coster, D.J. The corneal endothelium. *Eye* **1990**, *4*, 389–424.
9. Joyce, N.C. Proliferative capacity of corneal endothelial cells. *Exp. Eye Res.* **2012**, *95*, 16–23.
10. Nucci, P.; Brancato, R.; Mets, M.B.; Shevell, S.K. Normal endothelial cell density range in childhood. *Arch. Ophthalmol.* **1990**, *108*, 247–248.
11. Bourne, W.M.; Nelson, L.R.; Hodge, D.O. Central corneal endothelial cell changes over a ten-year period. *Investig. Ophthalmol. Vis. Sci.* **1997**, *38*, 779–782.
12. Steele, C. Corneal wound healing: A review. *Optom. Today* **1999**, *25*, 28–32.
13. Melles, G.R.; Ong, T.S.; Ververs, B.; van der Wees, J. Descemet membrane endothelial keratoplasty (DMEK). *Cornea* **2006**, *25*, 987–990.
14. Okumura, N.; Koizumi, N.; Kay, E.P.; Ueno, M.; Sakamoto, Y.; Nakamura, S.; Hamuro, J.; Kinoshita, S. The rock inhibitor eye drop accelerates corneal endothelium wound healing. *Investig. Ophthalmol. Vis. Sci.* **2013**, *54*, 2493–2502.
15. Mimura, T.; Shimomura, N.; Usui, T.; Noda, Y.; Kaji, Y.; Yamgami, S.; Amano, S.; Miyata, K.; Araie, M. Magnetic attraction of iron-endocytosed corneal endothelial cells to Descemet's membrane. *Exp. Eye Res.* **2003**, *76*, 745–751.
16. Okumura, N.; Koizumi, N.; Ueno, M.; Sakamoto, Y.; Takahashi, H.; Tsuchiya, H.; Hamuro, J.; Kinoshita, S. Rock inhibitor converts corneal endothelial cells into a phenotype capable of regenerating *in vivo* endothelial tissue. *Am. J. Pathol.* **2012**, *181*, 268–277.
17. Mimura, T.; Yamagami, S.; Usui, T.; Ishii, Y.; Ono, K.; Yokoo, S.; Funatsu, H.; Araie, M.; Amano, S. Long-term outcome of iron-endocytosing cultured corneal endothelial cell transplantation with magnetic attraction. *Exp. Eye Res.* **2005**, *80*, 149–157.
18. Moysidis, S.N.; Alvarez-Delfin, K.; Peschansky, V.J.; Salero, E.; Weisman, A.D.; Bartakova, A.; Raffa, G.A.; Merkhofer, R.M., Jr.; Kador, K.E.; Kunzevitzky, N.J.; *et al.* Magnetic field-guided cell delivery with nanoparticle-loaded human corneal endothelial cells. *Nanomed. Nanotechnol. Biol. Med.* **2015**, *11*, 499–509.

19. Hoppenreijs, V.P.T.; Pels, E.; Vrensen, G.F.J.M.; Treffers, W.F. Corneal endothelium and growth factors. *Surv. Ophthalmol.* **1996**, *41*, 155–164.

20. Pipparelli, A.; Arsenijevic, Y.; Thuret, G.; Gain, P.; Nicolas, M.; Majo, F. Rock inhibitor enhances adhesion and wound healing of human corneal endothelial cells. *PLoS ONE* **2013**, *8*.

21. Fan, T.J.; Zhao, J.; Hu, X.Z.; Ma, X.Y.; Zhang, W.B.; Yang, C.Z. Therapeutic efficiency of tissue-engineered human corneal endothelium transplants on rabbit primary corneal endotheliopathy. *J. Zhejiang Univ. Sci. B* **2011**, *12*, 492–498.

22. Fan, T.; Ma, X.; Zhao, J.; Wen, Q.; Hu, X.; Yu, H.; Shi, W. Transplantation of tissue-engineered human corneal endothelium in cat models. *Mol. Vis.* **2013**, *19*, 400–407.

23. Mohay, J.; Lange, T.M.; Soltau, J.B.; Wood, T.O.; McLaughlin, B.J. Transplantation of corneal endothelial cells using a cell carrier device. *Cornea* **1994**, *13*, 173–182.

24. Mohay, J.; Wood, T.O.; McLaughlin, B.J. Long-term evaluation of corneal endothelial cell transplantation. *Trans. Am. Ophthalmol. Soc.* **1997**, *95*, 131–151.

25. Hitani, K.; Yokoo, S.; Honda, N.; Usui, T.; Yamagami, S.; Amano, S. Transplantation of a sheet of human corneal endothelial cell in a rabbit model. *Mol. Vis.* **2008**, *14*, 1–9.

26. Jumblatt, M.M.; Maurice, D.M.; McCulley, J.P. Transplantation of tissue-cultured corneal endothelium. *Investig. Ophthalmol. Vis. Sci.* **1978**, *17*, 1135–1141.

27. Mimura, T.; Amano, S.; Usui, T.; Araie, M.; Ono, K.; Akihiro, H.; Yokoo, S.; Yamagami, S. Transplantation of corneas reconstructed with cultured adult human corneal endothelial cells in nude rats. *Exp. Eye Res.* **2004**, *79*, 231–237.

28. Mimura, T.; Yamagami, S.; Yokoo, S.; Usui, T.; Tanaka, K.; Hattori, S.; Irie, S.; Miyata, K.; Araie, M.; Amano, S. Cultured human corneal endothelial cell transplantation with a collagen sheet in a rabbit model. *Investig. Ophthalmol. Vis. Sci.* **2004**, *45*, 2992–2997.

29. Insler, M.S.; Lopez, J.G. Transplantation of cultured human neonatal corneal endothelium. *Curr. Eye Res.* **1986**, *5*, 967–972.

30. Insler, M.S.; Lopez, J.G. Heterologous transplantation *versus* enhancement of human corneal endothelium. *Cornea* **1991**, *10*, 136–148.

31. Insler, M.S.; Lopez, J.G. Extended incubation times improve corneal endothelial cell transplantation success. *Investig. Ophthalmol. Vis. Sci.* **1991**, *32*, 1828–1836.

32. Lai, J.Y.; Chen, K.H.; Hsiue, G.H. Tissue-engineered human corneal endothelial cell sheet transplantation in a rabbit model using functional biomaterials. *Transplantation* **2007**, *84*, 1222–1232.

33. Proulx, S.; Bensaoula, T.; Nada, O.; Audet, C.; d'Arc Uwamaliya, J.; Devaux, A.; Allaire, G.; Germain, L.; Brunette, I. Transplantation of a tissue-engineered corneal endothelium reconstructed on a devitalized carrier in the feline model. *Investig. Ophthalmol. Vis. Sci.* **2009**, *50*, 2686–2694.

34. Koizumi, N.; Sakamoto, Y.; Okumura, N.; Tsuchiya, H.; Torii, R.; Cooper, L.J.; Ban, Y.; Tanioka, H.; Kinoshita, S. Cultivated corneal endothelial transplantation in a primate: Possible future clinical application in corneal endothelial regenerative medicine. *Cornea* **2008**, *27*, S48–S55.

35. Koizumi, N.; Sakamoto, Y.; Okumura, N.; Okahara, N.; Tsuchiya, H.; Torii, R.; Cooper, L.J.; Ban, Y.; Tanioka, H.; Kinoshita, S. Cultivated corneal endothelial cell sheet transplantation in a primate model. *Investig. Ophthalmol. Vis. Sci.* **2007**, *48*, 4519–4526.

36. Tchah, H. Heterologous corneal endothelial cell transplantation—Human corneal endothelial cell transplantation in lewis rats. *J. Korean Med. Sci.* **1992**, *7*, 337–342.

37. Wilson, S.E.; Lloyd, S.A.; He, Y.G.; McCash, C.S. Extended life of human corneal endothelial cells transfected with the SV40 large T antigen. *Investig. Ophthalmol. Vis. Sci.* **1993**, *34*, 2112–2123.

38. Bednarz, J.; Teifel, M.; Friedl, P.; Engelmann, K. Immortalization of human corneal endothelial cells using electroporation protocol optimized for human corneal endothelial and human retinal pigment epithelial cells. *Acta Ophthalmol. Scand.* **2000**, *78*, 130–136.

39. Yokoi, T.; Seko, Y.; Yokoi, T.; Makino, H.; Hatou, S.; Yamada, M.; Kiyono, T.; Umezawa, A.; Nishina, H.; Azuma, N. Establishment of functioning human corneal endothelial cell line with high growth potential. *PLoS ONE* **2012**, *7*.

40. Kim, H.J.; Ryu, Y.H.; Ahn, J.I.; Park, J.K.; Kim, J.C. Characterization of immortalized human corneal endothelial cell line using HPV 16 E6/E7 on lyophilized human amniotic membrane. *Korean J. Ophthalmol.* **2006**, *20*, 47–54.

41. Takeuchi, M.; Takeuchi, K.; Ozawa, Y.; Kohara, A.; Mizusawa, H. Aneuploidy in immortalized human mesenchymal stem cells with non-random loss of chromosome 13 in culture. *Vitro Cell. Dev. Biol. Anim.* **2009**, *45*, 290–299.

42. Lin, Z.; Han, Y.; Wu, B.; Fang, W. Altered cytoskeletal structures in transformed cells exhibiting obviously metastatic capabilities. *Cell Res.* **1990**, *1*, 141–151.

43. Fan, T.; Zhao, J.; Ma, X.; Xu, X.; Zhao, W.; Xu, B. Establishment of a continuous untransfected human corneal endothelial cell line and its biocompatibility to denuded amniotic membrane. *Mol. Vis.* **2011**, *17*, 469–480.

44. Schmedt, T.; Chen, Y.; Nguyen, T.T.; Li, S.; Bonanno, J.A.; Jurkunas, U.V. Telomerase immortalization of human corneal endothelial cells yields functional hexagonal monolayers. *PLoS ONE* **2012**, *7*.

45. Mannagh, J.J.; Irving, A., Jr. Human corneal endothelium: Growth in tissue cultures. *Arch. Ophthalmol.* **1965**, *74*, 847–849.

46. Newsome, D.A.; Takasugi, M.; Kenyon, K.R.; Stark, W.F.; Opelz, G. Human corneal cells *in vitro*: Morphology and histocompatibility (HL-A) antigens of pure cell populations. *Investig. Ophthalmol.* **1974**, *13*, 23–32.

47. Baum, J.L.; Niedra, R.; Davis, C.; Yue, B.Y.J.T. Mass culture of human corneal endothelial cells. *Arch. Ophthalmol.* **1979**, *97*, 1136–1140.

48. Fabricant, R.N.; Alpar, A.J.; Centifanto, Y.M.; Kaufman, H.E. Epidermal growth factor receptors on corneal endothelium. *Arch. Ophthalmol.* **1981**, *99*, 305–308.

49. Tripathi, R.C.; Tripathi, B.J. Human trabecular endothelium, corneal endothelium, keratocytes, and scleral fibroblasts in primary cell culture. A comparative study of growth characteristics, morphology, and phagocytic activity by light and scanning electron microscopy. *Exp. Eye Res.* **1982**, *35*, 611–624.

50. Nayak, S.K.; Binder, P.S. The growth of endothelium from human corneal rims in tissue culture. *Investig. Ophthalmol. Vis. Sci.* **1984**, *25*, 1213–1216.

51. Engelmann, K.; Böhnke, M.; Friedl, P. Isolation and long-term cultivation of human corneal endothelial cells. *Investig. Ophthalmol. Vis. Sci.* **1988**, *29*, 1656–1662.

52. Yue, B.Y.; Sugar, J.; Gilboy, J.E.; Elvart, J.L. Growth of human corneal endothelial cells in culture. *Investig. Ophthalmol. Vis. Sci.* **1989**, *30*, 248–253.

53. Lass, J.H.; Reinhart, W.J.; Skelnik, D.L.; Bruner, W.E.; Shockley, R.P.; Park, J.Y.; Hom, D.L.; Lindstrom, R.L. An *in vitro* and clinical comparison of corneal storage with chondroitin sulfate corneal storage medium with and without dextran. *Ophthalmology* **1990**, *97*, 96–103.

54. Engelmann, K.; Friedl, P. Optimization of culture conditions for human corneal endothelial cells. *Vitro Cell. Dev. Biol.* **1989**, *25*, 1065–1072.

55. Samples, J.R.; Binder, P.S.; Nayak, S.K. Propagation of human corneal endothelium *in vitro* effect of growth factors *Exp. Eye Res.* **1991**, *52*, 121–128.

56. Insler, M.S.; Lopez, J.G. Microcarrier cell culture of neonatal human corneal endothelium. *Curr. Eye Res.* **1990**, *9*, 23–30.

57. Engelmann, K.; Friedl, P. Growth of human corneal endothelial cells in a serum-reduced medium. *Cornea* **1995**, *14*, 62–70.

58. Hoppenreijs, V.P.; Pels, E.; Vrensen, G.F.; Treffers, W.F. Basic fibroblast growth factor stimulates corneal endothelial cell growth and endothelial wound healing of human corneas. *Investig. Ophthalmol. Vis. Sci.* **1994**, *35*, 931–944.

59. Blake, D.A.; Yu, H.; Young, D.L.; Caldwell, D.R. Matrix stimulates the proliferation of human corneal endothelial cells in culture. *Investig. Ophthalmol. Vis. Sci.* **1997**, *38*, 1119–1129.

60. Bednarz, J.; Rodokanaki-von Schrenck, A.; Engelmann, K. Different characteristics of endothelial cells from central and peripheral human cornea in primary culture and after subculture. *Vitro Cell. Dev. Biol. Anim.* **1998**, *34*, 149–153.

61. Schonthal, A.H.; Hwang, J.J.; Stevenson, D.; Trousdale, M.D. Expression and activity of cell cycle-regulatory proteins in normal and transformed corneal endothelial cells. *Exp. Eye Res.* **1999**, *68*, 531–539.

62. Chen, K.H.; Azar, D.; Joyce, N.C. Transplantation of adult human corneal endothelium *ex vivo*: A morphologic study. *Cornea* **2001**, *20*, 731–737.

63. Engelmann, K.; Bednarz, J.; Schafer, H.J.; Friedl, P. Isolation and characterization of a mouse monoclonal antibody against human corneal endothelial cells. *Exp. Eye Res.* **2001**, *73*, 9–16.

64. Miyata, K.; Drake, J.; Osakabe, Y.; Hosokawa, Y.; Hwang, D.; Soya, K.; Oshika, T.; Amano, S. Effect of donor age on morphologic variation of cultured human corneal endothelial cells. *Cornea* **2001**, *20*, 59–63.

65. Orwin, E.J.; Hubel, A. *In vitro* culture characteristics of corneal epithelial, endothelial, and keratocyte cells in a native collagen matrix. *Tissue Eng.* **2000**, *6*, 307–319.

66. Mertens, S.; Bednarz, J.; Richard, G.; Engelmann, K. Effect of perfluorodecalin on human retinal pigment epithelium and human corneal endothelium *in vitro*. *Graefes Arch. Clin. Exp. Ophthalmol.* **2000**, *238*, 181–185.

67. Amano, S. Transplantation of cultured human corneal endothelial cells. *Cornea* **2003**, *22*, S66–S74.

68. Ishino Y, S.Y.; Nakamura, T.; Connon, C.J.; Rigby, H.; Fullwood, N.J.; Kinoshita, S. Amniotic membrane as a carrier for cultivated human corneal endothelial cell transplantation. *Investig. Ophthalmol. Vis. Sci.* **2004**, *45*, 800–806.

69. Zhu, C.; Joyce, N.C. Proliferative response of corneal endothelial cells from young and older donors. *Investig. Ophthalmol. Vis. Sci.* **2004**, *45*, 1743–1751.

70. Joyce, N.C.; Zhu, C.C. Human corneal endothelial cell proliferation: Potential for use in regenerative medicine. *Cornea* **2004**, *23*, S8–S19.

71. Amano, S.; Mimura, T.; Yamagami, S.; Osakabe, Y.; Miyata, K. Properties of corneas reconstructed with cultured human corneal endothelial cells and human corneal stroma. *Jpn. J. Ophthalmol.* **2005**, *49*, 448–452.

72. Yokoo, S.; Yamagami, S.; Yanagi, Y.; Uchida, S.; Mimura, T.; Usui, T.; Amano, S. Human corneal endothelial cell precursors isolated by sphere-forming assay. *Investig. Ophthalmol. Vis. Sci.* **2005**, *46*, 1626–1631.

73. Konomi, K.; Zhu, C.; Harris, D.; Joyce, N.C. Comparison of the proliferative capacity of human corneal endothelial cells from the central and peripheral areas. *Investig. Ophthalmol. Vis. Sci.* **2005**, *46*, 4086–4091.

74. Sumide, T.; Nishida, K.; Yamato, M.; Ide, T.; Hayashida, Y.; Watanabe, K.; Yang, J.; Kohno, C.; Kikuchi, A.; Maeda, N.; *et al.* Functional human corneal endothelial cell sheets harvested from temperature-responsive culture surfaces. *FASEB J.* **2006**, *20*, 392–394.

75. Enomoto, K.; Mimura, T.; Harris, D.L.; Joyce, N.C. Age differences in cyclin-dependent kinase inhibitor expression and RB hyperphosphorylation in human corneal endothelial cells. *Investig. Ophthalmol. Vis. Sci.* **2006**, *47*, 4330–4340.

76. Hsiue, G.H.; Lai, J.Y.; Chen, K.H.; Hsu, W.M. A novel strategy for corneal endothelial reconstruction with a bioengineered cell sheet. *Transplantation* **2006**, *81*, 473–476.

77. Lai, J.Y.; Lu, P.L.; Chen, K.H.; Tabata, Y.; Hsiue, G.H. Effect of charge and molecular weight on the functionality of gelatin carriers for corneal endothelial cell therapy. *Biomacromolecules* **2006**, *7*, 1836–1844.

78. Kikuchi, M.; Zhu, C.; Senoo, T.; Obara, Y.; Joyce, N.C. P27kip1 sirna induces proliferation in corneal endothelial cells from young but not older donors. *Investig. Ophthalmol. Vis. Sci.* **2006**, *47*, 4803–4809.

79. Joko, T.; Nanba, D.; Shiba, F.; Miyata, K.; Shiraishi, A.; Ohashi, Y.; Higashiyama, S. Effects of promyelocytic leukemia zinc finger protein on the proliferation of cultured human corneal endothelial cells. *Mol. Vis.* **2007**, *13*, 649–658.

80. Li, W.; Sabater, A.L.; Chen, Y.T.; Hayashida, Y.; Chen, S.Y.; He, H.; Tseng, S.C. A novel method of isolation, preservation, and expansion of human corneal endothelial cells. *Investig. Ophthalmol. Vis. Sci.* **2007**, *48*, 614–620.

81. Suh, L.H.; Zhang, C.; Chuck, R.S.; Stark, W.J.; Naylor, S.; Binley, K.; Chakravarti, S.; Jun, A.S. Cryopreservation and lentiviral-mediated genetic modification of human primary cultured corneal endothelial cells. *Investig. Ophthalmol. Vis. Sci.* **2007**, *48*, 3056–3061.

82. Yoeruek, E.; Spitzer, M.S.; Tatar, O.; Aisenbrey, S.; Bartz-Schmidt, K.U.; Szurman, P. Safety profile of bevacizumab on cultured human corneal cells. *Cornea* **2007**, *26*, 977–982.

83. Zhu, Y.T.; Hayashida, Y.; Kheirkhah, A.; He, H.; Chen, S.Y.; Tseng, S.C. Characterization and comparison of intercellular adherent junctions expressed by human corneal endothelial cells *in vivo* and *in vitro*. *Investig. Ophthalmol. Vis. Sci.* **2008**, *49*, 3879–3886.

84. Ishino, Y.; Zhu, C.; Harris, D.L.; Joyce, N.C. Protein tyrosine phosphatase-1B (PTP1B) helps regulate EGF-induced stimulation of S-phase entry in human corneal endothelial cells. *Mol. Vis.* **2008**, *14*, 61–70.

85. Miyai, T.; Maruyama, Y.; Osakabe, Y.; Nejima, R.; Miyata, K.; Amano, S. Karyotype changes in cultured human corneal endothelial cells. *Mol. Vis.* **2008**, *14*, 942–950.

86. Patel, S.V.; Bachman, L.A.; Hann, C.R.; Bahler, C.K.; Fautsch, M.P. Human corneal endothelial cell transplantation in a human *ex vivo* model. *Investig. Ophthalmol. Vis. Sci.* **2009**, *50*, 2123–2131.

87. Engler, C.; Kelliher, C.; Speck, C.L.; Jun, A.S. Assessment of attachment factors for primary cultured human corneal endothelial cells. *Cornea* **2009**, *28*, 1050–1054.

88. Yoeruek, E.; Saygili, O.; Spitzer, M.S.; Tatar, O.; Bartz-Schmidt, K.U.; Szurman, P. Human anterior lens capsule as carrier matrix for cultivated human corneal endothelial cells. *Cornea* **2009**, *28*, 416–420.

89. Joyce, N.C.; Harris, D.L. Decreasing expression of the G1-phase inhibitors, p21cip1 and p16INK4a, promotes division of corneal endothelial cells from older donors. *Mol. Vis.* **2010**, *16*, 897–906.

90. Choi, J.S.; Williams, J.K.; Greven, M.; Walter, K.A.; Laber, P.W.; Khang, G.; Soker, S. Bioengineering endothelialized neo-corneas using donor-derived corneal endothelial cells and decellularized corneal stroma. *Biomaterials* **2010**, *31*, 6738–6745.

91. He, Z.; Campolmi, N.; Ha Thi, B.M.; Dumollard, J.M.; Peoc'h, M.; Garraud, O.; Piselli, S.; Gain, P.; Thuret, G. Optimization of immunolocalization of cell cycle proteins in human corneal endothelial cells. *Mol. Vis.* **2011**, *17*, 3494–3511.

92. Shima, N.; Kimoto, M.; Yamaguchi, M.; Yamagami, S. Increased proliferation and replicative lifespan of isolated human corneal endothelial cells with l-ascorbic acid 2-phosphate. *Investig. Ophthalmol. Vis. Sci.* **2011**, *52*, 8711–8717.

93. Yamaguchi, M.; Ebihara, N.; Shima, N.; Kimoto, M.; Funaki, T.; Yokoo, S.; Murakami, A.; Yamagami, S. Adhesion, migration, and proliferation of cultured human corneal endothelial cells by laminin-5. *Investig. Ophthalmol. Vis. Sci.* **2011**, *52*, 679–684.

94. Watanabe, R.; Hayashi, R.; Kimura, Y.; Tanaka, Y.; Kageyama, T.; Hara, S.; Tabata, Y.; Nishida, K. A novel gelatin hydrogel carrier sheet for corneal endothelial transplantation. *Tissue Eng. A* **2011**, *17*, 2213–2219.

115

95. Lee, J.G.; Song, J.S.; Smith, R.E.; Kay, E.P. Human corneal endothelial cells employ phosphorylation of p27(Kip1) at both Ser10 and Thr187 sites for FGF-2-mediated cell proliferation via PI 3-kinase. *Investig. Ophthalmol. Vis. Sci.* **2011**, *52*, 8216–8223.

96. Hara, H.; Koike, N.; Long, C.; Piluek, J.; Roh, D.S.; SundarRaj, N.; Funderburgh, J.L.; Mizuguchi, Y.; Isse, K.; Phelps, C.J.; *et al.* Initial *in vitro* investigation of the human immune response to corneal cells from genetically engineered pigs. *Investig. Ophthalmol. Vis. Sci.* **2011**, *52*, 5278–5286.

97. Choi, J.S.; Kim, E.Y.; Kim, M.J.; Giegengack, M.; Khan, F.A.; Khang, G.; Soker, S. *In vitro* evaluation of the interactions between human corneal endothelial cells and extracellular matrix proteins. *Biomed. Mater.* **2013**, *8*.

98. Bi, Y.L.; Zhou, Q.; Du, F.; Wu, M.F.; Xu, G.T.; Sui, G.Q. Regulation of functional corneal endothelial cells isolated from sphere colonies by rho-associated protein kinase inhibitor. *Exp. Ther. Med.* **2013**, *5*, 433–437.

99. Kopsachilis, N.; Tsinopoulos, I.; Tourtas, T.; Kruse, F.E.; Luessen, U.W. Descemet's membrane substrate from human donor lens anterior capsule. *Clin. Exp. Ophthalmol.* **2012**, *40*, 187–194.

100. Kimoto, M.; Shima, N.; Yamaguchi, M.; Amano, S.; Yamagami, S. Role of hepatocyte growth factor in promoting the growth of human corneal endothelial cells stimulated by l-ascorbic acid 2-phosphate. *Investig. Ophthalmol. Vis. Sci.* **2012**, *53*, 7583–7589.

101. Levis, H.J.; Peh, G.S.; Toh, K.P.; Poh, R.; Shortt, A.J.; Drake, R.A.; Mehta, J.S.; Daniels, J.T. Plastic compressed collagen as a novel carrier for expanded human corneal endothelial cells for transplantation. *PLoS ONE* **2012**, *7*.

102. Okumura, N.; Hirano, H.; Numata, R.; Nakahara, M.; Ueno, M.; Hamuro, J.; Kinoshita, S.; Koizumi, N. Cell surface markers of functional phenotypic corneal endothelial cells. *Investig. Ophthalmol. Vis. Sci.* **2014**, *55*, 7610–7618.

103. Kopsachilis, N.; Tsaousis, K.T.; Tsinopoulos, I.T.; Welge-Luessen, U. Air toxicity for primary human-cultured corneal endothelial cells: An *in vitro* model. *Cornea* **2013**, *32*, e31–e35.

104. Peh, G.S.; Toh, K.P.; Ang, H.P.; Seah, X.Y.; George, B.L.; Mehta, J.S. Optimization of human corneal endothelial cell culture: Density dependency of successful cultures *in vitro*. *BMC Res. Notes* **2013**, *6*.

105. Cheong, Y.K.; Ngoh, Z.X.; Peh, G.S.; Ang, H.P.; Seah, X.Y.; Chng, Z.; Colman, A.; Mehta, J.S.; Sun, W. Identification of cell surface markers glypican-4 and CD200 that differentiate human corneal endothelium from stromal fibroblasts. *Investig. Ophthalmol. Vis. Sci.* **2013**, *54*, 4538–4547.

106. Fujita, M.; Mehra, R.; Lee, S.E.; Roh, D.S.; Long, C.; Funderburgh, J.L.; Ayares, D.L.; Cooper, D.K.; Hara, H. Comparison of proliferative capacity of genetically-engineered pig and human corneal endothelial cells. *Ophthalmic Res.* **2013**, *49*, 127–138.

107. Nakahara, M.; Okumura, N.; Kay, E.P.; Hagiya, M.; Imagawa, K.; Hosoda, Y.; Kinoshita, S.; Koizumi, N. Corneal endothelial expansion promoted by human bone marrow mesenchymal stem cell-derived conditioned medium. *PLoS ONE* **2013**, *8*.

108. Yoon, J.J.; Wang, E.F.; Ismail, S.; McGhee, J.J.; Sherwin, T. Sphere-forming cells from peripheral cornea demonstrate polarity and directed cell migration. *Cell Biol. Int.* **2013**, *37*, 949–960.

109. Chng, Z.; Peh, G.S.; Herath, W.B.; Cheng, T.Y.; Ang, H.P.; Toh, K.P.; Robson, P.; Mehta, J.S.; Colman, A. High throughput gene expression analysis identifies reliable expression markers of human corneal endothelial cells. *PLoS ONE* **2013**, *8*.

110. Numata, R.; Okumura, N.; Nakahara, M.; Ueno, M.; Kinoshita, S.; Kanematsu, D.; Kanemura, Y.; Sasai, Y.; Koizumi, N. Cultivation of corneal endothelial cells on a pericellular matrix prepared from human decidua-derived mesenchymal cells. *PLoS ONE* **2014**, *9*.

111. Giasson, C.J.; Deschambeault, A.; Carrier, P.; Germain, L. Adherens junction proteins are expressed in collagen corneal equivalents produced *in vitro* with human cells. *Mol. Vis.* **2014**, *20*, 386–394.

112. Choi, J.S.; Kim, E.Y.; Kim, M.J.; Khan, F.A.; Giegengack, M.; D'Agostino, R., Jr.; Criswell, T.; Khang, G.; Soker, S. Factors affecting successful isolation of human corneal endothelial cells for clinical use. *Cell Transplant.* **2014**, *23*, 845–854.

113. Niu, G.; Choi, J.S.; Wang, Z.; Skardal, A.; Giegengack, M.; Soker, S. Heparin-modified gelatin scaffolds for human corneal endothelial cell transplantation. *Biomaterials* **2014**, *35*, 4005–4014.

114. Ha Thi, B.M.; Campolmi, N.; He, Z.; Pipparelli, A.; Manissolle, C.; Thuret, J.Y.; Piselli, S.; Forest, F.; Peoc'h, M.; Garraud, O.; *et al.* Microarray analysis of cell cycle gene expression in adult human corneal endothelial cells. *PLoS ONE* **2014**, *9*.

115. Koo, S.; Muhammad, R.; Peh, G.S.; Mehta, J.S.; Yim, E.K. Micro- and nanotopography with extracellular matrix coating modulate human corneal endothelial cell behavior. *Acta Biomater.* **2014**, *10*, 1975–1984.

116. Muhammad, R.; Peh, G.S.; Adnan, K.; Law, J.B.; Mehta, J.S.; Yim, E.K. Micro- and nano-topography to enhance proliferation and sustain functional markers of donor-derived primary human corneal endothelial cells. *Acta Biomater.* **2015**, *19*, 138–148.

117. Okumura, N.; Kakutani, K.; Numata, R.; Nakahara, M.; Schlotzer-Schrehardt, U.; Kruse, F.; Kinoshita, S.; Koizumi, N. Laminin-511 and -521 enable efficient *in vitro* expansion of human corneal endothelial cells. *Investig. Ophthalmol. Vis. Sci.* **2015**, *56*, 2933–2942.

118. Vianna, L.M.; Kallay, L.; Toyono, T.; Belfort, R., Jr.; Holiman, J.D.; Jun, A.S. Use of human serum for human corneal endothelial cell culture. *Br. J. Ophthalmol.* **2015**, *99*, 267–271.

119. Peh, G.S.; Chng, Z.; Ang, H.P.; Cheng, T.Y.; Adnan, K.; Seah, X.Y.; George, B.L.; Toh, K.P.; Tan, D.T.; Yam, G.H.; *et al.* Propagation of human corneal endothelial cells: A novel dual media approach. *Cell Transplant.* **2015**, *24*, 287–304.

120. Hyldahl, L. Primary cell cultures from human embryonic corneas. *J. Cell Sci.* **1984**, *66*, 343–351.

121. Konomi, K.; Joyce, N.C. Age and topographical comparison of telomere lengths in human corneal endothelial cells. *Mol. Vis.* **2007**, *13*, 1251–1258.

122. Mimura, T.; Joyce, N.C. Replication competence and senescence in central and peripheral human corneal endothelium. *Investig. Ophthalmol. Vis. Sci.* **2006**, *47*, 1387–1396.

123. Song, Z.; Wang, Y.; Xie, L.; Zang, X.; Yin, H. Expression of senescence-related genes in human corneal endothelial cells. *Mol. Vis.* **2008**, *14*, 161–170.

124. Gao, Y.; Zhou, Q.; Qu, M.; Yang, L.; Wang, Y.; Shi, W. *In vitro* culture of human fetal corneal endothelial cells. *Graefe Arch. Clin. Exp. Ophthalmol.* **2011**, *249*, 663–669.

125. Noh, J.W.; Kim, J.J.; Hyon, J.Y.; Chung, E.S.; Chung, T.Y.; Yi, K.; Wee, W.R.; Shin, Y.J. Stemness characteristics of human corneal endothelial cells cultured in various media. *Eye Contact Lens* **2015**, *41*, 190–196.

126. Amano, S.; Yamagami, S.; Mimura, T.; Uchida, S.; Yokoo, S. Corneal stromal and endothelial cell precursors. *Cornea* **2006**, *25*, S73–S77.

127. Mimura, T.; Yamagami, S.; Yokoo, S.; Usui, T.; Amano, S. Selective isolation of young cells from human corneal endothelium by the sphere-forming assay. *Tissue Eng. Part C Methods* **2010**, *16*, 803–812.

128. Rajasekhar, V.K.; Begemann, M. Concise review: Roles of polycomb group proteins in development and disease: A stem cell perspective. *Stem Cells* **2007**, *25*, 2498–2510.

129. Yu, W.Y.; Sheridan, C.; Grierson, I.; Mason, S.; Kearns, V.; Lo, A.C.; Wong, D. Progenitors for the corneal endothelium and trabecular meshwork: A potential source for personalized stem cell therapy in corneal endothelial diseases and glaucoma. *J. Biomed. Biotechnol.* **2011**, *2011*.

130. Zhang, K.; Pang, K.; Wu, X. Isolation and transplantation of corneal endothelial cell-like cells derived from *in vitro*-differentiated human embryonic stem cells. *Stem Cells Dev.* **2014**, *23*, 1340–1354.

131. Hatou, S.; Yoshida, S.; Higa, K.; Miyashita, H.; Inagaki, E.; Okano, H.; Tsubota, K.; Shimmura, S. Functional corneal endothelium derived from corneal stroma stem cells of neural crest origin by retinoic acid and wnt/beta-catenin signaling. *Stem Cells Dev.* **2013**, *22*, 828–839.

132. Joyce, N.C.; Harris, D.L.; Markov, V.; Zhang, Z.; Saitta, B. Potential of human umbilical cord blood mesenchymal stem cells to heal damaged corneal endothelium. *Mol. Vis.* **2012**, *18*, 547–564.

133. Dai, Y.; Guo, Y.; Wang, C.; Liu, Q.; Yang, Y.; Li, S.; Guo, X.; Lian, R.; Yu, R.; Liu, H.; *et al.* Non-genetic direct reprogramming and biomimetic platforms in a preliminary study for adipose-derived stem cells into corneal endothelia-like cells. *PLoS ONE* **2014**, *9*.

134. Shao, C.; Fu, Y.; Lu, W.; Fan, X. Bone marrow-derived endothelial progenitor cells: A promising therapeutic alternative for corneal endothelial dysfunction. *Cells Tissues Organs* **2011**, *193*, 253–263.

135. Ju, C.; Zhang, K.; Wu, X. Derivation of corneal endothelial cell-like cells from rat neural crest cells *in vitro*. *PLoS ONE* **2012**, *7*.

136. Johnson, D.H.; Bourne, W.M.; Campbell, R.J. The ultrastructure of Descemet's membrane. I. Changes with age in normal corneas. *Arch. Ophthalmol.* **1982**, *100*, 1942–1947.

137. Bayyoud, T.; Thaler, S.; Hofmann, J.; Maurus, C.; Spitzer, M.S.; Bartz-Schmidt, K.U.; Szurman, P.; Yoeruek, E. Decellularized bovine corneal posterior lamellae as carrier matrix for cultivated human corneal endothelial cells. *Curr. Eye Res.* **2012**, *37*, 179–186.

138. Yoeruek, E.; Bayyoud, T.; Maurus, C.; Hofmann, J.; Spitzer, M.S.; Bartz-Schmidt, K.U.; Szurman, P. Decellularization of porcine corneas and repopulation with human corneal cells for tissue-engineered xenografts. *Acta Ophthalmol* **2012**, *90*, e125–e131.

139. Yoshida, J.; Oshikata-Miyazaki, A.; Yokoo, S.; Yamagami, S.; Takezawa, T.; Amano, S. Development and evaluation of porcine atelocollagen vitrigel membrane with a spherical curve and transplantable artificial corneal endothelial grafts. *Investig. Ophthalmol. Vis. Sci.* **2014**, *55*, 4975–4981.

140. Ide, T.; Nishida, K.; Yamato, M.; Sumide, T.; Utsumi, M.; Nozaki, T.; Kikuchi, A.; Okano, T.; Tano, Y. Structural characterization of bioengineered human corneal endothelial cell sheets fabricated on temperature-responsive culture dishes. *Biomaterials* **2006**, *27*, 607–614.

141. Lai, J.Y.; Chen, K.H.; Hsu, W.M.; Hsiue, G.H.; Lee, Y.H. Bioengineered human corneal endothelium for transplantation. *Arch. Ophthalmol.* **2006**, *124*, 1441–1448.

142. Fukuda, K.; Chikama, T.; Nakamura, M.; Nishida, T. Differential distribution of subchains of the basement membrane components type IV collagen and laminin among the amniotic membrane, cornea, and conjunctiva. *Cornea* **1999**, *18*, 73–79.

143. Chen, H.J.; Pires, R.T.; Tseng, S.C. Amniotic membrane transplantation for severe neurotrophic corneal ulcers. *Br. J. Ophthalmol.* **2000**, *84*, 826–833.

144. Kubo, M.; Sonoda, Y.; Muramatsu, R.; Usui, M. Immunogenicity of human amniotic membrane in experimental xenotransplantation. *Investig. Ophthalmol. Vis. Sci.* **2001**, *42*, 1539–1546.

145. Wencan, W.; Mao, Y.; Wentao, Y.; Fan, L.; Jia, Q.; Qinmei, W.; Xiangtian, Z. Using basement membrane of human amniotic membrane as a cell carrier for cultivated cat corneal endothelial cell transplantation. *Curr. Eye Res.* **2007**, *32*, 199–215.

146. Utheim, T.P.; Lyberg, T.; Raeder, S. The culture of limbal epithelial cells. *Methods Mol. Biol.* **2013**, *1014*, 103–129.

147. Proulx, S.; Audet, C.; Uwamaliya, J.; Deschambeault, A.; Carrier, P.; Giasson, C.J.; Brunette, I.; Germain, L. Tissue engineering of feline corneal endothelium using a devitalized human cornea as carrier. *Tissue Eng. Part A* **2009**, *15*, 1709–1718.

148. Hashimoto, Y.; Funamoto, S.; Sasaki, S.; Honda, T.; Hattori, S.; Nam, K.; Kimura, T.; Mochizuki, M.; Fujisato, T.; Kobayashi, H.; *et al.* Preparation and characterization of decellularized cornea using high-hydrostatic pressurization for corneal tissue engineering. *Biomaterials* **2010**, *31*, 3941–3948.

149. Madden, P.W.; Lai, J.N.; George, K.A.; Giovenco, T.; Harkin, D.G.; Chirila, T.V. Human corneal endothelial cell growth on a silk fibroin membrane. *Biomaterials* **2011**, *32*, 4076–4084.

150. Menashi, S.; Vlodavsky, I.; Ishai-Michaeli, R.; Legrand, Y.; Fridman, R. The extracellular matrix produced by bovine corneal endothelial cells contains progelatinase a. *FEBS Lett.* **1995**, *361*, 61–64.

151. Astete, C.E.; Sabliov, C.M. Synthesis and characterization of plga nanoparticles. *J. Biomater. Sci. Polym. Ed.* **2006**, *17*, 247–289.

152. Hadlock, T.; Singh, S.; Vacanti, J.P.; McLaughlin, B.J. Ocular cell monolayers cultured on biodegradable substrates. *Tissue Eng.* **1999**, *5*, 187–196.

153. Haghjou, N.; Soheilian, M.; Abdekhodaie, M.J. Sustained release intraocular drug delivery devices for treatment of uveitis. *J. Ophthal. Vis. Res.* **2011**, *6*, 317–329.

154. Wadood, A.C.; Armbrecht, A.M.; Aspinall, P.A.; Dhillon, B. Safety and efficacy of a dexamethasone anterior segment drug delivery system in patients after phacoemulsification. *J. Cataract Refract. Surg.* **2004**, *30*, 761–768.

155. Gao, X.; Liu, W.; Han, B.; Wei, X.; Yang, C. Preparation and properties of a chitosan-based carrier of corneal endothelial cells. *J. Mater. Sci. Mater. Med.* **2008**, *19*, 3611–3619.

156. Wang, T.J.; Wang, I.J.; Lu, J.N.; Young, T.H. Novel chitosan-polycaprolactone blends as potential scaffold and carrier for corneal endothelial transplantation. *Mol. Vis.* **2012**, *18*, 255–264.

Transcriptome Analysis of Cultured Limbal Epithelial Cells on an Intact Amniotic Membrane following Hypothermic Storage in Optisol-GS

Tor Paaske Utheim, Panagiotis Salvanos, Øygunn Aass Utheim, Sten Ræder, Ole Kristoffer Olstad, Maria Fideliz de la Paz and Amer Sehic

Abstract: The aim of the present study was to investigate the molecular mechanisms underlying activation of cell death pathways using genome-wide transcriptional analysis in human limbal epithelial cell (HLEC) cultures following conventional hypothermic storage in Optisol-GS. Three-week HLEC cultures were stored in Optisol-GS for 2, 4, and 7 days at 4 °C. Partek Genomics Suite software v.6.15.0422, (Partec Inc., St. Louis, MO, USA) was used to identify genes that showed significantly different ($P < 0.05$) levels of expression following hypothermic storage compared to non-stored cell sheets. There were few changes in gene expression after 2 days of storage, but several genes were differently regulated following 4 and 7 days of storage. The histone-coding genes HIST1H3A and HIST4H4 were among the most upregulated genes following 4 and 7 days of hypothermic storage. Bioinformatic analysis suggested that these two genes are involved in a functional network highly associated with cell death, necrosis, and transcription of RNA. HDAC1, encoding histone deacetylase 1, was the most downregulated gene after 7 days of storage. Together with other downregulated genes, it is suggested that HDAC1 is involved in a regulating network significantly associated with cellular function and maintenance, differentiation of cells, and DNA repair. Our data suggest that the upregulated expression of histone-coding genes together with downregulated genes affecting cell differentiation and DNA repair may be responsible for increased cell death following hypothermic storage of cultured HLEC. In summary, our results demonstrated that a higher number of genes changed with increasing storage time. Moreover, in general, larger differences in absolute gene expression values were observed with increasing storage time. Further understanding of these molecular mechanisms is important for optimization of storage technology for limbal epithelial sheets.

Reprinted from *J. Funct. Biomater.* Cite as: Utheim, T.P.; Salvanos, P.; Utheim, Ø.A.; Ræder, S.; Pasovic, L.; Olstad, O.K.; de la Paz, M.F.; Sehic, A. Transcriptome Analysis of Cultured Limbal Epithelial Cells on an Intact Amniotic Membrane following Hypothermic Storage in Optisol-GS. *J. Funct. Biomater.* **2016**, *7*, 4.

1. Introduction

Limbal epithelial stem cells exist in specialized niches in the limbus [1] where they function to maintain the corneal epithelium [2]. When this function is lost through disease or injury, limbal stem cell deficiency (LSCD) results. This is a potentially blinding and painful disease characterized by neovascularization and ingrowth of the conjunctiva over the cornea. Transplantation of *ex vivo* expanded human limbal epithelial cells (HLEC) has proved successful in treating LSCD [1].

As cell-based corneal regenerative therapies become more common, demand for access to culture laboratories is anticipated to increase [3]. Concurrently, there is a trend towards increased centralization of culture facilities to meet increasingly strict safety regulations. An effective, standardized transport strategy would therefore have widespread clinical impact, allowing widespread distribution of cell-based regenerative treatment to eye clinics from specialized culture facilities. Recent studies have illustrated the feasibility of this strategy. Oie *et al.* demonstrated that cultured oral mucosal cell sheets retained viability and phenotype following 12 hours transportation in Japan [4]. Moreover, cultured conjunctival epithelial cells were successfully used for treatment of pterygium in 23 patients following distribution to four eye clinics in India [5]. Advantages of a standardized short-term storage and transport method for cultured HLEC include provision of a window for sterility and quality assessment, improved surgery logistics, and wider access to treatment.

We have previously shown that storage temperature has a significant effect on the quality of cultured HLECs when stored in Optisol-GS for one week. Morphology and viability of cultured HLECs deteriorated significantly following storage of cultured HLECs at 5 °C [6,7] compared to storage at 23 °C. Hypothermic storage in serum-free media has been widely used [8]. Nonetheless, it has been shown that hypothermic storage can be injurious to a variety of cell types [9] and excised corneas [10]. Extended hypothermic preservation induces oxidative stress through increased reactive oxygen species production, resulting in a myriad of effects on cellular function, including DNA damage and impaired repair mechanisms [11]. If the production of repair proteins is insufficient to repair the injury, cell death occurs [11,12].

The aim of the present study was to investigate the molecular mechanisms underlying activation of cell death pathways using genome-wide transcriptional analysis in HLEC cultures following 2, 4, and 7 days of conventional hypothermic storage in Optisol-GS at 4 °C.

2. Results

2.1. Genes Exhibiting Higher Levels of Expression Following 2, 4, and 7 Days of Hypothermic Storage Compared to Control

Nine genes were upregulated (>2-fold) at at least one of the three time-points investigated (Table 1). Following 2 days of hypothermic storage, only 1 gene, GSTM2 encoding glutathione S-transferase 2, was upregulated (>2-fold). The most substantial increase in gene expression in HLEC cultures stored at 4 °C in Optisol-GS was displayed after 4 and 7 days, with 6 and 8 genes being upregulated (>2-fold), respectively (Table 1). GSTM2 was the only gene showing over 2-fold increase in expression across every time-point investigated, demonstrating 2.4, 3.3, and 5.3-fold upregulation after 2, 4, and 7 days of hypothermic storage, respectively (Table 1). Two histone-coding genes, HIST1H3A and HIST4H4, were upregulated (>2-fold) following both 4 and 7 days of hypothermic storage (Table 1). Following 4 days of storage, HIST1H3A and HIST4H4, were 2.5 and 2.7-fold upregulated compared to non-stored cell sheets, respectively. Furthermore, after 7 days of hypothermic storage, the same two genes were 4.7 and 2.2-fold upregulated, respectively. No histone genes exhibited significantly increased levels of expression in HLEC cultures compared to control following 2 days of storage (Table 1).

Table 1. Genes exhibiting higher levels of expression following hypothermic storage compared to control. Included genes exhibit over 2-fold upregulation at at least one of the time-points investigated.

Symbol	2 Days *vs.* Ctr		4 Days *vs.* Ctr		7 Days *vs.* Ctr	
	P-value	Fold Change	*P*-value	Fold Change	*P*-value	Fold Change
CD177	1.41×10^{-1}	1.981	5.31×10^{-2}	2.050	2.71×10^{-2}	3.053
FMO1	2.53×10^{-1}	1.368	3.75×10^{-1}	1.273	1.49×10^{-2}	2.108
GLUD1	3.61×10^{-2}	1.644	6.07×10^{-3}	1.975	3.32×10^{-4}	2.746
GSTM2	2.59×10^{-4}	2.387	6.38×10^{-6}	3.312	1.30×10^{-7}	5.251
HIST1H3A	3.84×10^{-2}	1.333	1.75×10^{-5}	2.528	2.80×10^{-8}	4.660
HIST4H4	9.95×10^{-2}	1.112	5.19×10^{-8}	2.655	2.49×10^{-10}	2.164
RNU11	4.65×10^{-5}	1.963	2.32×10^{-5}	2.045	5.64×10^{-6}	2.316
RNU4-1	1.02×10^{-1}	1.365	1.51×10^{-2}	1.624	6.75×10^{-4}	2.163
SLC27A2	4.71×10^{-1}	1.161	2.44×10^{-3}	2.033	7.55×10^{-1}	1.070

Ctr: non-stored cell sheets.

Bioinformatic analysis showed that the upregulated (>2-fold) genes after 4 and 7 days of hypothermic storage were involved in a functional network regulating molecular and cellular functions involved in "cell death," "necrosis," and "transcription of RNA" (Figure 1, Table 2). These results suggest these cellular functions to be more prominent in HLEC cultures after 4 and 7 days of storage in Optisol-GS at 4 °C compared to non-stored cell sheets. The resulting functional network consisted of a total of 26 genes with only 2 upregulated (>2-fold) genes (HIST1H3A and HIST4H4).

It was not suggested that the remaining 7 upregulated genes (CD177, FMO1, GLUD1, GSTM2, RNU11, RNU4-1, and SLC27A2) were involved in this network (Figure 1). These findings demonstrate that among 9 upregulated (>2-fold) genes, HIST1H3A and HIST4H4 may play the most important functional role.

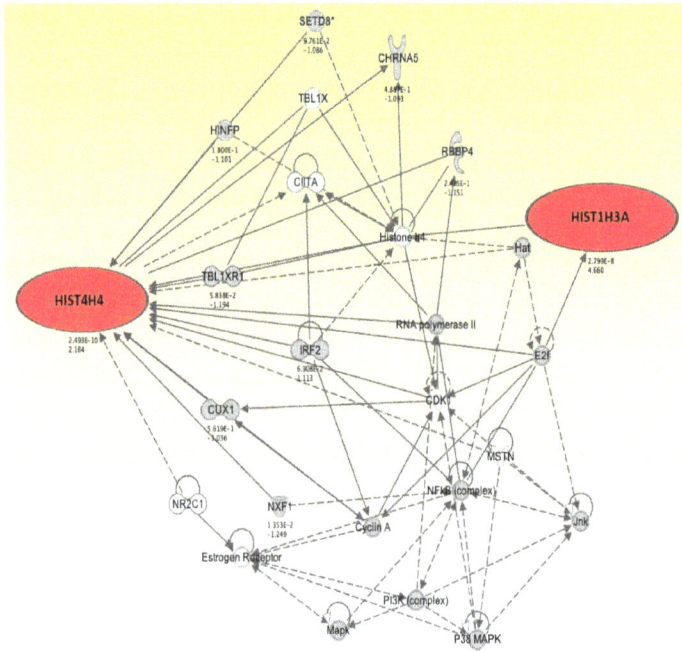

Figure 1. Functional network derived using upregulated (>2-fold) genes after 7 days of conventional hypothermic storage in Optisol-GS. Genes are represented as nodes and relationship between nodes are represented as lines. Expression ratios (7 days *vs.* control) are shown below the nodes. Red colored nodes represent upregulated (>2-fold) genes following 7 days of storage compared to non-stored cell sheets. The remaining nodes do not belong to the upregulated population of the genes, but are found as components of the network.

2.2. Genes Exhibiting Lower Levels of Expression Following 2, 4, and 7 Days of Hypothermic Storage Compared to Control

In total, 26 genes were downregulated (<-2-fold) at at least one of the three time-points investigated (Table 3). Seven of these genes were downregulated (<-2-fold) after 2 days of hypothermic storage, whereas 16 and 14 genes were downregulated (<-2-fold) following 4 and 7 days of storage, respectively (Table 3). Interestingly, only one gene, miR-21, showed an over 2-fold decrease in expression across every time-point investigated, with 2.7, 3.3, and 2.1-fold downregulation after

2, 4, and 7 days of hypothermic storage, respectively (Table 3). HDAC1, encoding histone deacetylase 1, was the most downregulated gene after 7 days of hypothermic storage, exhibiting 3.4-fold decrease in expression compared to control (Table 3).

Table 2. Top ten molecular and cellular functions significantly associated with the functional network (from Figure 1) derived using upregulated (>2-fold) genes after 7 days of hypothermic storage in Optisol-GS.

Functions	P-Value	No of Genes
Cell death	1.47×10^{-7}	20
Necrosis	3.06×10^{-7}	18
Transcription of RNA	4.12×10^{-4}	17
Binding of DNA	5.11×10^{-4}	17
Transcription of RNA	7.22×10^{-4}	12
Cellular assembly and organization	1.23×10^{-6}	9
Transcription of DNA	5.63×10^{-7}	8
Activation of DNA endogenous promotor	2.14×10^{-6}	5
Cell-cycle progression	2.27×10^{-3}	4
Gene expression	1.35×10^{-4}	2

Table 3. Genes Exhibiting Lower Levels of Expression Following Hypothermic Storage Compared to Control. Included genes exhibit over 2-fold downregulation at at least one of the time-points investigated.

Symbol	2 Days *vs.* Ctr		4 Days *vs.* Ctr		7 Days *vs.* Ctr	
	P-Value	Fold Change	P-Value	Fold Change	P-Value	Fold Change
ANKRD50	2.82×10^{-2}	−1.323	2.32×10^{-6}	−2.191	2.35×10^{-4}	−1.749
ANKRD36B	4.65×10^{-3}	−2.292	1.85×10^{-3}	−2.546	9.24×10^{-3}	−1.979
C9orf3	7.91×10^{-4}	−3.622	5.04×10^{-4}	−3.860	6.56×10^{-2}	−1.938
CCDC88C	1.82×10^{-1}	−1.363	1.51×10^{-1}	−1.398	6.70×10^{-3}	−2.041
CYP24A1	9.28×10^{-1}	−1.027	8.44×10^{-1}	−1.060	3.32×10^{-2}	−2.011
DGKH	3.48×10^{-5}	−1.619	8.80×10^{-8}	−2.114	2.51×10^{-5}	−1.681
DHFR	3.77×10^{-4}	−1.779	6.90×10^{-4}	−1.717	3.10×10^{-5}	−2.139
FAP	2.07×10^{-1}	−1.562	4.72×10^{-1}	−1.284	3.01×10^{-2}	−2.314
GTF2B	2.33×10^{-2}	−1.580	6.39×10^{-4}	−2.131	1.37×10^{-3}	−2.072
HDCA1	8.97×10^{-1}	−1.015	2.43×10^{-1}	−1.147	1.11×10^{-2}	−3.399
HAS2	6.50×10^{-2}	−1.375	7.49×10^{-4}	−1.920	8.18×10^{-5}	−2.342
LIF	4.97×10^{-2}	−1.906	2.85×10^{-1}	−1.403	1.23×10^{-2}	−2.440
LRRN1	2.41×10^{-1}	−1.422	1.78×10^{-1}	−1.501	1.71×10^{-2}	−2.217
mir-21	5.77×10^{-4}	−2.677	6.96×10^{-5}	−3.347	7.02×10^{-3}	−2.131
MPLKIP	2.86×10^{-4}	−1.829	1.59×10^{-4}	−1.983	4.89×10^{-6}	−2.618
NPIPL3	3.40×10^{-3}	−2.249	2.27×10^{-3}	−2.348	5.88×10^{-2}	−1.398
NRG1	1.57×10^{-3}	−1.666	4.65×10^{-5}	−2.067	1.12×10^{-4}	−2.022
PLA2G7	6.14×10^{-1}	−1.062	3.73×10^{-2}	−1.299	1.61×10^{-5}	−2.018
PSD3	1.43×10^{-1}	−1.141	1.78×10^{-5}	−2.162	9.70×10^{-5}	−2.012
RNF152	4.83×10^{-1}	−1.118	1.16×10^{-4}	−2.119	6.23×10^{-2}	−1.381
SESN3	6.85×10^{-2}	−1.206	5.50×10^{-7}	−2.044	6.02×10^{-5}	−1.684
SLC7A11	1.55×10^{-1}	−1.279	3.03×10^{-4}	−2.078	1.02×10^{-3}	−1.963
SMAD2	4.59×10^{-3}	−1.526	9.51×10^{-6}	−2.195	2.76×10^{-4}	−1.848
SMG1	1.43×10^{-4}	−2.030	2.45×10^{-5}	−2.503	4.67×10^{-4}	−1.395
TAF1D	7.99×10^{-3}	−1.961	2.85×10^{-3}	−2.177	2.71×10^{-2}	−1.770
TRA2A	1.38×10^{-3}	−2.456	3.45×10^{-3}	−2.229	6.73×10^{-4}	−1.357

Ctr: non-stored cell sheets.

Bioinformatic analysis demonstrated that downregulated (<-2-fold) genes following 4 and 7 days of hypothermic storage constituted an important part of a functional network involved in "cellular assembly and organization," "differentiation of cells," and "DNA repair" (Figure 2, Table 4). The network consisted of a total of 32 genes and included 9 downregulated (<-2-fold) genes, *i.e.* HDCA1, GTF2B, MiR-21, PLA2G7, LIF, NRG1, HAS2, DHFR, and CYP24A1 (Figure 2). Our results suggest that these functions are impaired in HLEC cultures after 4 and 7 days of conventional hypothermic storage compared to non-stored cell sheets.

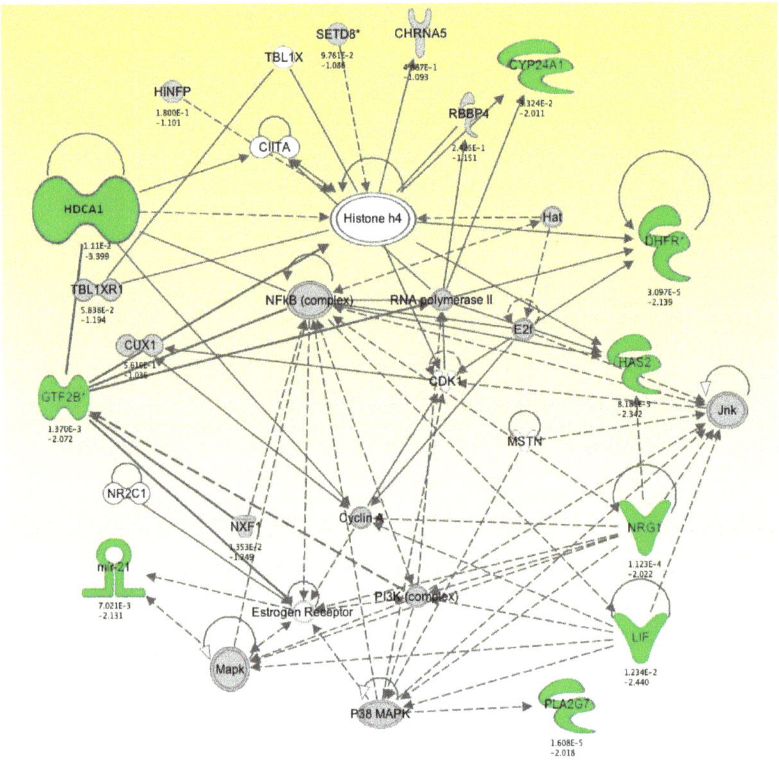

Figure 2. Functional network derived using downregulated (>2-fold) genes after 7 days of conventional hypothermic storage in Optisol-GS. Genes are represented as nodes and relationship between nodes are represented as lines. Expression ratios (7 days *vs.* control) are shown below the nodes. Green colored nodes represent downregulated (<-2-fold) genes following 7 days of storage compared to non-stored cell sheets. The remaining nodes do not belong to the downregulated population of the genes, but are found as components of the network.

Table 4. Top ten molecular and cellular functions significantly associated with the functional network (from Figure 2) derived using downregulated (<-2-fold) genes after 7 days of hypothermic storage in Optisol-GS.

Functions	P-Value	No of Genes
Cellular assembly and organization	4.28×10^{-5}	19
Differentiation of cells	1.51×10^{-2}	19
DNA repair	6.17×10^{-3}	16
Cellular function and maintenance	1.31×10^{-2}	13
Transactivation of RNA	4.60×10^{-4}	10
Binding of DNA	1.86×10^{-3}	9
Activation of DNA endogenous promotor	3.45×10^{-3}	8
G1/S phase transition	9.14×10^{-3}	6
Cell-cycle progression	3.23×10^{-3}	2
Transcription of DNA	1.12×10^{-2}	3

3. Discussion

Genes exhibiting higher levels of expression following 4 and 7 days of storage of cultured HLEC at 4 °C compared to non-stored cell sheets were significantly associated with cell death, necrosis, and transcription of RNA (Table 2). In contrast, downregulated genes following 4 and 7 days of hypothermic storage were associated with cellular assembly and organization, differentiation of cells, and DNA repair (Table 4). These results suggest enhanced apoptosis as a result of hypothermic storage, which is in line with previous studies showing both apoptosis and necrosis during corneal storage at 4 °C, with apoptosis appearing to predominate [10].

In a previous study on one-week storage in Optisol-GS of cultured LEC at 5 °C, few apoptotic cells were observed [6]. Interestingly, in contrast to ambient organ culture storage, storage in Optisol-GS at 5 °C induced dilated intercellular spaces, increased intracellular vacuoles, detachment of epithelial cells, and detachment of the epithelia from the amniotic membrane. Besides weak to moderate chromatin condensation, rupture of cell membranes and dissolution of organelles were frequently observed, indicative of necrosis [6].

Based on these results, the present study was designed to get insight into possible underlying mechanisms. Comparing cultured LEC subjected to one-week storage at 4 °C with cultured, non-stored cells (control cells) would be sufficient to meet this end. However, such a design would not give any insight into the effects of storage time on gene expression. Therefore, we extended the study to include three storage times to allow information on both the effects of storage time and 4 °C as a storage temperature. As cells are not cultured at 4 °C, we did not include such an experimental group. We did not perform gene analyses after 1 week of storage at 23 °C, as a previous study demonstrated excellent results at this temperature [7]. Moreover, our aim was to suggest possible mechanisms that deserve further studies to improve storage technology. In summary, our results demonstrated that a higher number of genes changed with increasing storage time. In general, larger differences in absolute gene expression values were observed with increasing storage time.

Explants from a total of four donors (two pairs) were distributed evenly between the four experimental groups. We made sure that explants from the superior region of the limbal ring was included in each of the experimental groups, as such explants are known to generate superior growth [13]. The superior region of the cornea was easily identified due to a suture at 12 o'clock position fastened by the surgeon at the time of enucleation.

After 4 days of hypothermic storage, a more than 2-fold increase in the expression of two histone-coding genes (HIST1H3A and HIST4H4) was observed. This upward trend was strengthened after 7 days of hypothermic storage. Transcriptome analysis of human corneal endothelium (HCE) has shown that HIST1H3A was among nine genes that displayed the most significant differential expression between pediatric and adult HCE [14]. The authors suggested that this gene was important for cell division in corneal endothelium. So far, there have not been any studies demonstrating the expression of HIST4H4 in human corneas; however, Zhang and colleagues have shown that transcriptional activation of histone H4 is important for adipocyte differentiation [15]. The histone gene transcription is cell-cycle dependent and rapidly induced by a chain of response effects at the transcriptional and translational levels when cells are subjected to diverse stress stimuli, independent of the type of stimulus [16–18]. A robust increase in unprocessed histone mRNA is observed upon activation of the DNA damage checkpoint [19]. Our findings suggest that the low viability after one week of hypothermic storage of HLEC [7] can be due to a histone-mediated mechanism and that failure to repair DNA damage may explain cell death and reduced viability of the transplants.

Allis and Turner proposed the "histone code" hypothesis where gene transcription is changed in response to the modification of histones, through altered access to promoter regions [20,21]. Specific histone modifications have also been linked to apoptotic chromatin changes, providing evidence for the existence of an apoptotic histone code [22]. Among the differentially downregulated genes, HDAC1 encoding histone deacetylase 1 exhibited the lowest levels of expression after 7 days of storage with 3.4-fold change compared to control. Acetylation and deacetylation of histones play an important role in transcription regulation of eukaryotic cells by decreasing histone-DNA interaction and promoting accessibility of the DNA for transcription activation [23,24]. In general, acetylation of histones promotes a more relaxed chromatin structure, allowing transcriptional activation [23]. HDACs can act as transcription repressors, due to histone deacetylation, and consequently promote chromatin condensation. HDAC inhibitors (HDACi) selectively alter gene transcription, in part, by chromatin remodeling and by changes in the structure of proteins in transcription factor complexes [25]. Further, the HDACs have many non-histone proteins substrates such as hormone receptors, chaperone proteins and cytoskeleton proteins, which regulate cell proliferation and cell death [25]. Thus, HDACi-induced cell death involves transcription-dependent and transcription-independent mechanisms [26–28]. It has also recently been postulated that histone deacetylase inhibitors could be used in the prevention and treatment of corneal haze and scar formation [29]. Decreased expression of HDAC1 in cultured HLEC following hypothermic storage compared to control may lead to increased acetylation of histones, which in turn results in enhanced transcription of RNA. This is in accordance with our findings suggesting that HIST1H13A and HIST4H4 and their related genes in the functional network (Figure 2) regulate transcription of RNA (Table 2).

Interestingly, only one gene, miR-21, was found among differentially expressed genes. MiR-21 exhibited significantly lower levels of expression across every time-point investigated after hypothermic storage (Table 3). MiRNAs play important functions in cell differentiation, cell proliferation, apoptosis, metabolism, and immune regulation by promoting the degradation of their target mRNA or inhibiting mRNA translation [30]. Overexpression of miR-21, an oncogenic miRNA, is associated with the progression, metastasis, and poor prognosis of many types of tumors [31–33]. MiR-21 is also known to be highly upregulated in malignant glioma, and inhibition of miR-21 activity was found to enhance cell death of malignant glioma cells [34]. Therefore, it may be speculated that increased cell death of HLEC following hypothermic storage is associated with decreased expression of miR-21.

Further research is warranted on the effect of different storage media and temperatures on gene expression. In conclusion, this study gives preliminary insight into the molecular mechanisms that may explain the low viability when HLEC are stored at 4 °C. Further investigations into time-dependent molecular mechanisms during storage of cultured cells may provide clues for optimization of storage medium for use in regenerative medicine technology.

4. Experimental Section

Dulbecco's minimal essential medium (DMEM), HEPES-buffered DMEM containing sodium bicarbonate and Ham's F12 (1:1), Dulbecco's modified Eagle's medium, Hanks' balanced salt solution, fetal bovine serum (FBS), insulin–transferrin–sodium selenite media supplement, human epidermal growth factor, dimethyl sulfoxide, hydrocortisone, gentamicin, and amphotericin B were purchased from Sigma-Aldrich (St. Louis, MO, USA). Dispase II was obtained from Roche Diagnostics (Basel, Switzerland), cholera toxin A subunit from Biomol (Exeter, UK), Ethicon Ethilon 6-0 C-2 monofilament suture from Johnson & Johnson (New Brunswick, NJ, USA), Netwell culture plate inserts from Costar Corning (New York,

NY, USA), vancomycin from Abbott Laboratories (Abbott Park, IL, USA), and the polypropylene containers from Plastiques Gosselin (Hazebrouck Cedex, France). Optisol-GS was purchased from Bausch&Lomb (Irvine, CA, USA). GeneChip HT One-Cycle cDNA Synthesis Kit, GeneChip HT IVT Labeling Kit, and GeneChip Human Gene 1.0 ST Arrays were from Affymetrix (Santa Clara, CA, USA).

4.1. Human Tissue Preparation

Human tissue was handled according to the Declaration of Helsinki. The experiment was conducted using four human corneas (two pairs) obtained from Centro de Oftalmologia Barraquer (Barcelona, Spain). Placing a suture in the superior scleral region prior to enucleation oriented the globes, and the corneoscleral tissue was excised using a 14 mm trephine. The limbal tissue was prepared in a class II safety cabinet as previously reported by Meller and colleagues [35]. The tissue was rinsed three times with DMEM (Sigma-Aldrich, St. Louis, MO, USA) containing 50 μg/mL gentamicin (Sigma-Aldrich) and 1.25 μg/mL amphotericin B (Sigma-Aldrich). After careful elimination of excessive sclera, conjunctiva, iris, and corneal endothelium, the remaining tissue was placed in a culture dish and exposed for 10 minutes to Dispase II (Roche Diagnostics) in Mg^{2+} and Ca^{2+} free Hanks' balanced salt solution (Sigma-Aldrich) at 37 °C under humidified 5% carbon dioxide and carefully rinsed with DMEM containing 10% FBS (Sigma-Aldrich). The central corneal button was eliminated using a KAI 6 mm trephine. The paired corneoscleral rims were divided into 24 explants, which were equally distributed between the four experimental groups with regard to limbal circumference origin.

4.2. Human Limbal Explant Cultures on Intact Amniotic Membranes

Human amniotic membrane was preserved in accordance with a method previously reported by Lee & Tseng and according to the Declaration of Helsinki. After thawing at room temperature, amniotic membrane with the epithelium intact and facing up was fastened to the polyester membrane of a Netwell culture plate insert (Costar Corning, New York, NY, USA) using Ethicon Ethilon 6-0 monofilament suture (Johnson & Johnson, New Brunswick, NJ, USA) as previously reported. On the center of each amniotic membrane insert, the explants were cultured in a supplemented hormonal epithelial medium with the epithelial side facing down as previously reported. The medium was made of HEPES-buffered DMEM (Sigma-Aldrich) containing sodium bicarbonate and Ham's F12 (1:1) and was supplemented with 5% FBS, 0.5% dimethyl sulfoxide (Sigma-Aldrich), 2 ng/m human epidermal growth factor (Sigma-Aldrich), 5 μg/mL insulin (Sigma-Aldrich), 5 μg/mL transferrin (Sigma-Aldrich), 5 ng/mL selenium (Sigma-Aldrich), 3 ng/mL hydrocortisone (Sigma-Aldrich), 30 ng/mL cholera toxin (Biomol), 50 μg/mL gentamicin, and 1.25 μg/mL amphotericin B. Cultures were incubated for 3 weeks at 37 °C in an atmosphere of humidified 5% carbon dioxide and 95% air, and the medium was changed every 2 to 3 days. Three-week HLEC cultures were prepared for eye bank storage ($n = 18$) and controls (nonstored tissue) ($n = 6$).

4.3. Hypothermic Storage of Cultured Human Limbal Epithelial Cells in Optisol-GS

Preparation for eye bank storage was performed in a class II safety cabinet. Twenty-five milliliters of Optisol-GS was added to radiation-sterilized 90-mL Plastiques Gosselin polypropylene containers (interior diameter 43 mm). Three-week HLEC cultures in polyester culture plate inserts were transferred by a disposable forceps to the storage containers (Figure 3). The hinged cap with septum was closed to establish a closed tissue storage system, and the containers were stored for 2 ($n = 6$), 4 ($n = 6$), and 7 days ($n = 6$) at 4 °C.

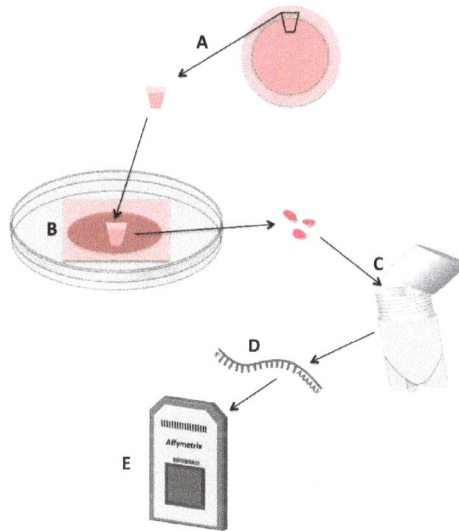

Figure 3. Experimental design of the study. The corneoscleral tissue was excised (**A**); HLECs were cultured for 3 weeks on intact amniotic membranes in supplemented hormonal epithelial medium (**B**); Disks of cultured epithelium were trephined with a 5-mm biopsy punch and stored in Optisol-GS at 4 °C (**C**); RNA was extracted (**D**); One hundred nanograms of total RNA was subjected to cDNA synthesis and labeling. Labeled and fragmented single stranded DNAs were hybridized to the gene microarray (**E**) before washing and staining.

4.4. RNA Isolation

Disks of cultured epithelium and amniotic membrane on polyester membranes were trephinated from the cultures using a 5-mm biopsy punch. Biopsies were stored in cryotubes at −80°C until needed. Total RNA was extracted from thawed biopsies using a Qiagen RNeasy Micro Kit (Hilden, Germany), according to the manufacturer's protocol. Three hundred and fifty microliters of RTL buffer containing beta-mercaptoethanol was added to the disks in microcentrifuge tubes and vortexed for 2 min. RNA concentration and purity was determined through measurement of A260/A280 ratios with the Nano Drop ND-1000 Spectrophotometer (Thermo Fisher Scientific, Wilmington, DE, USA). Confirmation of RNA quality was assessed by the Agilent BioAnalyzer 2100 System and RNA 6000 Nano Assay (Agilent Technologies, Santa Clara, CA, USA). RNA samples were immediately frozen and stored at $-80\,^\circ$C.

4.5. Microarray Analysis

The Affymetrix GeneChip Human Gene 1.0 ST Microarrays (Affymetrix, Santa Clara, CA, USA) used in this study included approximately 28,000 gene transcripts. Microarray analysis was carried out in triplicate using cultured HLEC stored in Optisol-GC at 4 $^\circ$C for 2, 4, and 7 days, and using non-stored control cultures. Preparation of complementary DNA (cDNA) was carried out using GeneChip HT One-Cycle cDNA Synthesis Kit (Affymetrix). Each of three microarrays was hybridized with cDNA prepared from 100 ng of total RNA from each resulting solution. Biotinylated and fragmented single stranded cDNAs were hybridized to the GeneChips. The arrays were washed and stained using FS-450 fluidics station (Affymetrix).

Signal intensities were detected by Hewlett Packard Gene Array Scanner 3000 7G (Hewlett Packard, Palo Alto, CA, USA). The scanned images were processed using the AGCC (Affymetrix GeneChip Command Console) software and the CEL files were imported into Partek Genomics Suite software (Partek Inc., St. Louis, MO, USA). The Robust Multichip Analysis (RMA) algorithm was applied to generate signal values and normalization. Gene transcripts with maximal signal values of less than 32 across all arrays were removed to filter for low and non-expressed genes. For expression comparisons of different groups, profiles were compared using a 1-way ANOVA model. The results were expressed as fold changes (FC) with corresponding P values.

4.6. Bioinformatic Analysis

Bioinformatic analysis using Ingenuity Pathways Analysis (IPA) (Ingenuity Inc., Redwood City, CA, USA) was carried out to find molecular and cellular functions and canonical pathways that were significantly associated with differentially expressed genes. Briefly, the data set containing gene identifiers and corresponding FCs and P values was uploaded onto the web-delivered application and each gene identifier was mapped to its corresponding gene object in the Ingenuity Pathways Knowledge Base (IPKB). Functional analysis identified the biological functions and/or diseases that were significantly associated with the data sets. Fisher's exact test was performed to calculate a P value determining the probability that each biological function and/or disease assigned to the data set was due to chance alone. The data sets were mined for significant pathways with the IPA library of canonical pathways, using IPA generated networks as graphical representations of the molecular relationships between genes and gene products.

Acknowledgments: Acknowledgments: We would like to thank Catherine Jackson and Astrid Østerud at the Department of Medical Biochemistry for their excellent assistance and support. Funding was obtained from the Department of Medical Biochemistry, Oslo University Hospital, Oslo, Norway and the South-Eastern Norway Regional Health Authority, Norway.

Author Contributions: Author Contributions: Tor Paaske Utheim, Øygunn Aass Utheim and Sten Ræder conceived and designed the experiments; Øygunn Aass Utheim, Tor Paaske Utheim and Sten Ræder performed the experiments; Panagiotis Salvanos, Lara Pasovic, Amer Sehic, Ole Kristoffer Olstad and Tor Paaske Utheim analyzed the data; Maria Fideliz de la Paz, Tor Paaske Utheim, Ole Kristoffer Olstad, Sten Ræder contributed reagents/materials/analysis tools; All wrote the paper.

Conflicts of Interest: Conflicts of interest: The authors declare no conflict of interest.

References

1. Schermer, A.; Galvin, S.; Sun, T.T. Differentiation-related expression of a major 64K corneal keratin *in vivo* and in culture suggests limbal location of corneal epithelial stem cells. *J. Cell Biol.* **1986**, *103*, 49–62.
2. Thoft, R.A.; Friend, J. The X, Y, Z hypothesis of corneal epithelial maintenance. *Invest. Ophthalmol. Vis. Sci.* **1983**, *24*, 1442–1443.
3. Ahmad, S.; Osei-Bempong, C.; Dana, R.; Jurkunas, U. The culture and transplantation of human limbal stem cells. *J. Cell Physiol.* **2010**, *225*, 15–19.
4. Oie, Y.; Nishida, K. Translational research on ocular surface reconstruction using oral mucosal epithelial cell sheets. *Cornea* **2014**, *33*, S47–S52.
5. Vasania, V.S.; Hari, A.; Tandon, R.; Shah, S.; Haldipurkar, S.; Shah, S.; Sachan, S.; Viswanathan, C. Transplantation of autologous *ex vivo* expanded human conjunctival epithelial cells for treatment of pterygia: A prospective open-label single arm multicentric clinical trial. *J. Ophthalmic Vis. Res.* **2014**, *9*, 407–416.
6. Raeder, S.; Utheim, T.P.; Utheim, O.A.; Nicolaissen, B.; Roald, B.; Cai, Y.; Haug, K.; Kvalheim, A.; Messelt, E.B.; Drolsum, L.; *et al.* Effects of organ culture and optisol-GS storage on structural integrity, phenotypes, and apoptosis in cultured corneal epithelium. *Invest. Ophthalmol. Vis. Sci.* **2007**, *48*, 5484–5493.
7. Utheim, T.P.; Utheim, T.P.; Raeder, S.; Eidet, J.; Stormo, C.; de la Paz, M.; Utheim, O.A. Storage of cultured human limbal epithelial cells in Optisol-GS at 23 °C *versus* 5 °C. *Invest. Ophthalmol. Vis. Sci.* **2009**, *50*, 1778–1779.
8. Pels, E.; Beele, H.; Claerhout, I. Eye bank issues: II. Preservation techniques: Warm *versus* cold storage. *Int. Ophthalmol.* **2008**, *28*, 155–163.

9. Abrahamse, S.L.; van Runnard Heimel, P.; Hartman, R.J.; Chamuleau, R.A.; van Gulik, T.M. Induction of necrosis and DNA fragmentation during hypothermic preservation of hepatocytes in UW, HTK, and Celsior solutions. *Cell Transplant.* **2003**, *12*, 59–68.

10. Komuro, A.; Hodge, D.O.; Gores, G.J.; Bourne, W.M. Cell death during corneal storage at 4 °C. *Investig. Ophthalmol. Vis. Sci.* **1999**, *40*, 2827–2832.

11. Rauen, U.; de Groot, H. Mammalian cell injury induced by hypothermia—The emerging role for reactive oxygen species. *Biol. Chem.* **2002**, *383*, 477–488.

12. Fitton, T.P.; Wei, C.; Lin, R.; Bethea, B.T.; Barreiro, C.J.; Amado, L.; Gage, F.; Hare, J.; Baumgartner, W.A.; Conte, J.V. Impact of 24 h continuous hypothermic perfusion on heart preservation by assessment of oxidative stress. *Clin. Transplant.* **2004**, *18*, 22–27.

13. Utheim, T.P.; Raeder, S.; Olstad, O.K.; Utheim, O.A.; de La Paz, M.; Cheng, R.; Huynh, T.T.; Messelt, E.; Roald, B.; Lyberg, T. Comparison of the histology, gene expression profile, and phenotype of cultured human limbal epithelial cells from different limbal regions. *Invest. Ophthalmol. Vis. Sci.* **2009**, *50*, 5165–5172.

14. Frausto, R.F.; Wang, C.; Aldave, A.J. Transcriptome analysis of the human corneal endothelium. *Investig. Ophthalmol. Vis. Sci.* **2014**, *55*, 7821–7830.

15. Zhang, Y.Y.; Li, X.; Qian, S.W.; Guo, L.; Huang, H.Y.; He, Q.; Liu, Y.; Ma, C.G.; Tang, Q.Q. Transcriptional activation of histone H4 by C/EBPβ during the mitotic clonal expansion of 3T3-L1 adipocyte differentiation. *Mol. Biol. Cell* **2011**, *22*, 2165–2174.

16. Bongiorno-Borbone, L.; de Cola, A.; Barcaroli, D.; Knight, R.A.; di Ilio, C.; Melino, G.; de Laurenzi, V. FLASH degradation in response to UV-C results in histone locus bodies disruption and cell-cycle arrest. *Oncogene* **2010**, *29*, 802–810.

17. Kratzmeier, M.; Albig, W.; Meergans, T.; Doenecke, D. Changes in the protein pattern of H1 histones associated with apoptotic DNA fragmentation. *Biochem. J.* **1999**, *337*, 319–327.

18. Sokol, A.; Kwiatkowska, A.; Jerzmanowski, A.; Prymakowska-Bosak, M. Up-regulation of stress-inducible genes in tobacco and arabidopsis cells in response to abiotic stresses and aba treatment correlates with dynamic changes in histone H3 and H4 modifications. *Planta* **2007**, *227*, 245–254.

19. Kaygun, H.; Marzluff, W.F. Translation termination is involved in histone mRNA degradation when DNA replication is inhibited. *Mol. Cell Biol.* **2005**, *25*, 6879–6888.

20. Strahl, B.D.; Allis, C.D. The language of covalent histone modifications. *Nature* **2000**, *403*, 41–45.

21. Turner, B.M. Histone acetylation and an epigenetic code. *Bioessays* **2000**, *22*, 836–845.

22. Fullgrabe, J.; Hajji, N.; Joseph, B. Cracking the death code: Apoptosis-related histone modifications. *Cell Death Differ.* **2010**, *17*, 1238–1243.

23. Lehrmann, H.; Pritchard, L.L.; Harel-Bellan, A. Histone acetyltransferases and deacetylases in the control of cell proliferation and differentiation. *Adv. Cancer Res.* **2002**, *86*, 41–65.

24. Mai, A.; Massa, S.; Rotili, D.; Cerbara, I.; Valente, S.; Pezzi, R.; Simeoni, S.; Ragno, R. Histone deacetylation in epigenetics: An attractive target for anticancer therapy. *Med. Res. Rev.* **2005**, *25*, 261–309.

25. Gui, C.Y.; Ngo, L.; Xu, W.S.; Richon, V.M.; Marks, P.A. Histone deacetylase (HDAC) inhibitor activation of $p21^{WAF1}$ involves changes in promoter-associated proteins, including HDAC1. *Proc. Natl. Acad. Sci. USA* **2004**, *101*, 1241–1246.

26. Bolden, J.E.; Peart, M.J.; Johnstone, R.W. Anticancer activities of histone deacetylase inhibitors. *Nat. Rev. Drug Discov.* **2006**, *5*, 769–784.

27. Marks, P.A.; Dokmanovic, M. Histone deacetylase inhibitors: Discovery and development as anticancer agents. *Expert. Opin. Investig. Drugs* **2005**, *14*, 1497–1511.

28. Rosato, R.R.; Grant, S. Histone deacetylase inhibitors: Insights into mechanisms of lethality. *Expert. Opin. Ther. Targets* **2005**, *9*, 809–824.

29. Zhou, Q.; Wang, Y.; Yang, L.; Wang, Y.; Chen, P.; Wang, Y.; Dong, X.; Xie, L. Histone deacetylase inhibitors blocked activation and caused senescence of corneal stromal cells. *Mol. Vis.* **2008**, *14*, 2556–2565.

30. Hammond, S.M. An overview of microRNAs. *Adv. Drug Deliv. Rev.* **2015**, *87*, 3–14.

31. Asangani, I.A.; Rasheed, S.A.; Nikolova, D.A.; Leupold, J.H.; Colburn, N.H.; Post, S.; Allgayer, H. MicroRNA-21 (miR-21) post-transcriptionally downregulates tumor suppressor Pdcd4 and stimulates invasion, intravasation and metastasis in colorectal cancer. *Oncogene* **2008**, *27*, 2128–2136.

32. Kadera, B.E.; Li, L.; Toste, P.A.; Wu, N.; Adams, C.; Dawson, D.W.; Donahue, T.R. MicroRNA-21 in pancreatic ductal adenocarcinoma tumor-associated fibroblasts promotes metastasis. *PLoS ONE* **2013**, *8*.

33. Petrovic, N.; Mandusic, V.; Stanojevic, B.; Lukic, S.; Todorovic, L.; Roganovic, J.; Dimitrijevic, B. The difference in miR-21 expression levels between invasive and non-invasive breast cancers emphasizes its role in breast cancer invasion. *Med. Oncol.* **2014**, *31*, 867.

34. Harmalkar, M.; Upraity, S.; Kazi, S.; Shirsat, N.V. Tamoxifen-induced cell death of malignant glioma cells is brought about by oxidative-stress-mediated alterations in the expression of BCL2 family members and is enhanced on miR-21 inhibition. *J. Mol. Neurosci.* **2015**, *57*, 197–202.

35. Meller, D.; Pires, R.T.; Tseng, S.C. *Ex vivo* preservation and expansion of human limbal epithelial stem cells on amniotic membrane cultures. *Br. J. Ophthalmol.* **2002**, *86*, 463–471.

Treatment of Silk Fibroin with Poly(ethylene glycol) for the Enhancement of Corneal Epithelial Cell Growth

Shuko Suzuki, Rebecca A. Dawson, Traian V. Chirila, Audra M. A. Shadforth, Thomas A. Hogerheyde, Grant A. Edwards and Damien G. Harkin

Abstract: A silk protein, fibroin, was isolated from the cocoons of the domesticated silkworm (*Bombyx mori*) and cast into membranes to serve as freestanding templates for tissue-engineered corneal cell constructs to be used in ocular surface reconstruction. In this study, we sought to enhance the attachment and proliferation of corneal epithelial cells by increasing the permeability of the fibroin membranes and the topographic roughness of their surface. By mixing the fibroin solution with poly(ethylene glycol) (PEG) of molecular weight 300 Da, membranes were produced with increased permeability and with topographic patterns generated on their surface. In order to enhance their mechanical stability, some PEG-treated membranes were also crosslinked with genipin. The resulting membranes were thoroughly characterized and compared to the non-treated membranes. The PEG-treated membranes were similar in tensile strength to the non-treated ones, but their elastic modulus was higher and elongation lower, indicating enhanced rigidity. The crosslinking with genipin did not induce a significant improvement in mechanical properties. In cultures of a human-derived corneal epithelial cell line (HCE-T), the PEG treatment of the substratum did not improve the attachment of cells and it enhanced only slightly the cell proliferation in the longer term. Likewise, primary cultures of human limbal epithelial cells grew equally well on both non-treated and PEG-treated membranes, and the stratification of cultures was consistently improved in the presence of an underlying culture of irradiated 3T3 feeder cells, irrespectively of PEG-treatment. Nevertheless, the cultures grown on the PEG-treated membranes in the presence of feeder cells did display a higher nuclear-to-cytoplasmic ratio suggesting a more proliferative phenotype. We concluded that while the treatment with PEG had a significant effect on some structural properties of the *B. mori* silk fibroin (BMSF) membranes, there were minimal gains in the performance of these materials as a substratum for corneal epithelial cell growth. The reduced mechanical stability of freestanding PEG-treated membranes makes them a less viable choice than the non-treated membranes.

Reprinted from *J. Funct. Biomater.* Cite as: Suzuki, S.; Dawson, R.A.; Chirila, T.V.; Shadforth, A.M.A.; Hogerheyde, T.A.; Edwards, G.A.; Harkin, D.G. Treatment of Silk Fibroin with Poly(ethylene glycol) for the Enhancement of Corneal Epithelial Cell Growth. *J. Funct. Biomater.* **2015**, *6*, 345–366.

1. Introduction

The silk produced by the larvae of domesticated silkmoth (*Bombyx mori*) or some wild silkmoths have been known in the textile manufacturing for millennia [1–4]. In medicine, the use of silk threads as surgical sutures can be traced back to the beginning of the Common Era, when it was suggested by Galen of Pergamon [5,6]. With the increasing availability of *B. mori* silk throughout the subsequent centuries, the silk sutures became steadily used and, starting with the 19th century [7], they dominated the surgical field owing to some remarkable properties [8–11]. In 1866, Williams used for the first time silk sutures in the eye surgery in cataract operations [12], and Kuhnt followed his example in corneoscleral surgery [13]. Relatively slowly, silk became the suture material of choice in ophthalmic surgery [14–17]. Today, although the silk sutures are still available on the market and in clinical use, the sutures made of synthetic polymers (such as polyamides, polyesters, lactone-based polymers, and polyolefins) are generally preferred by surgeons. However, the medical applications of *B. mori* silk have not stopped at sutures. With the significant progress over the last few decades in understanding the complex structure and composition of silk and with the advent of methods enabling the isolation of its polypeptidic components, new applications emerged for the two main constitutive proteins of silk, fibroin and sericin [18–20]. Due to an array of desirable properties (they can be processed into various forms; do not elicit toxic or traumatic effects to living tissues; elicit low immune response; are permeable for oxygen, fluids and biomolecules; degrade protractedly in physiologic media and the resulting products do not accumulate in the body; and fibroin, in particular, also displays suitable mechanical strength), the silk proteins have been extensively investigated as biomaterials for tissue engineering, regenerative medicine and sustained drug delivery [21–33].

The feasibility of utilizing silk proteins as biomaterials for reconstructing tissue of clinical significance in the human eye was first reported by our group when we demonstrated that primary human corneal limbal epithelial cells could attach and proliferate on membranes of *B. mori* silk fibroin (BMSF) at levels comparable to those observed on tissue culture plastic (TCP) substrata, both in serum-supplemented and serum-free media [34,35]. Subsequent work has established BMSF as a functional substratum of significant potential in ocular tissue engineering [36–39]. Our investigations extended also to *B. mori* sericin [40], and to the fibroin produced by a wild species of silkmoth, *Antheraea pernyi* [41,42]. We have reported extensively on the evaluation of silk proteins as substrata for corneal cells (epithelial, limbal epithelial, limbal mesenchymal stromal, endothelial) [34–36,40–46], and retinal pigment epithelial cells [37,47].

For ocular tissue-engineered constructs, the templates should ideally be thin (2–10 μm), transparent, flexible, strong enough for surgical manipulation, permeable to solutes, and should promote adequate levels of cell attachment and growth. While most of these prerequisites are fulfilled by the membranes made of BMSF, there is still a need to optimize some properties. Indeed, it can be said that the attachment of cells to BMSF substrata is generally weak when compared to other materials. The enhancement of substratum's transport properties and of the adhesion and growth of cells would be important for the development of better tissue-engineered constructs, and strategies to achieve it have been actively pursued by some dedicated research groups. To this aim, methods for creating surface topographic features and/or rendering the substratum porous were investigated in order to improve colonization by corneal cells of the BMSF templates. One of strategies consists of mixing poly(ethylene glycol) (PEG), a water-soluble polymer, into the solutions of BMSF prior to stabilizing the structure by conversion to the conformation "Silk II" that makes the membrane insoluble in water. Subsequent washing in water removes PEG, which thus fulfills its role as a porogen. NOTE: The nomenclature for PEG needs, perhaps, some clarification. Poly(ethylene oxide) (PEO) is frequently used as an alternative name, usually when the molecular weight (MW) of the polymer is over 20 kDa, although this is rather a non-abiding convention. Equivalent names, such as "polyoxyethylene" or "polyoxirane", are seldom used, while the official IUPAC-recommended name, "poly(oxyethane-1,2-diyl)", is never seen in literature. In this report, we will use exclusively the acronym PEG regardless of MW.

The first use of PEG to modify the properties of BMSF, with the explicit aim of generating porosity, has been reported by Asakura and coworkers [48,49]. Their objectives have been either to study the interaction between metal ions trapped within the porous structure of BMSF [48] or to enhance the permeability of the BMSF membranes used for enzyme immobilization [49]. PEG with a MW of 300 Da was used, which probably explains why no microscopic evidence for pores could be obtained, as the size (more precisely the diameter of an equivalent sphere) of this particular PEG molecule is only about 1 nm [50]. However, the roughness of the membrane surface and the permeability of membranes were both enhanced significantly as the weight ratio PEG/BMSF increased. For instance, at a weight ratio PEG/BMSF of 3, the permeability to glucose or to salt increased 20 times. As a drawback, the mechanical strength and elasticity were drastically reduced with increasing PEG content [49]. Nevertheless, Asakura's studies have revealed that the incorporation of PEGs, at least of those with low MWs, into BMSF led not only to an increase of the permeability but also to changes in the surface topography.

Following the recognition of BMSF as a potential biomaterial, its blending with relatively low amounts of PEG with a much higher MW (900 kDa, which corresponds to a molecular size in the region of 100 nm [50]) has been investigated as a method either to reduce the brittleness of BMSF templates (as fibrous scaffolds or membranes) [51,52], or to induce porosity [53]. PEG blending also served in fundamental studies to create a model mimicking the behavior of natural silk proteins *in vivo* [54]. In the field of ocular tissue engineering, PEG with a MW of 900 kDa has been used to induce porosity in the BMSF membranes as substrata for corneal cells [44,55] or retinal cells [47], while PEG with a MW of 300 Da has been used with the same aim of improving the growth of corneal epithelial cells [56], the latter study being in fact a continuation of Asakura's work applied in ophthalmic tissue engineering. The effects upon corneal cells' growth of differing surface topographic patterns, created by lithographic techniques on the surface of BMSF membranes, have been also investigated on both porous [55] and non-porous membranes [57].

By using a PEG with high MW (900 kDa) as a porogen, well defined and microscopically detectable porous features were achieved in the BMSF membranes, but their performance as substrata for corneal cells was inferior to that of non-porous membranes [44,47,55]. The use of a PEG with a much lower MW (300 Da = 0.3 kDa) led to BMSF films (coated on cell culture inserts that are porous) with increased permeability and roughness of the surface [56]. While the rough topography was evident under the microscope, it appears that no pores could be seen inside the material. Remarkably, the cultures of primary rabbit corneal limbal epithelial cells on the PEG-treated substrata resulted in stratified epithelial layers, while only monolayers were noticed on the original BMSF substrata [56]. This finding could be indeed a consequence of favorable combined effects of higher permeability and rougher surface topography. The use of an underlying layer of 3T3 murine fibroblasts as feeder cells in this study almost certainly contributed to the improved growth of cells of the BMSF membranes with higher permeability. Nevertheless, the authors did not compare the growth in the presence and absence of the feeder cells. The precise mechanism of PEG action remains therefore somewhat unclear.

In the present report, we compared the attachment and proliferation of human corneal epithelial cells (HCECs as a cell line) and of human corneal limbal epithelial cells (HCLECs) on BMSF membranes that either were treated with PEG (MW 300 Da) or were not treated. Although the processing of substrata was similar to that described by Higa *et al.* [56], our study was different in many respects, including: human-derived cells instead of animal cells; freestanding BMSF membranes instead of porous culture membranes coated with BMSF films; and crosslinked membranes for enhanced mechanical stability. Moreover, we compared the growth of primary cell cultures both in the presence and absence of the feeder cells. Other differences will be discussed in the next section of this report. The aim of this study was to investigate whether the treatment of BMSF substrata with a PEG of low MW is of benefit to corneal epithelial cellular growth due to the potential synergism of higher permeability and irregular patterning of the surface.

2. Results and Discussion

2.1. Background

Being associated inherently with an enhancement of permeability, the presence of pores in the templates for cellular constructs is beneficial for the cells' growth due to increased diffusion of oxygen, nutrients and biomolecules that must be supplied to the cells and regenerating tissue, and to improved diffusion-based waste transport. Porosity also has favorable effects on the intercellular communication and signaling, and on the spatiotemporal control of the regions where the cells are expected to operate [58,59]. Validity of these general principles for the system BMSF/ocular cells (corneal or retinal) has been investigated in some studies [44,47,55,56]. It has been found [55] that immortalized human corneal stromal fibroblasts were able to colonize stacked BMSF layers (each 2 μm thick), where pores of size from 0.5 to 5 μm were created by treatment with PEG (900 kDa), but no comparative quantitative evaluation of cellular growth was provided. Our group has previously reported [44] BMSF membranes (thickness 2.3 ± 1 μm), where pores (2.9 ± 1.5 μm) were made by the use of the same PEG (900 kDa), which were evaluated *in vitro* as substrata for cultures of human corneal limbal epithelial cells (HCLECs). The relatively larger number of cells attached on the porous BMSF as compared to non-porous BMSF substrata or TCP was not statistically significant. On the non-porous substratum, cultivation of HCLECs for two weeks resulted in stratified layers of cells with a basal cuboidal layer. In contrast, cells on the porous substratum formed flattened and squamous monolayers. The same porous BMSF membranes have also been used as substrata for the growth of retinal pigment epithelial (RPE) cells (line ARPE-19) [47]. It was found that the attachment of cells was inferior to that on TCP, but no experimental comparison was carried out against a non-porous BMSF substratum. Based on the above results, porous morphologies induced by using a PEG of high MW appear to offer no advantages for cell growth, perhaps due to the large size of the pores (see further).

The ability of corneal cells to respond to the topography of the template has been demonstrated on a variety of materials and involving a range of topographic features. For instance, employing bovine corneal epithelial tissue explants or primary corneal epithelial cells, has shown [60–62] that both tissue outgrowth and cell proliferation were strongly affected by the size and number of the surface pores. These studies have been carried out on various commercially available membranes such as polycarbonate, cellulose, or polyester (Mylar®), over the pore size range 0.1 to 3 μm. Continuous cell layers were seen on the surfaces with the smallest pore size. At pore sizes over 0.9 μm the outgrowth and proliferation were almost halted. Comparing the growth on the same material (polycarbonate), regular hemidesmosomal adhesive structures occurred only on the surface with pores of 0.1 μm, while at higher pore size these structures were restricted, and they did not occur at all at the highest pore size or on the smooth surface. In a series of reports [63–67], surface topographic patterns consisting of features such as grooves and ridges were created on the surface of silicon wafers (by lithography) or polyurethane membranes (by moulding) with a pitch range between 400 and 4000 nm, the pitch being the distance between the centres of two consecutive holes. The levels of adhesion and proliferation of primary human corneal epithelial cells (HCECs) [63,67] or SV40-immortalized HCECs [64–67] were systematically investigated. While on the substrata with smooth surfaces the cells were mostly round, on the patterned surfaces they were elongated and tending to adopt a stellate morphology, as well as aligned along the grooves and ridges. Following normal incubation, the cells proliferated better on silicon wafers when the features had high pitch values, and also on the smooth surface. On the contrary, when the cells were exposed to shear stress in a laminar flow chamber, the features with lowest pitch value induced the highest level of adherent cells; at the highest pitch, the effect of topography was lost. On the patterned polyurethane substrata, however, the proliferation of both types of cells decreased as the dimensions of topographic features became smaller [67].

In a study involving BMSF [55], rabbit corneal stromal fibroblasts and immortalized human corneal stromal fibroblasts were seeded on membranes patterned with concentric circular or linear grooves. While the alignment of cells during growth was evident on the patterned surfaces, the amount of adherent cells was lower than on the smooth BMSF or TCP surfaces. In a more recent study from the same laboratory [56], the initial attachment of an immortalized HCLEC line on BMSF substrata patterned with linear grooves was greater than that on glass, smooth BMSF or BMSF surfaces with circular grooves. After eight days of culture, the situation reversed and the glass and smooth BMSF substrata supported the highest levels of cellular growth. Significant improvement in the attachment and proliferation of pig vascular endothelial cells has been reported on fibrous BMSF substrata fabricated by electrospinning [68]. However, it is problematic to ascertain whether this result is due to porosity, to surface topography, or to their combined effect.

The findings in all these studies, sometimes contradictory or difficult to interpret, illustrate the complexity of the mechanochemical signalling mechanisms governing the response of corneal cells to surface topographic cues. Notwithstanding such complexity, there might be a distinct possibility of harnessing the cells' response for the purpose of enhancing the biocompatibility of the cell/template systems, resulting in more extensive cellular colonization of the BMSF templates and, ultimately, to functional and stable constructs for the restoration of ocular surface.

Considering the rather ambiguous results reported with a PEG of high MW [44,47,55], and the promising results reported [56] using a PEG of low MW, we developed freestanding BMSF membranes that were modified with PEG of

MW 300 Da, with the expectation of increasing permeability and also of generating topographic features on the surface of the membranes. However, our approach was somewhat different from that adopted in the mentioned report [56]. Table 1 presents the experimental differences between the two studies. Critically, our studies were performed using freestanding membranes (as the substrata for clinical applications would be required), and growth of primary cultures was compared in the presence and absence of feeder cells.

2.2. Characterization of Silk Fibroin Membranes

BMSF membranes of *ca.* 3 μm or *ca.* 6 μm in thickness were produced on a casting table. Upon addition of PEG with MW of 300 Da, at a PEG/fibroin weight ratio of 2, the thickness of the resulting membranes almost doubled. While the non-treated fibroin membranes were easy to peel off from the casting plate and to handle (Figure 1a), the PEG-treated membranes were fragile and difficult to remove without breaking them (Figure 1d).

Table 1. Comparison between experimental designs: reference [56] *vs.* this report.

Aspect	Reference [56]	This report
Cells	Primary rabbit CLECs	Primary human CLECs; SV40-immortalized HCECs
Feeder cells	Always present in cultures	Growth of primary cultures compared in the presence and absence of feeder cells
Maximum duration of cultures	7 days	12 days
Substrata	BMSF films coated onto porous cell culture membranes	Freestanding BMSF membranes
Control substrata	Non-treated BMSF film; AM	Non-treated BMSF membrane; TCP
Ratio PEG/BMSF (by wt.)	0 to 38 (assessed); 2 (recommended)	2
Mol. wt. of molecules assessed for permeability	0.376 to 15 kDa	26–28 kDa
Modification of membranes	No	Yes (by chemical crosslinking)
In vivo evaluation	Yes (animals)	No

CLECs: corneal limbal epithelial cells; HCECs: human corneal epithelial cells; BMSF: *Bombyx mori* silk fibroin; AM: amniotic membrane; TCP: tissue culture plastic; PEG: poly(ethylene glycol).

To improve mechanical stability, crosslinking of PEG-treated fibroin with genipin was performed before mixing with PEG, using a previously established protocol [40]. Although the resulting membranes were thicker (10 to 15 μm), they remained more fragile than the non-treated BMSF membranes (Figure 1h). Their handling, however, became somewhat easier than of the uncrosslinked PEG-treated membranes. With care, therefore, a sufficient number of genipin-crosslinked PEG-treated membranes of suitable size could be generated for the next experiments.

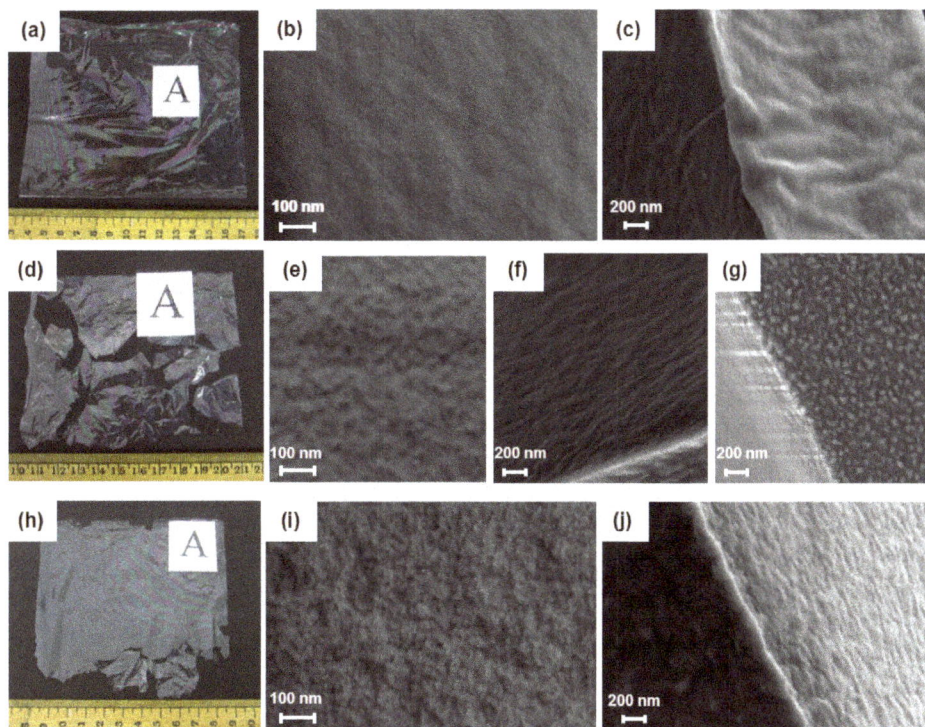

Figure 1. Scanning electron micrographs of the *B. mori* silk fibroin (BMSF) membranes. Physical appearance of non-treated (**a–c**), PEG-treated (**d–g**), and genipin-crosslinked PEG-treated (**h–j**) fibroin membranes. (**a,d,h**) Gross appearance of dried membranes after removal from the casting plate. Images of surfaces (**b,e,i**), cross-sections (**c,f,j**) and the edge of the PEG-treated membrane (**g**).

Scanning electron microscopy revealed that the surfaces of PEG-treated membranes, either uncrosslinked (Figure 1e) or crosslinked (Figure 1i) were rougher than that of non-treated membranes (Figure 1b), and no pores were noticeable. These findings are in agreement with previous reports [49,56]. In cross-section, the PEG-treated membranes also showed rough morphologies (Figure 1f,j), whereas the fractured surface of the non-treated membranes was smoother and denser (Figure 1c).

In the case of the uncrosslinked PEG-treated membrane, nanoscale fibroin globules were observed mainly in a region close to the edge of the membrane (Figure 1g), which has been a general occurrence on the BMSF substrata reported previously [56]. The surface roughness of membranes was further investigated by contact mode atomic force microscopy (AFM) (Figure 2). The roughness average (R_a) values measured from these images are given in Table 2. It is obvious that the treatment with PEG induced a significant increase in the value of R_a, very likely due to phase separation induced through its presence.

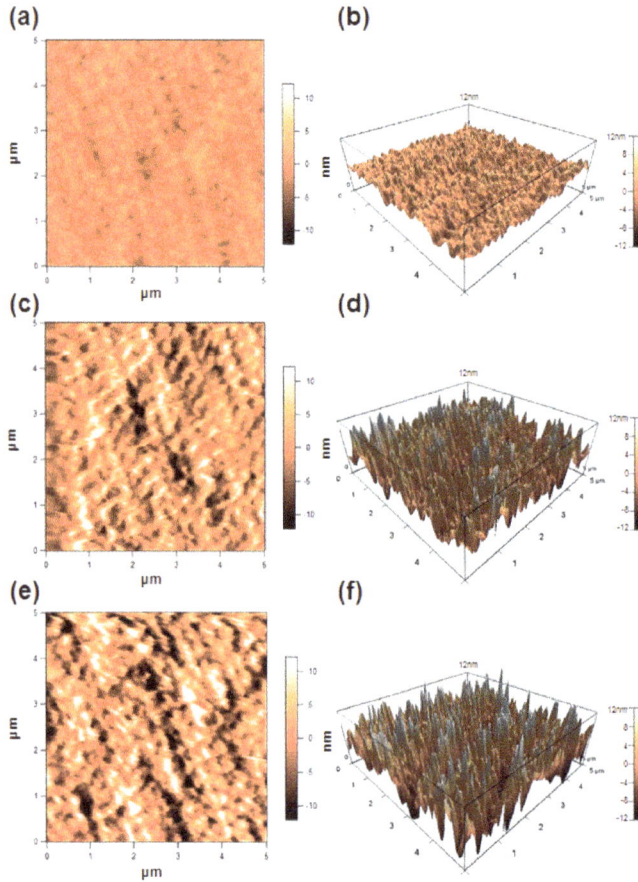

Figure 2. 2-D and 3-D AFM images of the surfaces of non-treated (**a,b**), PEG-treated (**c,d**) and genipin-crosslinked PEG-treated (**e,f**) fibroin membranes. Analysed area: 5 μm × 5 μm.

Table 2. Roughness average of membranes estimated by AFM.

Fibroin membrane	R_a (nm)
Non-treated	1.3
PEG-treated	4.4
Genipin-crosslinked PEG-treated	5.9

Infrared spectroscopy was employed to characterize the secondary structure of BMSF in the membranes [69,70]. The spectra in the Amide I region (1590–1720 cm^{-1}) of the annealed non-treated membrane and of the PEG-treated membranes are shown in Figure 3. The spectrum of the non-treated membrane displayed a broad absorption band with a peak at 1640 cm^{-1}, indicating a substantial amount of random-coil conformation (Figure 3a).

The broad shape of this band with a shoulder at 1619 cm^{-1} indicates a small, but significant, amount of β-sheet component in the non-treated BMSF. The Amide I band spectra of the PEG-treated membranes, either crosslinked or not, showed strong peaks at 1619 cm^{-1} and shoulders at 1700 cm^{-1}, respectively, indicating a significant proportion of β-sheet conformations (Figure 3b,c) in both materials, and suggesting a negligible effect of the crosslinking upon the secondary structure of fibroin. More important here, the high content in β-sheet conformations proves that the PEG-treated membranes do not need to be water-annealed in order to induce the conformational conversion responsible for rendering the fibroin insoluble in water, as this process is accomplished due to the presence of PEG as a polar agent able to induce conversion to the β-sheet conformation more effectively than water.

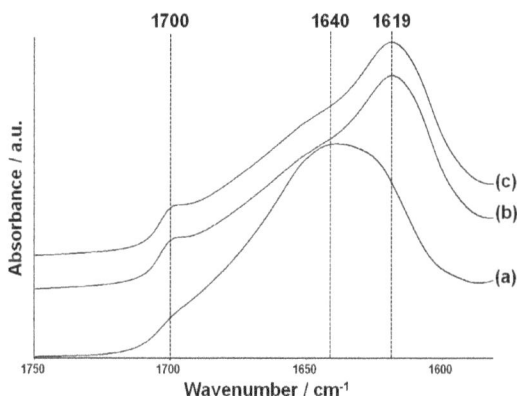

Figure 3. Fourier-transform infrared spectroscopy-ATR spectra of non-treated (**a**), PEG-treated (**b**), and genipin-crosslinked PEG-treated (**c**) fibroin membranes.

The results of mechanical testing (Figure 4) indicated important differences between certain tensile characteristics of the three types of membranes. Although the ultimate strength values were similar for all samples, the elastic moduli of the PEG-treated membranes were significantly higher than those of the non-treated membranes, while their elongation at break was significantly lower. This can be a consequence of increased rigidity due to higher proportion of β-sheet conformations induced by the treatment with PEG, an assumption strongly suggested by the infrared spectrometric analysis. Rather unexpectedly, the crosslinking did not improve the tensile strength of the PEG-treated membranes.

To estimate the permeability of the BMSF membranes to biomolecules, the growth factor VEGF (vascular endothelial growth factor) was chosen as the permeant molecule, and a method was designed for the purpose (Figure 5a). VEGF has a MW of 26–28 kDa, and plays an important role in certain pathophysiological processes in the eye. In this study, we determined the relative permeability of the non-treated and of the crosslinked PEG-treated membranes. As shown in Figure 5b, the PEG-treated membranes were relatively more permeable to VEGF molecule as compared to the non-treated membranes. This clearly indicates that by blending BMSF with PEG (MW 300 Da), the permeability is enhanced, thus supporting the observations of Higa *et al.* [56]. Interestingly, approximately 50% and 70% of VEGF (*i.e.*, 7.5 and 10.5 ng) for the non-treated and PEG-treated membranes, respectively, were lost as shown by comparing the total amounts of protein in the apical and basal compartments after 24 h to the initial amounts. This could be due to the trapping of VEGF within BMSF due to electrostatic interactions of positively charged VEGF and negatively charged fibroin molecules.

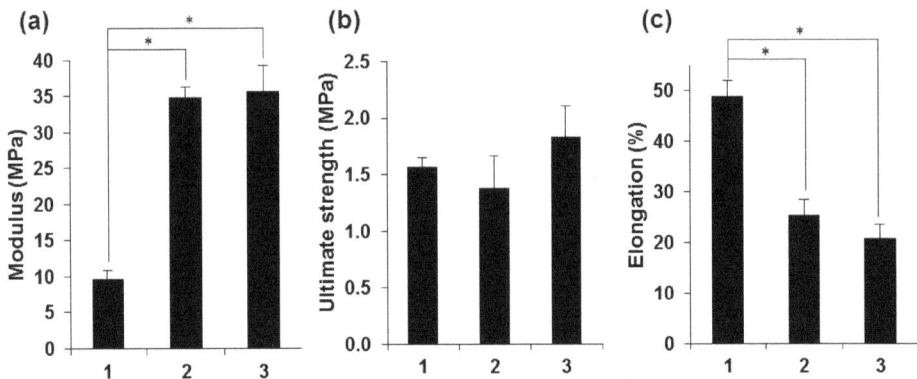

Figure 4. Quantitative comparison of the tensile characteristics of non-treated (1), PEG-treated (2) and genipin-crosslinked PEG-treated (3) fibroin membranes. (a) Young's modulus; (b) Ultimate tensile strength; (c) Elongation at break. Bars represents mean $\pm$ standard error of the mean ($n = 6$). An asterisk indicates that the difference is statistically significant ($p < 0.05$).

(a)   (b)

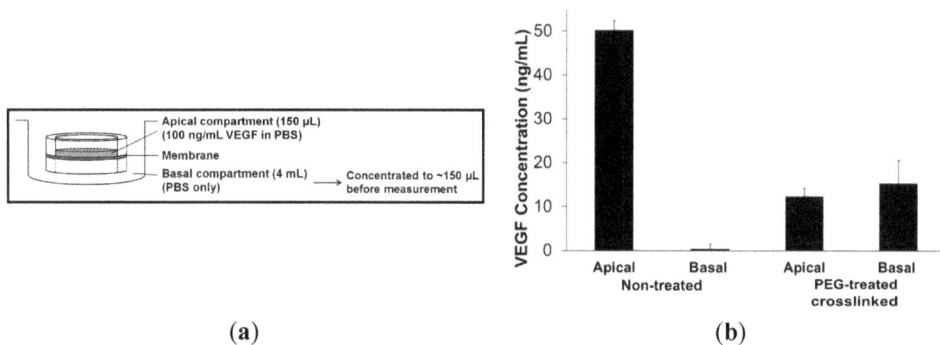

Figure 5. Relative permeability of BMSF membranes to VEGF. (**a**) Schematic representation of the permeability experimental setup. (**b**) Comparison of VEGF concentrations after 24 h in the apical and basal compartments delimiting the membranes.

2.3. Attachment and Proliferation of HCE-T Cell Line

The attachment and proliferation of an SV40-immortalized cell line (HCE-T) was examined on membranes (*ca.* 6 μm in thickness) placed at the bottom of the culture-plate wells. Since these cells can be serially propagated in the absence of feeder cells, they provided a useful model of the human corneal epithelial cells' growth in the absence of any accessory cells. The numbers of adherent cells was expressed as the total DNA content with the PicoGreen® assay (Figure 6). In a short-term attachment assay (over a period of 90 min), no quantitative difference between the numbers of cells attached to the genipin-crosslinked PEG-treated and those attached to non-treated membranes in serum-free conditions was noticed (Figure 6a), but they were significantly lower than the number of cells attached to the TCP control. In longer-term cultures (up to seven days), cell growth on the PEG-treated membrane was higher than that on non-treated membrane in serum-supplemented growth medium, albeit the differences were not statistically significant (Figure 6b). However, the level of cell growth on the non-treated membranes was found to be significantly lower than that on TCP substrata.

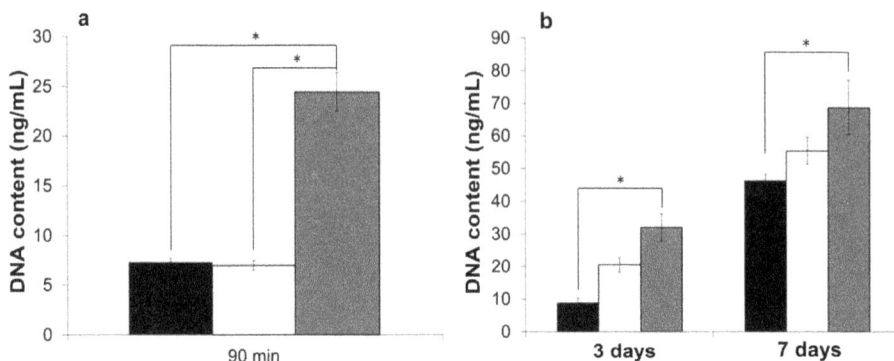

Figure 6. Attachment and proliferation of cells of HCE-T line on BMSF membranes. (**a**) Cellular attachment in serum-free medium; (**b**) Proliferation in serum-supplemented medium on non-treated fibroin membrane (black), genipin-crosslinked PEG-treated fibroin membrane (white) and TCP (grey). Numbers of cells were measured *via* quantification of DNA content (PicoGreen[®] assay). Bars represent mean ± standard error of the mean. The asterisk indicates that the difference is statistically significant ($p < 0.05$).

2.4. Growth of Primary Human Corneal Limbal Epithelial Cells (CLECs)

Primary cultures of human CLECs were cultivated for up to 12 days on freestanding BMSF membranes (*ca.* 6 µm) that had been mounted in Teflon[®] cell culture chambers. The design of these chambers facilitates separation of culture medium between the upper and lower membrane surfaces. The growth of cells on genipin-crosslinked PEG-treated membranes (10 to 15 µm in thickness) was compared to that observed on non-treated membranes. Moreover, both membrane types were tested in both the presence and absence of irradiated 3T3 cells grown on the lower membrane surface.

When examined by phase contrast microscopy after five days of growth (Figure 7), a marked difference in culture morphology was observed in the presence of 3T3 cells. In short, in the presence of feeder cells, the cultures displayed a more confluent and compact morphology, which is indicative of a more proliferative phenotype, and this effect of the feeder cells was observed irrespectively of PEG-treatment.

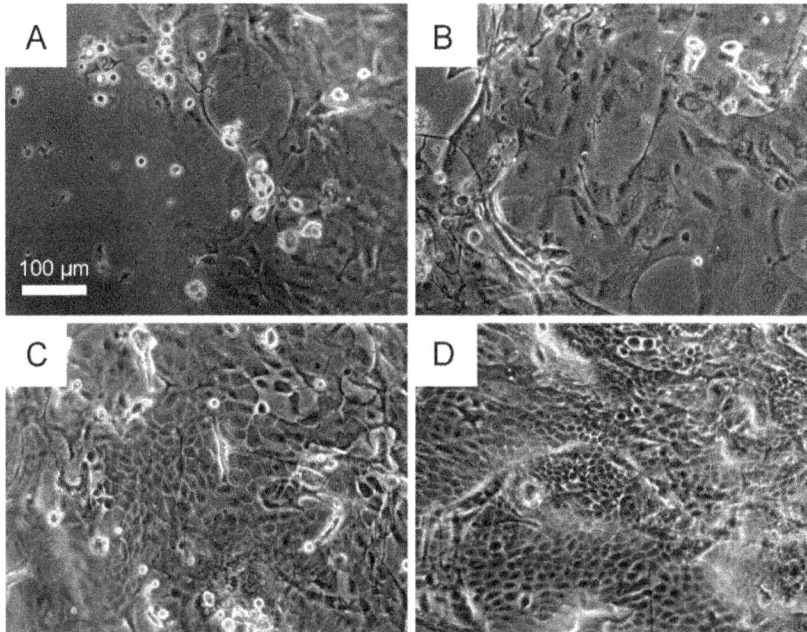

Figure 7. Phase contrast micrographs of primary cultures of human CLECs after five days of growth on either non-treated BMSF membranes (**A,C**) or genipin-crosslinked PEG-treated BMSF membranes (**B,D**), in either the absence (**A,B**) or presence (**C,D**) of an underlying culture of feeder cells (irradiated 3T3 murine fibroblasts).

After 12 days of growth, all cultures were fixed and subsequently stained with rhodamine phalloidin (to display F-actin filaments) and Hoechst nuclear dye (to display cell nuclei). Using confocal fluorescence microscopy, a high-resolution optical cross-section was obtained through each culture when folded and mounted in glycerol under a glass coverslip (Figure 8). This technique revealed that human CLEC cultures grown on fibroin membranes are consistently more stratified when an underlying layer of irradiated 3T3 cells is present, and the stratification was observed irrespectively of treatment with PEG. Nevertheless, the cells present within the cultures grown on PEG-treated membranes, in the presence of feeder cells, displayed a higher nuclear-to-cytoplasmic ratio suggesting a more proliferative phenotype. This observation tends to support the conclusions of Higa *et al.* [56] that superior growth is seen using PEG-treated membranes. Logically, this enhanced growth is due at least in part to the presence of feeder cells, but since cultures grown on non-treated membranes also displayed increased stratification, we cannot discount the potential role of changes in membrane topography created by PEG in conjunction with effects mediated by the feeder cells.

Ultimately, both membranes may well support the manufacture of human CLEC cultures of sufficient quality to enable therapeutic applications. Nevertheless, the difficulties that we encountered in producing freestanding PEG-treated membranes suggest that any potential benefits bestowed by this material are insufficient to warrant changing our clinical strategy.

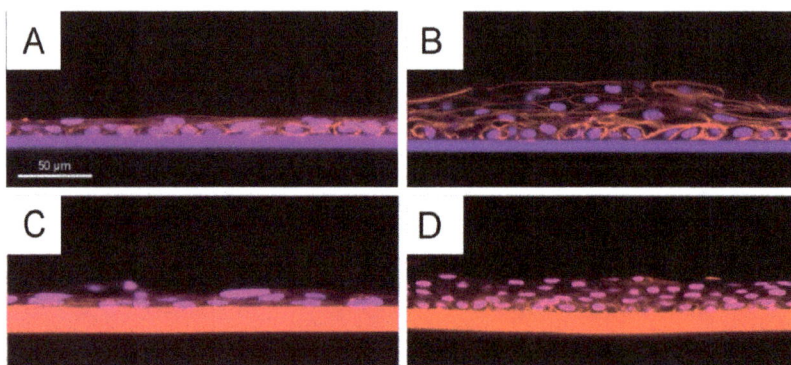

Figure 8. Histology by confocal microscopy after cultivation of primary human CLECs for 12 days on non-treated (**A,B**) and genipin-crosslinked PEG-treated (**C,D**) BMSF membranes: without feeder cells (**A,C**); co-cultured with feeder cells (irradiated 3T3 murine fibroblasts) (**B,D**). The feeder cells have become dislodged during culture and subsequent preparation of samples for confocal microscopy. The genipin-crosslinked PEG-treated membranes were thicker than the non-treated membranes and displayed intense auto-fluorescence, as seen in (**C**) and (**D**).

2.5. Summary

As this study has duplicated experiments previously reported [56], albeit with some important differences (see Table 1), we expected that our results would confirm those findings. While the PEG-treated membranes were called "porous" [56], there was no microscopic evidence for pores in the bulk of material, *i.e.*, in the cross-sectioned membranes. The only micrographs provided (see "Fig. 2A,B" in ref. [56]) were those of the membrane surfaces, which showed that the surfaces became rougher following treatment with PEG (MW 300 Da). Our study confirms this observation, and also confirmed that the permeability increased after treatment with PEG.

No difference in cell attachment and growth was observed in the experiments using a transformed human corneal epithelial cell line. However, the results obtained using primary cultures of human corneal limbal epithelial cells suggested that an enhanced permeability of PEG-treated membranes does lead to subtle changes in cell behaviour that could be of a clinical value (a more proliferative cell phenotype).

Nevertheless, since an underlying culture of irradiated 3T3 murine fibroblasts (as feeder cells) can also influence positively the growth of HCECs when cultivated on non-treated membranes, it would appear that even without PEG-treatment the BMSF substrata are perhaps sufficiently permeable to support the manufacture of clinically suitable cultures. Ultimately, further studies in a pre-clinical model of ocular surface disease will be needed to resolve this issue. From our perspectives, we believe that the technical difficulties associated with the routine manufacture of freestanding PEG-treated membranes will outweigh any potential benefits arising from the apparent increase in permeability of the substratum.

A rather tenuous cell adhesion is a known drawback of the BMSF templates [71]. However, the biomaterial characteristics of fibroin make these templates attractive for tissue engineering applications, as proved by the growing number of the published studies, and the enhancement of cell attachment to BMSF remains therefore a topic of great interest. Rather than cell adhesion based on non-specific interactions, which likely govern this process on BMSF, there is a need to promote physicochemical characteristics in the substratum's surface that will be able to mediate the cell-surface anchorage in a specific way. Ideally, the surface shall comprise structural elements leading to its recognition by the cells' integrin receptors and thus generating true focal adhesions between cells and surface. Whether the modification of BMSF surface through covalent binding of extracellular matrix proteins or/and through topographic patterning are sufficient for introducing specific interactions in order to facilitate stronger cell attachment is yet to be determined, notwithstanding the volume of research dedicated to this topic. The contribution of higher porosity and/or permeability is limited to improving intercellular communication, however without promoting specific interactions.

3. Experimental Section

3.1. Materials

Bombyx mori silkworm cocoons (with pupae removed) were purchased from Tajima Shoji Co. Ltd. (Yokohama, Japan). Genipin (98% purity) was supplied by Erica Co. Ltd. (Shenzhen, China). Topas® 8007S-04 (olefin copolymer) was supplied by Topas Advanced Polymers (Frankfurt, Germany). Minisart®-GF pre-filters (0.7 μm) and Minisart® filters (0.2 μm) were supplied by Sartorius Stedim Biotech (Göttingen, Germany), and the dialysis cassettes Slid-A-Lyzer® (MWCO 3.5 kDa) by Thermo Scientific (Rockford, IL, USA). Sodium carbonate, lithium bromide, PEG (MW 300 Da) and 10% formaldehyde solution were supplied by Sigma-Aldrich. High purity water (Milli-Q) was used in all experiments. The human vascular endothelial growth factor (VEGF, #3045-VE-025/CF) and its enzyme-linked immunosorbent assay (ELISA) kit (#DY3045) were purchased from R&D Systems (Melbourne, Australia). Amicon

Ultra-4 centrifugal filters (#UFC801024, 10 kDa MWCO) were supplied from Merck Millipore Ltd. (Darmstadt, Germany). Foetal bovine serum (FBS) was supplied by Thermo Scientific (USA). All other cell culture reagents and supplements, as well as Quant-iT™ PicoGreen® dsDNA assay kit were purchased from Life Technologies (Melbourne, Australia).

3.2. Preparation of Fibroin Membranes

Silk fibroin solution was prepared according to a previously established protocol [40]. The concentration of solution used in experiments was 1.78% (as determined by gravimetric analysis). The standard BMSF membranes were prepared by casting the fibroin solution in a custom-made casting table where the supporting glass plate was pre-coated with a polyolefin polymer (Topas®) film [45]. The blade height was set in order to generate an approximate dry thickness of either 3 μm or 6 μm for the resulting BMSF membranes. After drying, the membranes were water-annealed in a vacuum chamber at −80 kPa for 6 h at room temperature in the presence of a container filled with water, followed by peeling off from the supporting Topas® film.

The PEG-treated BMSF membranes were prepared according to a published protocol [56], with some modifications. In brief, PEG was slowly blended into the 1.78-% fibroin solution at a PEG/fibroin ratio of 2 (by weight). The solution was cast as described above. After drying, the membranes were soaked in 2 L of water for 3 days with two water exchanges per day to remove PEG. The dried membrane was then peeled off from the underlying Topas® film.

In order to crosslink the PEG-treated membranes, an amount of genipin equivalent to 12 wt% of fibroin was mixed with fibroin solution and stirred slowly for 5 h at 40 °C [40]. The mixture acquired a light blue hue, which is indicative of a reaction taking place between genipin and amino acids [72]. Subsequently, the membranes were processed following the method described above for the PEG-treated BMSF membrane.

3.3. Scanning Electron Microscopy (SEM)

Small pieces of 3-μm thick membranes or freeze-fractured fragments (the latter for the examination in cross-section), were placed on specimen stubs using double-sided adhesive tapes and coated with a layer of iridium using a sputter coater. Field-emission scanning electron microscopy (FE-SEM Sigma, Zeiss, Germany) was employed to examine the surface and internal morphologies of various membranes.

3.4. Atomic Force Microscopy (AFM)

Fibroin films were cast on clean glass slides. A MultiScan AFM (BMT, Ettlingen, Germany) in contact mode was employed using a silicon cantilever (ContAl-G,

BudgetSensors, Bulgaria), with a tip radius of less than 10 nm and a scan rate of 0.50 Hz. The R_a value was obtained from the total area of $5.0 \times 5.0 \ \mu m^2$ of the AFM image.

3.5. Fourier-Transform Infrared Spectroscopy (FTIR)

A Nicolet FTIR spectrometer (Thermo Electron Corp., Waltham, MA, USA), equipped with a diamond attenuated total reflectance (ATR) sampling accessory, was used to analyse the secondary structure of each type of BMSF membrane. Each spectrum was obtained by co-adding 64 scans over the range 4000 to 525 cm^{-1} at a resolution of 8 cm^{-1}. The OMNIC 7 software package (Thermo Electron Corp., Waltham, MA, USA) was used to analyse and plot the spectra.

3.6. Tensile Testing

Strips (1 cm $\times$ 3 cm) cut out from each of the 3-μm thick membranes were subjected to tensile measurements using an Instron 5848 microtester (Instron, UK), equipped with a 5 N load cell, at a crosshead speed of 14 mm/min. The stripes were loaded by pneumatic grips, which were set to a gauge distance of 14 mm, and soaked in phosphate buffered saline (PBS) (pre-heated to 37 $\pm$ 3 °C) in a BioPuls™ unit for 5 min prior to stretching. Stress-strain plots were recorded and Young's modulus was determined from the slope of the linear region of the curve. The mean values were calculated from six measurements of each membrane.

3.7. Permeability of the Membranes

A permeability test was designed to quantify the movement of biomolecules across the fibroin membranes. Custom-designed Teflon® chambers (Figure 5a) were used to suspend the BMSF membrane, which creates separate upper and lower compartments. This assay uses a known concentration of vascular endothelial growth factor (VEGF) in the upper compartment, and at a set time point the movement of the VEGF molecules into the lower compartment (through the membrane) can be quantified by ELISA (enzyme-linked immunosorbent assay). VEGF is a basic protein with a MW of 26–28 kDa and an isoelectric point of 8.5.

Discs cut out of each membrane were assembled in chambers and sterilized by immersion in a 70% ethanol solution for 1 h, air-dried in a biohazard hood and rinsed 3 times with PBS. Chambers were inserted into a 6-well plate well with 4 mL fresh PBS. This volume creates the lower compartment below the membrane. A VEGF solution (100 ng/mL) was prepared and 150 μL was added to the upper compartment, above the membranes. The plate was incubated at 37 °C, and upper and lower compartment volumes were collected after 24 h. Samples were frozen at -40 °C immediately after collection. ELISA assay was performed to examine each sample volume. The purpose of the assay is to determine if a particular protein is present in a sample and, if so, how much is present. We performed the assay using a commercially available VEGF sandwich ELISA kit. Briefly, the assay TCP plates (96-well) were prepared by coating with the capture antibody, and incubated at room temperature overnight. The plates were then washed and blocked with bovine serum albumin. The upper and lower compartment volumes were thawed and added to the corresponding assay plate/wells. Lower compartment volumes were concentrated using Amicon Ultra-4 centrifugal filters (MWCO 10 kDa) to be equivalent to the upper compartment volume (about 150 μL). A VEGF standard curve was also included on each assay plate. All samples and standards were tested in duplicate. After incubation at room temperature for 2 h, each plate was washed, and the detection antibody was added, and incubated.

In order to quantify the interaction of the detection antibody, the plate was washed, and a secondary antibody coupled to horseradish peroxidase (HRP) was added and incubated for 20 min. After a final wash step, the HRP enzyme was activated using a Substrate System, incubated for 20 min, which initiated the visible colour reaction. The enzyme reaction was completed with the addition of the Stop Solution, and each plate was placed into an absorbance microplate reader. The optical density was determined for each well at 450 nm and at 540 nm. The intensity of the colour reaction is proportional to the amount of VEGF protein in the original sample volumes, *i.e.* bound to the capture antibody on the bottom of the wells.

3.8. Culture and Growth of Transformed Human CECs on BMSF Substrata

A SV40-immortalized cell line (HCE-T) derived from human corneal epithelial cells (CECs) was used for assaying the initial cell attachment and growth. HCE-T cells were cultured in Dulbecco's modified Eagle medium supplemented with 10% FBS, glutamine and 1% v/v penicillin/ streptomycin. The cells were cultured in a humidified atmosphere of 5% CO_2 at 37 °C, and passaged using Versene and TrypLE®.

Silk fibroin membranes were cut using a trephine blade to produce circular pieces of approximately 14 mm in diameter, and placed into individual wells of a 24-well TCP plate using rubber O-rings. They were sterilized in 70% ethanol for 30 min followed by washing three times with PBS. The HCE-T cells (20,000/cm^2) were seeded into each well with 0.5 mL/well medium, and 0.25 mL medium was exchanged every third day. Serum-free medium was used in the case of the short-term attachment assay (90 min). For proliferation assay, the cultures were assayed after incubation for 3 or 7 days in serum-supplemented media. At the end of each time point, the O-rings were removed and the cultures were rinsed three times with PBS. The DNA content of adhered cells was quantified using the PicoGreen® assay as previously described [40]. All experiments were conducted in triplicate for each series of three assessments.

3.9. Culture and Growth of Primary Human CLECs on BMSF Substrata

Cadaveric human eye tissue was obtained with human research ethics committee approval and donor consent from the Queensland Eye Bank, Brisbane, Australia. The primary cultures of human corneal limbal epithelial cells were established from the corneal limbus as described previously [40], and cultured in Green's medium [73]. Freshly isolated human CLECs were seeded into 25 cm^2 flasks containing 1×10^6 irradiated 3T3 murine fibroblasts (i3T3) as feeder cells. Membranes, *ca.* 6 μm in thickness, were cut into circular pieces of approximately 14 mm in diameter and mounted in sterile Teflon® cell culture chambers as described previously [43]. After sterilizing in 70% ethanol followed by rinsing with PBS, 3×10^4 i3T3 cells were seeded on to the underside of the membrane and allowed to attach for 24 hours. The chambers were then inverted before 1×10^4 human CLECs were seeded all in Green's medium. The cells were cultured for 12 days with chamber re-feeds every two to three days. After 12 days, the cells were fixed by immersing the chambers in 3.7% formaldehyde and stained with rhodamine phalloidin and Hoechst nuclear dye to highlight the actin fibres and the nuclei in the cells for examination by confocal fluorescence microscopy on a Nikon A1 confocal system.

3.10. Statistical Analysis

The results of mechanical testing and cell culture were statistically processed by the one-way analysis of variance (ANOVA) in conjunction with Tukey-Kramer multiple comparisons, using the GraphPad Prism® version 6.0.

4. Conclusions

The characteristics of BMSF as a substratum for the growth of corneal epithelial cells can be modified by blending with a PEG of low molecular weight such as 300 Da. Both permeability and surface topography are indeed changed in ways that are expected to be beneficial to the process of cell attachment and proliferation. In practice, however, the treatment with PEG enhances the fragility of membranes. This effect appears to negate any potential benefits to cell growth.

Acknowledgments: This work was supported by Queensland Eye Institute Foundation (formerly Prevent Blindness Foundation), Australia, and by funding from the National Health and Medical Research Council of Australia (research grants 553038 and APP1049050). R.A.D., A.M.A.S. and D.G.H. obtained supplementary funding from Queensland University of Technology (QUT), Brisbane, Australia. We thank Sanjileena Singh and Rachel Hancock at QUT's Central Analytical Research Facility, and Ron Rasch at the Centre for Microscopy and Microanalysis of the University of Queensland for help with microscopic analyses. Finally, we wish to acknowledge the generous support offered by the staff from the Queensland Eye Bank and Queensland Eye Hospital for facilitating access to donor eye tissue.

Author Contributions: S.S. contributed to the design of study and interpretation of results, carried out the production and analysis of silk materials, drafted sections of the manuscript, and prepared the graphic matter. R.A.D. carried out the cell culture experiments, interpreted the results and drafted sections of the manuscript. T.V.C. contributed to the design of study and to the interpretation of results, carried out literature search, and wrote the manuscript. A.M.A.S. carried out the permeability measurements, interpreted the results, and drafted a section of the manuscript. T.A.H. assisted to cell culture work and interpretation of results. G.A.E. coordinated mechanical testing and assisted to interpretation of results. D.G.H. proposed the study, contributed to the design of experiments, coordinated cell culture experiments and the interpretation of results, performed the confocal microscopic examination, and drafted sections of the manuscript. All authors read the manuscript and approved its final version.

Conflicts of Interest: The authors declare no conflict of interest.

References

1. Gervers, V. *Studies in Textile History*; Royal Ontario Museum: Toronto, Canada, 1977.
2. Barber, E.J.W. *Prehistoric Textiles*; Princeton University Press: Princeton, NJ, USA, 1991.
3. Good, I.L.; Kenoyer, J.M.; Meadow, R.H. New evidence for early silk in the Indus Civilization. *Archaeometry* **2009**, *51*, 457–466.
4. Hansen, V. *The Silk Road: A New History*; Oxford University Press: New York, NY, USA, 2012; pp. 2–23.
5. Mackenzie, D. The history of sutures. *Med. Hist.* **1973**, *17*, 158–168.
6. Muffly, T.M.; Tizzano, A.P.; Walters, M.D. The history and evolution of sutures in pelvic surgery. *J. R. Soc. Med.* **2011**, *104*, 107–112.
7. Edgar, I.I. Modern surgery and Lord Lister. *J. Hist. Med. Allied Sci.* **1961**, *16*, 145–160.
8. Kiliani, O.G.T. On traumatic keloid of the median nerve, with observations upon the absorption of silk sutures. *Ann. Surg.* **1901**, *33*, 13–22.
9. Halsted, W.S. Ligature and suture material. The employment of fine silk in preference to catgut and the advantages of transfixion of tissues and vessels in control of hemorrhage. *J. Am. Med. Assoc.* **1913**, *60*, 1119–1126.
10. Smit, I.B.; Witte, E.; Brand, R.; Trimbos, J.B. Tissue reaction to suture materials revisited: Is there argument to change our views? *Eur. Surg. Res.* **1991**, *23*, 347–354.
11. Deveikis, J.P.; Manz, H.J.; Luessenhop, A.J.; Caputy, A.J.; Kobrine, A.I.; Schellinger, D.; Patronas, N. A clinical and neuropathologic study of silk suture as an embolic agent for brain arteriovenous malformations. *Am. J. Neuroradiol.* **1994**, *15*, 263–271.
12. Williams, H.W. *Recent Advances in Ophthalmic Science*; Ticknor & Fields: Boston, MA, USA, 1866; pp. 90–92.
13. Kuhnt, H. *Beiträge zur operativen Augenheilkunde*; Verlag Gustav Fischer: Jena, Germany, 1883; pp. 69–97.
14. McLean, J.M. A new corneoscleral suture. *Arch. Ophthalmol.* **1940**, *23*, 554–559.
15. Hughes, W.L.; Guy, L.P.; Romaine, H.H. Use of absorbable sutures in cataract surgery. *Arch. Ophthalmol.* **1944**, *32*, 362–367.
16. Larmi, T. Sutures in eye surgery. *Acta Ophthalmol.* **1961**, *39(S63)*, 15–19.
17. Salthouse, T.N.; Matlaga, B.F.; Wykoff, M.H. Comparative tissue response to six suture materials in rabbit cornea, sclera, and ocular muscle. *Am. J. Ophthalmol.* **1977**, *84*, 224–233.

18. Minoura, N.; Tsukada, M.; Nagura, M. Physico-chemical properties of silk fibroin membrane as a biomaterial. *Biomaterials* **1990**, *11*, 430–434.

19. Minoura, N.; Aiba, S.; Higuchi, M.; Gotoh, Y.; Tsukada, M.; Imai, Y. Attachment and growth of fibroblast cells on silk fibroin. *Biochem. Biophys. Res. Commun.* **1995**, *208*, 511–516.

20. Minoura, N.; Aiba, S.; Gotoh, Y.; Tsukada, M.; Imai, Y. Attachment and growth of cultured fibroblast cells on silk protein matrices. *J. Biomed. Mater. Res.* **1995**, *29*, 1215–1221.

21. Altman, G.H.; Diaz, F.; Jakuba, C.; Calabro, T.; Horan, R.L.; Chen, J.; Lu, H.; Richmond, J.; Kaplan, D.L. Silk-based biomaterials. *Biomaterials* **2003**, *24*, 401–416.

22. Wang, Y.; Kim, H.-J.; Vunjak-Novakovic, G.; Kaplan, D.L. Stem cell-based tissue engineering with silk biomaterials. *Biomaterials* **2006**, *27*, 6064–6082.

23. Vepari, C.; Kaplan, D.L. Silk as a biomaterial. *Prog. Polym. Sci.* **2007**, *32*, 991–1007.

24. Hakimi, O.; Knight, D.P.; Vollrath, F.; Vadgama, P. Spider and mulberry silkworm silks as compatible biomaterials. *Composites B* **2007**, *38*, 324–337.

25. Kundu, S.C.; Dash, B.C.; Dash, R.; Kaplan, D.L. Natural protective glue protein, sericin bioengineered by silkworms: potential for biomedical and biotechnological applications. *Prog. Polym. Sci.* **2008**, *33*, 998–1012.

26. Wang, X.; Cebe, P.; Kaplan, D.L. Silk proteins—Biomaterials and bioengineering. In *Protein Engineering Handbook*; Lutz, S., Bornscheuer, U.T., Eds.; Wiley-VCH Verlag: Weinheim, Germany, 2009; pp. 939–959.

27. Murphy, A.R.; Kaplan, D.L. Biomedical applications of chemically-modified silk fibroin. *J. Mater. Chem.* **2009**, *19*, 6443–6450.

28. Hardy, J.G.; Scheibel, T.R. Composite materials based on silk proteins. *Prog. Polym. Sci.* **2010**, *35*, 1093–1115.

29. Wenk, E.; Merkle, H.P.; Meinel, L. Silk fibroin as a vehicle for drug delivery applications. *J. Control. Rel.* **2011**, *150*, 128–141.

30. Kasoju, N.; Bora, U. Silk fibroin in tissue engineering. *Adv. Healthc. Mater.* **2012**, *1*, 393–412.

31. Gil, E.S.; Panilaitis, B.; Bellas, E.; Kaplan, D.L. Functionalized silk biomaterials for wound healing. *Adv. Healthc. Mater.* **2013**, *2*, 206–217.

32. Kundu, B.; Rajkhowa, R.; Kundu, S.C.; Wang, X. Silk fibroin biomaterials for tissue regenerations. *Adv. Drug Deliv. Rev.* **2013**, *65*, 457–470.

33. Pereira, R.F.P.; Silva, M.M.; de Zea Bermudez, V. *Bombyx mori* silk fibers: An outstanding family of materials. *Macromol. Mater. Eng.* **2014**.

34. Chirila, T.; Barnard, Z.; Zainuddin; Harkin, D. Silk as substratum for cell attachment and proliferation. *Mater. Sci. Forum* **2007**, *561–565*, 1549–1552.

35. Chirila, T.V.; Barnard, Z.; Zainuddin; Harkin, D.G.; Schwab, I.R.; Hirst, L.W. *Bombyx mori* silk fibroin membranes as potential substrata for epithelial constructs used in the management of ocular surface disorders. *Tissue Eng. Part A* **2008**, *14*, 1203–1211.

36. Chirila, T.V.; Hirst, L.W.; Barnard, Z.; Harkin, D.G.; Schwab, I.R. Reconstruction of the ocular surface using biomaterials. In *Biomaterials and Regenerative Medicine in Ophthalmology*; Chirila, T., Ed.; Woodhead Publishing Ltd.: Cambridge, UK, 2010; pp. 213–242.

37. Kwan, A.S.L.; Chirila, T.V.; Cheng, S. Development of tissue-engineered membranes for the culture and transplantation of retinal pigment epithelial cells. In *Biomaterials and Regenerative Medicine in Ophthalmology*; Chirila, T., Ed.; Woodhead Publishing Ltd.: Cambridge, UK, 2010; pp. 390–408.

38. Harkin, D.G.; George, K.A.; Madden, P.W.; Schwab, I.R.; Hutmacher, D.W.; Chirila, T.V. Silk fibroin in ocular tissue reconstruction. *Biomaterials* **2011**, *32*, 2445–2458.

39. Harkin, D.G.; Chirila, T.V. Silk fibroin in ocular surface reconstruction: What is its potential as a biomaterial in ophthalmics? *Future Med. Chem.* **2012**, *4*, 2145–2147.

40. Chirila, T.V.; Suzuki, S.; Bray, L.J.; Barnett, N.L.; Harkin, D.G. Evaluation of silk sericin as a biomaterial: *in vitro* growth of human corneal limbal epithelial cells on Bombyx mori sericin membranes. *Prog. Biomater.* **2013**, *2*.

41. Bray, L.J.; Suzuki, S.; Harkin, D.G.; Chirila, T.V. Incorporation of exogenous RGD peptide and inter-species blending as strategies for enhancing human corneal limbal epithelial cell growth on bombyx mori silk fibroin membrane. *J. Funct. Biomater.* **2013**, *4*, 74–88.

42. Hogerheyde, T.A.; Suzuki, S.; Stephenson, S.A.; Richardson, N.A.; Chirila, T.V.; Harkin, D.G.; Bray, L.J. Assessment of freestanding membranes prepared from Antheraea pernyi silk fibroin as a potential vehicle for corneal epithelial cell transplantation. *Biomed. Mater.* **2014**, *9*.

43. Bray, L.J.; George, G.A.; Ainscough, S.L.; Hutmacher, D.W.; Chirila, T.V.; Harkin, D.G. Human corneal epithelial equivalents constructed on *Bombyx mori* silk fibroin membranes. *Biomaterials* **2011**, *32*, 5086–5091.

44. Bray, L.J.; George, K.A.; Hutmacher, D.W.; Chirila, T.V.; Harkin, D.G. A dual-layer silk fibroin scaffold for reconstructing the human corneal limbus. *Biomaterials* **2012**, *33*, 3529–3538.

45. Bray, L.J.; George, K.A.; Suzuki, S.; Chirila, T.V.; Harkin, D.G. Fabrication of a corneal-limbal tissue substitute using silk fibroin. In *Corneal Regenerative Medicine. Methods and Protocols*; Wright, B., Connon, C.J., Eds.; Humana Press: New York, NY, USA, 2013; pp. 165–178.

46. Madden, P.W.; Lai, J.N.X.; George, K.A.; Giovenco, T.; Harkin, D.G.; Chirila, T.V. Human corneal endothelial cell growth on a silk fibroin membrane. *Biomaterials* **2011**, *32*, 4076–4084.

47. Shadforth, A.M.; George, K.A.; Kwan, A.S.; Chirila, T.V.; Harkin, D.G. The cultivation of human retinal pigment epithelial cells on Bombyx mori silk fibroin. *Biomaterials* **2012**, *33*, 4110–4117.

48. Asakura, T.; Demura, M.; Tsutsumi, M. ^{23}Na and ^{27}Al NMR studies of the interaction between Bombyx mori silk fibroin and metal ions trapped in the porous silk fibroin membrane. *Makromol. Chem. Rapid Commun.* **1988**, *9*, 835–839.

49. Demura, M.; Asakura, T. Porous membrane of Bombyx mori silk fibroin: structure characterization, physical properties and application to glucose oxidase immobilization. *J. Membr. Sci.* **1991**, *59*, 39–52.

50. Kuga, S. Pore size distribution analysis of gel substances by size exclusion chromatography. *J. Chromatog.* **1981**, *206*, 449–461.

51. Jin, H.-J.; Fridrich, S.V.; Rutledge, G.C.; Kaplan, D.L. Electrospinning Bombyx mori silk with poly(ethylene oxide). *Biomacromolecules* **2002**, *3*, 1233–1239.

52. Jin, H.-J.; Park, J.; Valluzzi, R.; Cebe, P.; Kaplan, D.L. Biomaterial films of *Bombyx mori* silk fibroin with poly(ethylene oxide). *Biomacromolecules* **2004**, *5*, 711–717.

53. Lawrence, B.D.; Omenetto, F.; Chui, K.; Kaplan, D.L. Processing methods to control silk fibroin film biomaterial features. *J. Mater. Sci.* **2008**, 6967–6985.

54. Jin, H.-J.; Kaplan, D.L. Mechanism of silk processing in insect and spiders. *Nature* **2003**, 1057–1061.

55. Lawrence, B.D.; Marchant, J.K.; Pindrus, M.A.; Omenetto, F.G.; Kaplan, D.L. Silk film biomaterials for cornea tissue engineering. *Biomaterials* **2009**, *30*, 1299–1308.

56. Higa, K.; Takeshima, N.; Moro, F.; Kawakita, T.; Kawashima, M.; Demura, M.; Shimazaki, J.; Asakura, T.; Tsubota, K.; Shimmura, S. Porous silk fibroin film as a transparent carrier for cultivated corneal epithelial sheets. *J. Biomater. Sci. Polym. Edn.* **2010**, *22*, 2261–2276.

57. Lawrence, B.D.; Pan, Z.; Liu, A.; Kaplan, D.L.; Rosenblatt, M.I. Human corneal limbal epithelial cell response to varying silk film geometric topography *in vitro*. *Acta Biomater.* **2012**, *8*, 3732–3743.

58. Harley, B.A.; Yannas, I.V. *In vivo* synthesis of tissues and organs. In *Principles of Tissue Engineering*, 3rd ed.; Lanza, R., Langer, R., Vacanti, J., Eds.; Elsevier Academic Press: Burlington, MA, USA, 2007; pp. 219–238.

59. Williams, D. *Essential Biomaterials Science*; Cambridge University Press: Cambridge, UK, 2014; pp. 340–355.

60. Fitton, J.H.; Dalton, B.A.; Beumer, G.; Johnson, G.; Griesser, H.J.; Steele, J.G. Surface topography can interfere with epithelial tissue migration. *J. Biomed. Mater. Res.* **1998**, *42*, 245–257.

61. Evans, M.D.; Dalton, B.A.; Steele, J.G. Persistent adhesion of epithelial tissue is sensitive to polymer topography. *J. Biomed. Mater. Res.* **1999**, *46*, 485–493.

62. Steele, J.G.; Johnson, G.; McLean, K.M.; Beumer, G.; Griesser, H.J. Effect of porosity and surface hydrophilicity on migration of epithelial tissue over synthetic polymer. *J. Biomed. Mater. Res.* **2000**, *50*, 475–482.

63. Teixeira, A.I.; Abrams, G.A.; Bertics, P.J.; Murphy, C.J.; Nealey, P.F. Epithelial contact guidances on well-defined micro- and nanostructured substrates. *J. Cell Sci.* **2003**, *116*, 1881–1892.

64. Karuri, N.W.; Liliensiek, S.; Teixeira, A.I.; Abrams, G.; Campbell, S.; Nealey, P.F.; Murphy, C.J. Biological length scale topography enhances cell-substratum adhesion of human corneal epithelial cells. *J. Cell Sci.* **2004**, *117*, 3153–3164.

65. Diehl, K.A.; Foley, J.D.; Nealey, P.F.; Murphy, C.J. Nanoscale topography modulates corneal epithelial cell migration. *J. Biomed. Mater. Res. A* **2005**, *75*, 603–611.

66. Karuri, N.W.; Porri, T.J.; Albrecht, R.M.; Murphy, C.J.; Nealey, P.F. Nano- and microscale holes modulate cell-substrate adhesion, cytoskeletal organization, and -β1 integrin localization in Sv40 human corneal epithelial cells. *IEEE Trans. Nanobiosci.* **2006**, *5*, 273–280.

67. Liliensiek, S.J.; Campbell, S.; Nealey, P.F.; Murphy, C.J. The scale of substratum topographic features modulates proliferation of corneal epithelial cells and corneal fibroblasts. *J. Biomed. Mater. Res. A* **2006**, *79*, 185–192.

68. Zhang, K.; Mo, X.; Huang, C.; He, C.; Wang, H. Electrospun scaffolds from silk fibroin and their cellular compatibility. *J. Biomed. Mater. Res. A* **2010**, *93*, 976–983.

69. Hu, X.; Kaplan, D.; Cebe, P. Determining beta-sheet crystallinity in fibrous proteins by thermal analysis and infrared spectroscopy. *Macromolecules* **2006**, *39*, 6161–6170.

70. Hu, X.; Shmelev, K.; Sun, L.; Gil, E.-S.; Park, S.-H.; Cebe, P.; Kaplan, D.L. Regulation of silk material structure by temperature-controlled water vapour annealing. *Biomacromolecules* **2011**, *12*, 1686–1696.

71. Leal-Egaña, A.; Scheibel, T. Interactions of cells with silk surfaces. *J. Mater. Chem.* **2012**, *22*, 14330–14336.

72. Djerassi, C.; Gray, J.D.; Kincl, F.A. Naturally occurring oxygen heterocycles. IX. Isolation and characterization of genipin. *J. Org. Chem.* **1960**, *25*, 2174–2177.

73. Dawson, R.A.; Upton, Z.; Malda, J.; Harkin, D.G. Preparation of cultured skin for transplantation using insulin-like growth factor I in conjunction with insulin-like growth factor binding protein 5, epidermal growth factor, and vitronectin. *Transplantation* **2006**, *81*, 1668–1676.

Culture of Oral Mucosal Epithelial Cells for the Purpose of Treating Limbal Stem Cell Deficiency

Tor Paaske Utheim, Øygunn Aass Utheim, Qalb-E-Saleem Khan and Amer Sehic

Abstract: The cornea is critical for normal vision as it allows allowing light transmission to the retina. The corneal epithelium is renewed by limbal epithelial cells (LEC), which are located in the periphery of the cornea, the limbus. Damage or disease involving LEC may lead to various clinical presentations of limbal stem cell deficiency (LSCD). Both severe pain and blindness may result. Transplantation of cultured autologous oral mucosal epithelial cell sheet (CAOMECS) represents the first use of a cultured non-limbal autologous cell type to treat this disease. Among non-limbal cell types, CAOMECS and conjunctival epithelial cells are the only laboratory cultured cell sources that have been explored in humans. Thus far, the expression of p63 is the only predictor of clinical outcome following transplantation to correct LSCD. The optimal culture method and substrate for CAOMECS is not established. The present review focuses on cell culture methods, with particular emphasis on substrates. Most culture protocols for CAOMECS used amniotic membrane as a substrate and included the xenogeneic components fetal bovine serum and murine 3T3 fibroblasts. However, it has been demonstrated that tissue-engineered epithelial cell sheet grafts can be successfully fabricated using temperature-responsive culture surfaces and autologous serum. In the studies using different substrates for culture of CAOMECS, the quantitative expression of p63 was generally poorly reported; thus, more research is warranted with quantification of phenotypic data. Further research is required to develop a culture system for CAOMECS that mimics the natural environment of oral/limbal/corneal epithelial cells without the need for undefined foreign materials such as serum and feeder cells.

Reprinted from *J. Funct. Biomater.* Cite as: Utheim, T.P.; Utheim, Ø.A.; Khan, Q.-E.-S.; Sehic, A. Culture of Oral Mucosal Epithelial Cells for the Purpose of Treating Limbal Stem Cell Deficiency. *J. Funct. Biomater.* **2016**, *7*, 5.

1. Introduction

1.1. Limbal Stem Cell Deficiency

The regenerating organs in the body (e.g., cornea, skin, and gut) harbor tissue-specific stem cells, which are responsible for tissue homeostasis and efficient healing in case of injury. The ocular surface is composed of corneal and conjunctival epithelium [1]. The corneal epithelium in particular plays a crucial role in maintaining the cornea's avascularity and transparency [2]. The self-renewal of the corneal surface is a multistep process dependent on a small population of limbal stem cells [3,4] located in structures referred to as limbal crypts [5] or limbal epithelial crypts [6].

Numerous external factors and disorders (e.g., chemical or thermal injuries, microbial infections, surgeries involving the limbus, cicatricial pemphigoid, and aniridia) can lead to dysfunction or loss of limbal epithelial cells (LEC), resulting in either partial or total limbal stem cell deficiency (LSCD) [2]. The condition can be painful and may lead to reduced vision, or even blindness, by causing persistent epithelial defects, fibrovascular pannus, conjunctivalization, and superficial and deep vascularization of the cornea. The persistence of epithelial defects may result in ulceration, scarring, and corneal perforation [2]. Limbal stem cell deficiency is most often bilateral.

1.2. Treatment Strategies for Limbal Stem Cell Deficiency

Treatment approaches for LSCD can be categorized as follows: (a) transplantation of cultured cells [2]; (b) transplantation of non-cultured cells [2]; and (c) approaches that do not involve transplantation of cells, for example keratoprostheses [7]. A great variety of cell-based therapeutic strategies have been suggested for LSCD [8]. The stem cells of the corneal epithelium are believed to be located in the limbus [3,4]. In 1989, limbal grafts were transplanted to eyes suffering from LSCD to restore the corneal surface [9]. The results were promising. However, the procedure carries a risk of inducing LSCD in the healthy eye because of large limbal cell withdrawal [10], and the therapy is not possible in cases of bilateral LSCD. This led to a novel therapeutic strategy with *ex vivo* expansion of LEC first reported by Pellegrini and colleagues in 1997 [11]. In their study, successful ocular surface reconstruction was achieved using autologous cultivated LEC isolated from small biopsies in two patients, both affected with severe unilateral ocular surface disease. Since then, more than 1000 patients suffering from LSCD have been treated with *ex vivo* cultured LEC [11–18]. Since 2003, nine cultured non-limbal cell sources have been successfully used to reconstruct the corneal epithelium in bilateral LSCD, in which limbal tissue is not recommended for harvest [8]. The sources include oral mucosal epithelial cells [19], embryonic stem cells [20], conjunctival epithelial cells [21], epidermal stem cells [22], dental pulp stem cells [23], bone marrow-derived mesenchymal stem cells [24], hair follicle bulge-derived stem cells [25], umbilical cord lining stem cells [26], and orbital fat-derived stem cells [27]. Among non-limbal cell types, cultured autologous oral mucosal epithelial cell sheet (CAOMECS) and conjunctival epithelial cells are the only laboratory cultured cell sources that have been explored in humans.

2. Cultured Autologous Oral Mucosal Epithelial Cell Sheet

A significant advantage of CAOMECS as a cell source is easy isolation from biopsies that heal quickly without residual scarring. As the CAOMECS are autologous, there is no risk of immune rejection, thus making immunosuppression unnecessary. However, a disadvantage of transplantation of CAOMECS is the development of peripheral neovascularization [28–31]. Studies have demonstrated that angiogenesis related factors were expressed in corneas after transplantation [32–35]. Anti-angiogenic therapy has been proposed as a method to prevent corneal neovascularization and improve the outcomes after transplantation with CAOMECS [36]. Thus far, 242 patients with LSCD have been reported as treated, with a success rate of 72% and a follow-up time of between one and 7.5 years [19,28–32,37–51].

An ideal substrate is easily available, transparent, and easy to manipulate; it permits cells to proliferate and retain high viability. Though transplant success has been demonstrated using various culture methods, the optimal culture method for CAOMECS for use in corneal regeneration has not been established. The determination of appropriate substrates and culture protocols for CAOMECS may contribute to the development of standardized, safe, and effective regenerative therapy for LSCD. The present review focuses on the current state of knowledge of the culture methods and substrates used for CAOMECS in ocular regeneration. The review was prepared by searching the National Library of Medicine database using the search term "oral mucosal epithelial cells" in an attempt not to leave out any relevant publications. In total, the search resulted in 4897 studies, of which 41 studies, published from 2003 to 2015, were related directly to the core topic of the present review.

3. Characteristics of the Culture Protocol for Cultured Autologous Oral Mucosal Epithelial Cell Sheet

The standard culture conditions used for production of transplantable epithelial cell sheets, including CAOMECS, typically requires fetal bovine serum (FBS) and murine 3T3 feeder layers [52]. The epithelial progenitor or stem cells isolated from small biopsies can, under these conditions, be expanded *in vitro* to create stratified epithelial layers that closely resemble native tissues [53]. However, these constructs are classified as xenogeneic products, with the inherent possibility of infection or pathogen transmission from animal-derived materials [54]. In addition, xeno-contamination may result in immunogenicity [55]. The use of feeder layers and foreign serum is, therefore, a concern in regenerative medicine. Furthermore, dispase, a bacteria-derived protease, is commonly used to enable cell isolation [53].

Treatment of LSCD based on various methods using CAOMECS is presented in Figure 1. The following culture methods and substrates have been used in order to produce transplantable CAOMECS: (1) amniotic

163

membrane [28–30,32,35,37,39,40,42,43,45–47,49,51,56–63] (Table 1); (2) temperature-responsive cell-culture surfaces [31,38,64–70] (Table 2); (3) fibrin-coated culture plates [41,48] (Table 3); (4) fibrin gel [71] (Table 3); (5) collagen IV-coated culture plates [72] (Table 3); and (6) culture plates without any substrate [33,34,73,74] (Table 3).

The possibility of pathogen transmission cannot be excluded from xenogeneic or allogeneic materials, such as human amniotic membrane obtained following elective Caesarean operations [17,18,75], collagen isolated from porcine or bovine skin [76], and hydrated gels made from fibrin derived from human donor blood [77–79]. Therefore, the establishment of culture conditions avoiding animal-derived products and foreign undefined components is warranted.

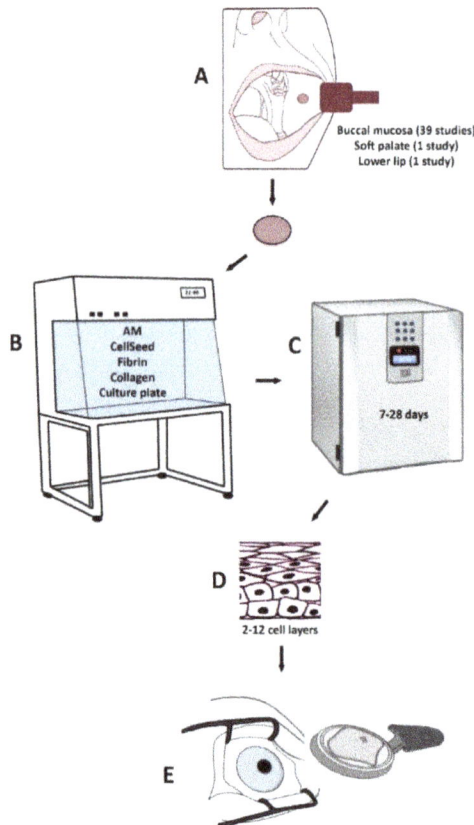

Figure 1. Treatment of LSCD based on various methods using CAOMECS. A biopsy from the mucosa is harvested from the oral cavity (**A**). The biopsy is cultured in the laboratory on different substrates (**B**) for 7–28 days (**C**). A stratified cultured tissue is produced (**D**) and is transplanted to the diseased eye (**E**).

Table 1. Culture of oral mucosal epithelial cells on amniotic membrane.

Author, Year	Type of Study	Cell Suspension/Explant	Substrate	Air-Lifting	Serum	3T3	Culture Medium	Culture Time (Days)	Morphology	Phenotype
Shimazaki et al., 2009 [62]	Animal	Cell suspension	Denuded AM	Yes	FBS	Yes	SHEM (aprotinin)	7–10	Multilayered stratified epithelium; Tight junctions	Expression of K3, ZO-1, and occludin
Sekiyama et al., 2006 [35]	in Vitro	Cell suspension	Denuded AM	Yes	–	Yes	DMEM:F12 (penicillin, streptomycin, insulin, cholera toxin, EGF)	7–14	–	Expression of VEGF and Flt-1; Low expression of PEDF
Sotozono et al., 2013 [50]	Clinical	Cell suspension	Denuded AM	Yes	HAS	Yes	DMEM:F12 (penicillin, streptomycin, insulin, cholera toxin, EGF)	8–9	–	–
Sotozono et al., 2014 [49]	Clinical	Cell suspension	Denuded AM	Yes	HAS	Yes	DMEM:F12 (penicillin, streptomycin, insulin, cholera toxin, EGF)	8–9	–	–
Gaddipati et al., 2014 [40]	Clinical	Explant	Denuded AM	–	–	No	DMEM:F12 (penicillin, streptomycin, insulin, cholera toxin, EGF)	9	5–6 cell layers; Stratified epithelium	Expression of K3, K12, K19, Ki-67, p75, and PAX6; p63 expression in most of the basal and supra basal cells
Sen et al., 2011 [60]	in Vitro	Explant	Denuded AM	Yes	FCS	Yes	DMEM:F12 (penicillin, streptomycin, amphotericin, EGF, insulin)	14	Stratified epithelium; Desmosomes; Abundant mucin granules	Expression of K3, K4, K13, connexin 43, p63, p75, β₁-integrin, CD29, ABCG2, and MUC 1, 5B, 6, 13, 15 and 16
Satake et al., 2008 [47]	Clinical	Cell suspension	Denuded AM	Yes	FBS	Yes	DMEM:F12 (gentamycin, streptomycin, penicillin, amphotericin, EGF, insulin)	>14	Non-keratinized, squamous, polygonal, cells with a low nuclear to cytoplasmatic ratio	–
Takeda et al., 2011 [51]	Clinical	Cell suspension	Denuded AM	Yes	–	Yes	DMEM:F12 (penicillin, streptomycin, insulin, cholera toxin, EGF)	14–16	–	–
Chen et al., 2009 [39]	Clinical	Cell suspension	Denuded AM	No	FCS	Yes	DMEM:F12(penicillin, streptomycin, insulin, cholera toxin, EGF)	14–21	2–5 cell layers; Elongated cell nuclei	Expression of K3, K4, K13, p63, p75, and ABCG2
Chen et al., 2012 [32]	Clinical	Cell suspension	Denuded AM	No	FCS	Yes	SHEM (penicillin, streptomycin, insulin, cholera toxin, EGF)	14–21	5–10 cell layers; Stratified epithelium	Expression of FGF2, K8, VEGF, endostatin, PEDF, and IL-1ra

Table 1. *Cont.*

Author, Year	Type of Study	Cell Suspension/Explant	Substrate	Air-Lifting	Serum	3T3	Culture Medium	Culture Time (Days)	Morphology	Phenotype
Ma et al., 2009 [29]	Clinical	Suspension	Denuded AM	No	FBS	Yes	DMEM:F12 (penicillin, streptomycin, insulin, cholera toxin, EGF)	14–21	2–5 cell layers; Elongated cell nuclei	Expression of K3, K13, p63, p75, and ABCG2
Nakamura et al., 2004 [30]	Clinical	Cell suspension	Denuded AM	Yes	FBS	Yes	DMEM:F12 (penicillin, streptomycin, insulin, cholera toxin, EGF)	14–21	5–6 cell layers; Desmosomes and hemidesmosomes	Expression of K3, K4, and K13
Ang et al., 2006 [37]	Clinical	Cell suspension	Denuded AM	Yes	HAS/FBS	Yes	KGM (penicillin, streptomycin, insulin, EGF)	15–16	4–6 cell layers; Cuboidal cells, More flattened cells superficially	Expression of K3, K4, K13, ZO-1, desmoplakin, integrin-α_1, laminin 5, and collagen IV
Ang et al., 2006 [37]	Clinical	Cell suspension	Denuded AM	Yes	HAS/FBS	Yes	KGM (penicillin, streptomycin, insulin, EGF)	15–16	4–6 cell layers; Cuboidal cells, More flattened cells superficially	Expression of K3, K4, K13, ZO-1, desmoplakin, integrin-α_1, laminin 5, and collagen IV
Inatomi et al., 2006 [28]	Clinical	Cell suspension	Denuded AM	Yes	FCS	Yes	DMEM:F12 (penicillin, streptomycin, insulin, cholera toxin, EGF)	15–16	5–6 cell layers; Cuboidal cells, several suprabasal cell layers, and flat apical cell layers	Expression of VEGF, FGF, and thrombospondin 1
Inatomi et al., 2006 [28]	Clinical	Cell suspension	Denuded AM	Yes	FCS	Yes	DMEM:F12 (penicillin, streptomycin, insulin, cholera toxin, EGF)	15–16	5–6 cell layers; Cuboidal cells, several suprabasal cell layers, and flat apical cell layers	Expression of VEGF, FGF, and thrombospondin 1
Inatomi et al., 2006 [42]	Clinical	Cell suspension	Denuded AM	Yes	HAS/FCS	Yes	DMEM:F12 (penicillin, streptomycin, insulin, cholera toxin, EGF)	15–16	5–6 cell layers; Cuboidal cells, several suprabasal cell layers, and flat apical cell layers	
Nakamura et al., 2011 [45]	Clinical	Cell suspension	Denuded AM	Yes	HAS	Yes	KGM (penicillin, streptomycin, insulin, cholera toxin, EGF)	15–16	–	–
Priya et al., 2011 [46]	Clinical	Cell suspension	Denuded AM	No	AS	Yes	DMEM:F12 (PI, mouse IgG1/IgG2a, mitomycin C, EGF, insulin, penicillin, streptomycin)	18–21	Flat and uniformly distributed epithelial cells	Low expression of p63 (3.0% ± 1.7% of cells); Negative expression of K12

Table 1. *Cont.*

Author, Year	Type of Study	Cell Suspension/Explant	Substrate	Air-Lifting	Serum	3T3	Culture Medium	Culture Time (Days)	Morphology	Phenotype
Sharma et al., 2011 [61]	In vitro	Cell suspension	Denuded AM	–	FBS	Yes	DMEM:F12 (penicillin, streptomycin, insulin, cholera toxin, EGF)	21	3–5 cell layers; Stratified epithelium	Expression of K3 and β1-integrin; High expression of p63
Promprasit et al., 2014 [59]	in Vitro	Explant	Denuded AM	Yes	FBS	Yes	DMEM:F12 (penicillin, streptomycin, insulin, EGF)	21	2–5 cell layers; Stratified epithelium; Cuboidal cells in basal layer, flat superficial cells	Expression of K3 and connexin 43; High expression of p63
Nakamura et al., 2003 [57]	Animal	Cell suspension	Denuded AM	Yes	FBS	Yes	DMEM:F12 (penicillin, streptomycin, insulin, cholera toxin, EGF)	21	3–5 cell layers; Stratified epithelium;	Expression of K3, K4, and K13
Nakamura et al., 2003 [58]	Animal	Cell suspension	Denuded AM	Yes	FBS	Yes	DMEM:F12 (penicillin, streptomycin, insulin, cholera toxin, EGF)	21	5–6 cell layers; Stratified epithelium;	Expression of K3, K4, and K13
Nakamura et al., 2003 [58]	Animal	Cell suspension	Denuded AM	Yes	FBS	Yes	DMEM:F12 (penicillin, streptomycin, insulin, cholera toxin, EGF)	21	5–6 cell layers; Stratified epithelium;	Expression of K3, K4, and K13
Kolli et al., 2014 [43]	Clinical	Explant	Intact AM	Yes	HAS	No	DMEM:F12 (penicillin, streptomycin, insulin, cholera toxin, EGF, hydrocortisone, triiodothyronine, adenine)	21	3–7 cell layers, firmly attached to each other; High nucleus to cytoplasm ratio	Expression of K3, ABCG2, and C/EBPδ; High expression of ΔNp63α; Negative for K12 and PAX6
Madhira et al., 2008 [56]	in Vitro	Cell suspension	Denuded AM	No	FCS	No	DMEM:F12 (penicillin, streptomycin, amphotericin, gentamycin, insulin, cholera toxin, EGF)	21–28	2–3 cell layers; Stratified epithelium; Gap junctions and desmosomes	Expression of K3, K4, K15, and connexin 43; Negative for K12 and PAX6
Yokoo et al., 2006 [63]	in Vitro	Cell suspension	Denuded AM	Yes	FBS	Yes	DMEM/F12 (penicillin, streptomycin, amphotericin)	28	3–5 cell layers; Stratified epithelium	–

ABCG2, ATP binding cassette subfamily G member; AM, amniotic membrane; AS, autologous serum; DMEM, Dulbecco's modified eagle medium; EGF, epidermal growth factor; FBS, fetal bovine serum; FCS, fetal calf serum; FGF2, fibroblast growth factor 2; Flt-1, Fms-like tyrosine kinase 1; HAS, human autologous serum; IgG2a, immunoglobulin G2a; IL-1ra, interleukin 1ra; KGM, keratinocyte growth medium; MUC, mucin; PAX6, paired box 6; PEDF, pigment epithelium derived factor; PI, propidium iodide; SHEM, supplemented hormonal epithelial medium; VEGF, vascular endothelial growth factor; ZO-1, zona occludens protein 1; –, indicates not reported.

Table 2. Culture of oral mucosal epithelial cells on temperature-responsive surfaces.

Author, Year	Type of Study	Cell Suspension/Explant	Substrate	Air-Lifting	Serum	3T3	Culture Medium	CultureTime (Days)	Morphology	Phenotype
Burillon et al., 2012 [38]	Clinical	Cell suspension	CellSeed [a]	No	–	Yes	–	–	Similar characteristics to normal corneal epithelium; Basal membrane	Expression of K3/76, p63, laminin 5, and β_1-integrin
Soma et al., 2014 [69]	Animal	Cell suspension	CellSeed [a]	–	FBS	Yes	DMEM:F12 (insulin, triiodthyronine, hydrocortisone)	10–12	3–4 cell layers; Stratified epithelium; Cobble stone-like cell morphology	Expression of K14 and p63
Sugiyama et al., 2014 [70]	Animal	Cell suspension	CellSeed [a]	–	FBS	Yes	DMEM:F12 (penicillin, streptomycin, insulin, cholera toxin, EGF, hydrocortisone, triiodothyronine)	14	3–5 cell layers; Stratified epithelium; Cuboidal cells in the basal layer, squamous epithelium on the apical side	Expression of K4, K13, MUC5
Nishida et al., 2004 [31]	Clinical	Cell suspension	CellSeed [a]	No	–	Yes	–	14	Multilayered cell sheets; Microvilli, desmosomes, basement membrane	Expression of β_1-integrin, K3, and p63
Bardag-Gorce et al., 2015 [64]	Animal	Cell suspension	CellSeed [a]	–	FBS	Yes	–	14	Multilayered stratified epithelium	Expression of K4, ΔNp63, TIMP-1, TIMP-3, and connexin 43
Hayashida et al., 2005 [66]	Animal	Cell suspension	CellSeed [a]	–	FBS	Yes	–	14	3–5 cell layers; Stratified epithelium;	Expression of K3, K4, K13, p63, ΔNp63, and β_1-integrin
Murakami et al., 2006 [67]	in Vitro	Cell suspension	CellSeed [a]	–	HAS	No	DMEM/F12 (penicillin, streptomycin, fungizone, transferrin, EGF, cholera toxin, hydrocortisone, triiodothyronine)	14	3–5 cell layers; Cuboidal basal cells, flattened middle cells, and polygonal flattened superficial cells	Expression of p63 and Ki67
Oie et al., 2010 [68]	Clinical	Cell suspension	CellSeed [a]	–	HAS	Yes	–	14–17	4–5 cell layers; Small basal cells, flattened middle cells, and polygonal flattened superficial cells	Expression of K1, K3/76, K4, K10, K12, K13, K15, ZO-1, and MUC16; Moderate expression of p63 (30.7% ± 7.6% of cells)

168

Table 3. Culture of oral mucosal epithelial cells on other substrates.

Author, Year	Type of Study	Cell Suspension/Explant	Substrate	Air-Lifting	Serum	3T3	Culture Medium	Culture Time (Days)	Morphology	Phenotype
Satake et al., 2011 [48]	Clinical	Cell suspension	Fibrin-coated cell culture inserts	Yes	HAS	Yes	DMEM:F12 (penicillin, streptomycin, transferrin, EGF, hydrocortisone, triiodothyronine)	–	5–6 cell layers;	–
Hirayama et al., 2012 [41]	Clinical	Cell suspension	Fibrin-coated cell culture inserts	Yes	HAS	Yes	DMEM:F12 (penicillin, streptomycin, insulin, EGF, hydrocortisone)	–	5–6 cell layers;	–
Sheth et al., 2014 [71]	in Vitro	Explant	Fibrin gel	–	FCS	No	DMEM:F12 (penicillin, cholera toxin, insulin, EGF, hydrocortisone)	–	Multilayered epithelium; Cobblestone morphology	Expression of K3, K13, and K19; High expression of p63
Ilmarinen et al., 2012 [72]	in Vitro	Cell suspension	Collagen IV-coated cell culture inserts	Yes	No	No	Serum-free oral PCT epithelium medium (EGF)	13–17	4–12 cell layers; Stratified epithelium; Cuboidal basal cells and flat intermediate and superficial cells	Expression of K3/12, K4, K13, Ki67, and p63
Kanayama et al., 2007 [34]	in Vitro	Cell suspension	Culture plate	–	FBS	Yes	DMEM (Supplements not reported)	–	Multilayered cells; Normal epithelial morphology	Expression of FGF2, VEGF, Ang1, and TGF-β1
Kanayama et al., 2009 [33]	in Vitro	Cell suspension	Culture plate	–	FBS	Yes	DMEM (Supplements not reported)	14	Multilayered cells	Expression of VEGFr-1
Hyun et al., 2014 [74]	Animal	Cell suspension	Culture plate	–	FBS	Yes	DMEM:F12 (penicillin, streptomycin, gentamycin, amphotericin)	14	2–6 cell layers; Stratified epithelium	Expression of K3, K4, and Ki67; High expression of p63
Krishnan et al., 2010 [73]	in Vitro	Explant	Culture plate	–	FBS	–	DMEM:F12 (streptomycin, amphotericin, EGF, insulin, transferrin, selenium, hydrocortisone)	21	Multilayered cells; Normal epithelial morphology	Expression of ABCG2, K3, MUC1/4/16, hBD1/2/3; High expression of p63 and ΔNp63

ABCG2, ATP binding cassette subfamily G member; Ang1, angiopoietin; DMEM, Dulbecco's modified eagle medium; EGF, epidermal growth factor; FBS, fetal bovine serum; FCS, fetal calf serum; FGF2, fibroblast growth factor 2; HAS, human autologous serum; hBD, human beta defensing; MUC, mucin; PCT, progenitor cell-targeted; TGF-β1, transforming growth factor beta 1; VEGF, vascular endothelial growth factor; –, indicates not reported.

4. Culture of Oral Mucosal Epithelial Cells on Amniotic Membrane

Amniotic membrane has been used on the ocular surface since 1940 [80], and for the first time in treatment of LSCD in 1946 [81]. In cases of partial LSCD, amniotic membrane can be applied to the affected eye and provide a suitable substrate for corneal epithelial repopulation [82,83]. The amniotic membrane secretes several growth factors such as hepatocyte growth factor, basic fibroblast growth factor, and transforming growth factor β [84,85]. Amniotic membrane is suggested to exert its effects by suppressing inflammation and scarring [86]. There is currently a discussion over whether amniotic membrane should be deepithelialized/denuded prior to culture, or if this substrate should remain intact. It has been reported that native, intact amniotic membrane comprise higher levels of growth factors compared to denuded amniotic membrane [87].

Amniotic membrane is the most common culture substrate for CAOMECS, and has been used in 15 clinical, three animal, and six *in vitro* studies (Table 1). With one exception [43], the amniotic membrane was denuded, *i.e.*, the single layer of epithelial cells on the amniotic membrane was removed (Table 1). In the studies using amniotic membrane as a substrate for cultured CAOMECS, cell suspension [28–30,32,35,37,39,42,45–47,49–51,56–58,61–63] was applied in all studies, except four using the explant method [40,43,59,60]. The number of fabricated, stratified epithelial cell layers varied from two [56] to 10 [32]. Oral mucosal epithelial cells were normally cultivated between two to three weeks; however, the culture time varied between seven [62] and 28 [63] days. The most frequently used culture medium with added supplements was Dulbecco's Modified Eagle Medium (DMEM:F12) [28–30,35,39,40,42,43,46,47,49–51,56–61,63], followed by keratinocyte growth medium (KGM) [37,45] and supplemented hormonal epithelial medium (SHEM) [32,62] (Table 1). Murine fibroblasts (3T3 strain) were used in all but three studies [40,43,56]. Most of the culture protocols exposed the cells to air-lifting (lowering the level of the culture medium to allow the cells to be cultured at the air–liquid interface), including clinical [28,30,37,42,43,45,47,49–51], animal [57,58], and *in vitro* studies [35,59,60,63] (Table 1). Fetal bovine serum (FBS) [29,30,37,47,57–59,61–63] and fetal calf serum (FCS) [28,32,39,42,56,60] were broadly used; however, six studies used human autologous serum (HAS) [28,43,45,46,49,50] in an attempt to minimize/avoid the use of animal derived components (Table 1).

Oral mucosal epithelial cells cultivated on amniotic membrane exhibited multilayered, stratified epithelium and appeared very similar to a normal corneal epithelium (Table 1). The presence of non-keratinized, stratified-specific keratins K3 and K4/K13 was detected by immunohistochemistry, reverse transcription polymerase chain reaction, and Western blotting (Table 1). The expression of p63, a marker for undifferentiated cells, was reported in 33% (8/24) of the studies (Table 4).

Using transmission electron microscopy it was demonstrated that the cultivated oral epithelial sheet had junctional contacts, such as desmosomes, hemidesmosomes, and tight junctions, which were almost identical to those of normal corneal epithelial cells [30,56,60,62].

5. Culture of Oral Mucosal Epithelial Cells on Temperature-Responsive Surfaces

In order to avoid the use of allogenic bacteria [53,67] and animal derived [52] components in the cornea-engineered constructs, carrier-free epithelial cell sheets using temperature-responsive culture dishes have been developed [31,88,89]. The modified surfaces transition between hydrophilic and hydrophobic states—depending on the temperature—by covalently immobilizing the temperature-responsive polymer poly(N-isopropylacrylamide) onto commercially available tissue culture wells. Under *in vitro* culture conditions at 37 °C, numerous cell types adhere and proliferate similarly to those of normal tissue culture polystyrene. By reducing the temperature to 20 °C, the cultured cells spontaneously detach along with their deposited extracellular matrix (ECM) without the need for proteolytic enzymes such as dispase [89,90]. Therefore, with temperature-responsive culture surfaces the undesirable factors inherent to some substrates can be excluded from transplantable constructs.

Nine studies (three clinical, four animal, and two *in vitro*) have utilized the temperature-responsive cell-culture surfaces as a substrate for CAOMECS. In all studies the cells were applied as a cell suspension and DMEM:F12 with added supplements was used as a culture medium (Table 2). The culture time for CAOMECS in these studies ranged from 10 [69] to 28 [65] days, but was most often two weeks [31, 64,66,67,70]. The most common nutrient used was FBS [64–66,69,70]; however, two studies utilized HAS [67,68]. None of the studies exposed the cells to air-lifting (Table 2). The number of fabricated cell layers varied from three [69] to eight [65]. Only one study did not use 3T3 murine fibroblasts [67]. Two studies reported the cell viability of the cultured sheets to be 86% [68] and 93% [65]. The presence of p63 in the fabricated cell sheets was reported in 78% (7/9) of the studies (Table 4).

Table 4. Expression of p63 in cultured autologous oral mucosal epithelial cell sheet cultivated on different substrates.

Substrate	Total Number of Studies	Expression of p63 Not Reported	Non-Quantitative Expression of p63 Reported	Quantitative Expression of p63 Reported
Amniotic membrane	24	16 studies	4 studies: p63 expressed; 1 study: high expression of ΔNp63; 2 studies: high expression of p63	1 study: 3.0% ± 1.7% of cells
Temperature-responsivecell-culture inserts	9	2 studies	6 studies: p63 expressed	1 study: 30.7% ± 7.6% of cells
Fibrin-coated culture plate	2	2 studies	–	–
Fibrin gel	1	–	1 study: high expression of p63	–
Collagen IV-coated culture plate	1	–	1 study: p63 expressed	–
Culture plate	4	2 studies	2 studies: high expression of p63	–

6. Culture of Oral Mucosal Epithelial Cells on Fibrin Substrates

Fibrin has been broadly used as a substrate in regenerative medicine and for wound-healing [91,92]. It is easily available, assists epithelial cell growth, and its degradation can be controlled by addition of fibrinolytic components. Rama and colleagues first established the use of fibrin gels as a substrate for ocular surface reconstruction in 2001 [78]. Fibrin gel is a hemostatic compound of thrombin, fibrinogen, and calcium chloride [93]. The mixture of these components fabricates a gel that is similar to the physiological lump formed at the last stage of the coagulation cascade [94]. The gel produced by this reaction is biodegradable, non-toxic, and inhibits fibrosis, tissue necrosis, and inflammation [94–96]. *In vivo*, the gel is completely resorbed and ultimately replaced by matrix components such as collagen [95]. A major disadvantage with fibrin as a substrate is that it encourages angiogenesis [97]. The gel, however, is resorbed within days to weeks after transplantation [94], minimizing the effects. Sheth *et al.* have demonstrated that CAOMECS cultivated on fibrin gel results in production of multilayered epithelium *in vitro*. The fabricated cell sheets expressed keratins K3, K4, and K13 [71]. The putative epithelial progenitor cell marker p63 [98] was also highly expressed (Table 3). Sheth and associates modified the pre-existing methodology to produce a reproducible, robust gel that supports the expansion and transplantation of CAOMECS, without the need for murine 3T3 fibroblasts. Fibrin-coated culture plates have also been used as a substrate for CAOMECS [41,48] (Table 3). Both studies utilized murine 3T3 fibroblasts and DMEM:F12 with added supplements as a culture medium. Human autologous serum was used as nutrient, and the cells were exposed to air-lifting [41,48].

7. Culture of Oral Mucosal Epithelial Cells on Collagen Substrates

All of the previously reported culture protocols for CAOMECS use serum, and most also use feeder cells to support the stratification of the epithelial cells. Due to the risk of infections associated with murine feeder cells and non-autologous serum in the cultivation of cell sheets, Ilmarinen and colleagues sought other options to support the stratification of isolated CAOMECS [72]. In their *in vitro* study, stratified epithelium was generated on collagen IV-coated culture plates in serum-free culture conditions without using 3T3 feeder cells. The authors analyzed the functional properties of the cell sheets by transepithelial electrical resistance measurements, in addition to morphology, differentiation, and regenerative capacity. This study is the only report of a successful stratification of oral mucosal epithelium for ocular surface regeneration in the absence of serum. The results showed that, in serum-free conditions, oral mucosal epithelial cells attached to and proliferated on collagen IV–coated inserts more readily than on amniotic membrane [72]. Ilmarinen and colleagues also studied the effects of increased epidermal growth factor (EGF) concentration, as EGF is known to stimulate the growth and differentiation of a variety of epithelial tissues [99,100]. However, they detected no major effects on the phenotype of the cell sheets using additional EGF.

8. Culture of Oral Mucosal Epithelial Cells on Non-Coated Culture Plates

Four studies (three *in vitro* and one animal) have used non-coated, substrate-free culture plates in order to fabricate transplantable CAOMECS [33,34,73,74] (Table 3). All of the studies used DMEM:F12 with added supplements as a culture medium and FBS as a nutrient, without including air-lifting. In three studies, murine 3T3 feeder cells were included [33,34,74]. The authors reported formation of a multilayered epithelium [33,34,73], one study specifying the number of cell layers [74]. Two of the four studies confirmed the expression of K3 and high expression of p63 [73,74].

9. Challenges and Future Perspectives

Recently, a meta-analytic concise review about transplantation of CAOMECS for treating LSCD has reported a success rate of 72% [19]. In this review, the focus was on clinical features of transplants of CAOMECS over the past 10 years, including surgery and pre- and postoperative considerations. In contrast, herein we focus on cell culture methods, with particular emphasis on substrates. Moreover, in the present review we expand on both *in vitro* and animal studies.

A complete xenobiotic-free culture protocol has become a goal in regenerative medicine; this is to avoid the risk of transferring known and unknown microorganisms and to standardize the culture conditions. The properties of epithelial cells are dependent upon extracellular signals supplied by the cell–cell

173

and cell–substratum interactions. Further research is warranted to develop a culture system for CAOMECS that mimics the natural environment of oral/limbal/corneal epithelial cells without the need for undefined foreign materials such as serum and feeder cells.

It is likely that the phenotype of CAOMECS affects clinical success following transplantation. Thus far, p63 is the only predictor of clinical outcome following transplantation to correct LSCD [12]. Recently, Rama *et al.* demonstrated that the phenotype of cultured LEC is critical to ensure successful reconstruction of the ocular surface following LSCD [12]. The authors showed that successful transplantation was achieved in 78% of patients when using cell cultures in which p63-bright cells constituted more than 3% of the total number of clonogenic cells. In contrast, successful transplantation was only seen in 11% of patients when p63-bright cells made up 3% or less of the total number of cells. In the studies using different substrates for culture of CAOMECS, the expression of p63 varied considerably (Table 4). Few studies reported the expression of p63 when using fibrin-coated culture plates, fibrin gels, collagen-coated culture plates, and culture plates without substrate (Table 4). When comparing amniotic membrane and temperature-responsive inserts, 33% (8/24) and 78% (7/9) of the studies showed the expression of p63, respectively (Table 4). The quantitative expression of p63 was generally poorly reported; thus, more research is warranted with quantification of phenotypic data.

The use of culture inserts with autologous serum has also been shown to facilitate the stratification of oral mucosal epithelial cells in the absence of 3T3 feeders [67]. Kolli *et al.* found that autologous serum was superior to FCS in generating an undifferentiated epithelium [43], and in another study the porcine trypsin was replaced with xeno-free trypsin with successful outcomes [61]. Hirayama *et al.* [41] showed that transplantation of a substrate-free cell sheet resulted in significantly better results than engrafting oral mucosal cells cultured on an amniotic membrane. The improvements were significantly higher graft survival rate, better visual acuity (1, 3, 6, and 12 months postoperatively), and reduction of neovascularization (12 months postoperatively) [41]. Furthermore, except collagen IV-coated culture plate, this review demonstrates that the use of different methods and substrates for culture of CAOMECS did not appear to have any effect on the number of cell layers generated (Table 5).

Table 5. Overall Effect of Different Culture Methods and Substrates for Cultured Autologous Oral Mucosal Epithelial Cell Sheet.

Substrate/Method	Air-lifting	Animal-derived Nutrient	Use of 3T3	Serum-free Medium	Viability	Morphology	Phenotype (Expression of p63)
Amniotic membrane	17/24	16/24	21/24	0/24	>98% (1)	4.2 cell layers (15)	++
Temperature-responsive cell-culture inserts	0/9	5/9	8/9	0/9	86%–93% (2)	4.3 cell layers (6)	++
Fibrin-coated culture plate	2/2	0/2	2/2	0/2		5–6 cell layers (2)	–
Fibrin gel	0/1	1/1	0/1	0/1	–	–	+++
Collagen IV-coated culture plate	1/1	0/3	0/1	1/1	–	4–12 cell layers (1)	+
Culture plate	0/4	4/4	3/4	0/4	–	2–6 cell layers (1)	+++

Number of studies using different culture parameters is presented in the Table; –, indicates not reported; +, low expression of p63; ++, moderate expression of p63; +++, high expression of p63.

Due to the lack of mechanical strength provided by various culture substrates, transplantation of substrate-free cell sheets can be challenging. Hence, methods to enhance the strength and durability of the epithelial cell sheets should be further explored. Using the air-lifting technique, originally developed to formulate skin cell culture sheets for transplantation, the mechanical strength of epithelial cell sheets can be increased. The present review reveals that only 48.8% of the studies applied the air-lifting method (Tables 1–3). Interestingly, the majority of studies using amniotic membrane (71%) did utilize air-lifting, while none of the studies with temperature-responsive surfaces applied this method (Table 5). Arguments for air-lifting include the promotion of migration [101], proliferation [101], epithelial stratification [101], and increased barrier function of LEC [102]. Arguments against air-lifting include induction of squamous metaplasia [103], gradual loss of stem cells [104], and differentiation of LEC [104,105]. Until 2010, the clinical implications of increased differentiation of transplanted cells in corneal reconstruction were unknown. This changed when Rama and colleagues demonstrated the critical importance for clinical success of a substantial, putative stem cell population within the cultured cells [12]. It is yet to be investigated whether the potential advantages of air-lifting outweigh the disadvantages in corneal regeneration using CAOMECS.

10. Conclusions

Most culture protocols for CAOMECS used amniotic membrane as a substrate and included the xenogeneic components FBS and murine 3T3 fibroblasts. However, it has been demonstrated that tissue-engineered epithelial cell sheet grafts can be successfully fabricated using temperature-responsive culture surfaces and autologous serum. More studies on how various substrates and other culture parameters affect the cell sheet, with special emphasis on the phenotype, are warranted. Furthermore, it is important to focus on cell culture methods using xenobiotic-free conditions.

Acknowledgments: The authors would like to thank Astrid Østerud, Department of Medical Biochemistry, Oslo University Hospital, Oslo, and Catherine Jackson at Department of Medical Biochemistry. Funding received from Department of Oral Biology, Faculty of Dentistry, University of Oslo and Department of Medical Biochemistry, Oslo University Hospital, Oslo, Norway is acknowledged.

Author Contributions: Tor Paaske Utheim, Øygunn Aass Utheim and Amer Sehic searched and identified the literature; Tor Paaske Utheim and Amer Sehic analyzed the data; All wrote the paper.

Conflicts of Interest: The authors declare no conflict of interest.

References

1. Land, M.F.; Fernald, R.D. The evolution of eyes. *Ann. Rev. Neurosci.* **1992**, *15*, 1–29.
2. Utheim, T.P. Limbal epithelial cell therapy: Past, present, and future. *Methods Mol. Biol.* **2013**, *1014*, 3–43.
3. Cotsarelis, G.; Cheng, S.Z.; Dong, G.; Sun, T.T.; Lavker, R.M. Existence of slow-cycling limbal epithelial basal cells that can be preferentially stimulated to proliferate: Implications on epithelial stem cells. *Cell* **1989**, *57*, 201–209.
4. Davanger, M.; Evensen, A. Role of the pericorneal papillary structure in renewal of corneal epithelium. *Nature* **1971**, *229*, 560–561.
5. Shortt, A.J.; Secker, G.A.; Munro, P.M.; Khaw, P.T.; Tuft, S.J.; Daniels, J.T. Characterization of the limbal epithelial stem cell niche: Novel imaging techniques permit *in vivo* observation and targeted biopsy of limbal epithelial stem cells. *Stem cells* **2007**, *25*, 1402–1409.
6. Dua, H.S.; Shanmuganathan, V.A.; Powell-Richards, A.O.; Tighe, P.J.; Joseph, A. Limbal epithelial crypts: A novel anatomical structure and a putative limbal stem cell niche. *Br. J. Ophthalmol.* **2005**, *89*, 529–532.
7. Modjtahedi, B.S.; Eliott, D. Vitreoretinal complications of the boston keratoprosthesis. *Semin. Ophthalmol.* **2014**, *29*, 338–348.
8. Sehic, A.; Utheim, O.A.; Ommundsen, K.; Utheim, T.P. Pre-clinical cell-based therapy for limbal stem cell deficiency. *J. Funct. Biomater.* **2015**, *6*, 863–888.
9. Kenyon, K.R.; Tseng, S.C. Limbal autograft transplantation for ocular surface disorders. *Ophthalmol.* **1989**, *96*, 709–722.
10. Jenkins, C.; Tuft, S.; Liu, C.; Buckley, R. Limbal transplantation in the management of chronic contact-lens-associated epitheliopathy. *Eye* **1993**, *7*, 629–633.
11. Pellegrini, G.; Traverso, C.E.; Franzi, A.T.; Zingirian, M.; Cancedda, R.; de Luca, M. Long-term restoration of damaged corneal surfaces with autologous cultivated corneal epithelium. *Lancet* **1997**, *349*, 990–993.

12. Rama, P.; Matuska, S.; Paganoni, G.; Spinelli, A.; de Luca, M.; Pellegrini, G. Limbal stem-cell therapy and long-term corneal regeneration. *New Engl. J. Med.* **2010**, *363*, 147–155.

13. Nakamura, T.; Sotozono, C.; Bentley, A.J.; Mano, S.; Inatomi, T.; Koizumi, N.; Fullwood, N.J.; Kinoshita, S. Long-term phenotypic study after allogeneic cultivated corneal limbal epithelial transplantation for severe ocular surface diseases. *Ophthalmology* **2010**, *117*, 2247–2254.

14. Kolli, S.; Ahmad, S.; Lako, M.; Figueiredo, F. Successful clinical implementation of corneal epithelial stem cell therapy for treatment of unilateral limbal stem cell deficiency. *Stem Cells* **2010**, *28*, 597–610.

15. Shortt, A.J.; Secker, G.A.; Rajan, M.S.; Meligonis, G.; Dart, J.K.; Tuft, S.J.; Daniels, J.T. *Ex vivo* expansion and transplantation of limbal epithelial stem cells. *Ophthalmology* **2008**, *115*, 1989–1997.

16. Nakamura, T.; Inatomi, T.; Sotozono, C.; Ang, L.P.; Koizumi, N.; Yokoi, N.; Kinoshita, S. Transplantation of autologous serum-derived cultivated corneal epithelial equivalents for the treatment of severe ocular surface disease. *Ophthalmology* **2006**, *113*, 1765–1772.

17. Koizumi, N.; Inatomi, T.; Suzuki, T.; Sotozono, C.; Kinoshita, S. Cultivated corneal epithelial stem cell transplantation in ocular surface disorders. *Ophthalmology* **2001**, *108*, 1569–1574.

18. Tsai, R.J.; Li, L.M.; Chen, J.K. Reconstruction of damaged corneas by transplantation of autologous limbal epithelial cells. *New Engl. J. Med.* **2000**, *343*, 86–93.

19. Utheim, T.P. Concise review: Transplantation of cultured oral mucosal epithelial cells for treating limbal stem cell deficiency-current status and future perspectives. *Stem Cells* **2015**, *33*, 1685–1695.

20. Homma, R.; Yoshikawa, H.; Takeno, M.; Kurokawa, M.S.; Masuda, C.; Takada, E.; Tsubota, K.; Ueno, S.; Suzuki, N. Induction of epithelial progenitors *in vitro* from mouse embryonic stem cells and application for reconstruction of damaged cornea in mice. *Investig. Ophthalmol. Vis. Sci.* **2004**, *45*, 4320–4326.

21. Tanioka, H.; Kawasaki, S.; Yamasaki, K.; Ang, L.P.; Koizumi, N.; Nakamura, T.; Yokoi, N.; Komuro, A.; Inatomi, T.; Kinoshita, S. Establishment of a cultivated human conjunctival epithelium as an alternative tissue source for autologous corneal epithelial transplantation. *Investig. Ophthalmol. Vis. Sci.* **2006**, *47*, 3820–3827.

22. Yang, X.; Qu, L.; Wang, X.; Zhao, M.; Li, W.; Hua, J.; Shi, M.; Moldovan, N.; Wang, H.; Dou, Z. Plasticity of epidermal adult stem cells derived from adult goat ear skin. *Mol. Reprod. Dev.* **2007**, *74*, 386–396.

23. Monteiro, B.G.; Serafim, R.C.; Melo, G.B.; Silva, M.C.; Lizier, N.F.; Maranduba, C.M.; Smith, R.L.; Kerkis, A.; Cerruti, H.; Gomes, J.A.; *et al.* Human immature dental pulp stem cells share key characteristic features with limbal stem cells. *Cell Prolif.* **2009**, *42*, 587–594.

24. Ma, Y.; Xu, Y.; Xiao, Z.; Yang, W.; Zhang, C.; Song, E.; Du, Y.; Li, L. Reconstruction of chemically burned rat corneal surface by bone marrow-derived human mesenchymal stem cells. *Stem Cells* **2006**, *24*, 315–321.

25. Meyer-Blazejewska, E.A.; Call, M.K.; Yamanaka, O.; Liu, H.; Schlotzer-Schrehardt, U.; Kruse, F.E.; Kao, W.W. From hair to cornea: Towards the therapeutic use of hair follicle-derived stem cells in the treatment of limbal stem cell deficiency. *Stem Cells* **2010**, *29*, 57–66.

26. Reza, H.M.; Ng, B.Y.; Gimeno, F.L.; Phan, T.T.; Ang, L.P. Umbilical cord lining stem cells as a novel and promising source for ocular surface regeneration. *Stem Cell Rev.* **2011**, *7*, 935–947.

27. Lin, K.J.; Loi, M.X.; Lien, G.S.; Cheng, C.F.; Pao, H.Y.; Chang, Y.C.; Ji, A.T.; Ho, J.H. Topical administration of orbital fat-derived stem cells promotes corneal tissue regeneration. *Stem Cell Res. Ther.* **2013**, *4*.

28. Inatomi, T.; Nakamura, T.; Koizumi, N.; Sotozono, C.; Yokoi, N.; Kinoshita, S. Midterm results on ocular surface reconstruction using cultivated autologous oral mucosal epithelial transplantation. *Am. J. Ophthalmol.* **2006**, *141*, 267–275.

29. Ma, D.H.; Kuo, M.T.; Tsai, Y.J.; Chen, H.C.; Chen, X.L.; Wang, S.F.; Li, L.; Hsiao, C.H.; Lin, K.K. Transplantation of cultivated oral mucosal epithelial cells for severe corneal burn. *Eye* **2009**, *23*, 1442–1450.

30. Nakamura, T.; Inatomi, T.; Sotozono, C.; Amemiya, T.; Kanamura, N.; Kinoshita, S. Transplantation of cultivated autologous oral mucosal epithelial cells in patients with severe ocular surface disorders. *Br. J. Ophthalmol.* **2004**, *88*, 1280–1284.

31. Nishida, K.; Yamato, M.; Hayashida, Y.; Watanabe, K.; Yamamoto, K.; Adachi, E.; Nagai, S.; Kikuchi, A.; Maeda, N.; Watanabe, H.; *et al.* Corneal reconstruction with tissue-engineered cell sheets composed of autologous oral mucosal epithelium. *Engl. J. Med.* **2004**, *351*, 1187–1196.

32. Chen, H.C.; Yeh, L.K.; Tsai, Y.J.; Lai, C.H.; Chen, C.C.; Lai, J.Y.; Sun, C.C.; Chang, G.; Hwang, T.L.; Chen, J.K.; *et al.* Expression of angiogenesis-related factors in human corneas after cultivated oral mucosal epithelial transplantation. *Investig. Ophthalmol. Vis. Sci.* **2012**, *53*, 5615–5623.

33. Kanayama, S.; Nishida, K.; Yamato, M.; Hayashi, R.; Maeda, N.; Okano, T.; Tano, Y. Analysis of soluble vascular endothelial growth factor receptor-1 secreted from cultured corneal and oral mucosal epithelial cell sheets *in vitro*. *Br. J. Ophthalmol.* **2009**, *93*, 263–267.

34. Kanayama, S.; Nishida, K.; Yamato, M.; Hayashi, R.; Sugiyama, H.; Soma, T.; Maeda, N.; Okano, T.; Tano, Y. Analysis of angiogenesis induced by cultured corneal and oral mucosal epithelial cell sheets *in vitro*. *Exp. Eye Res.* **2007**, *85*, 772–781.

35. Sekiyama, E.; Nakamura, T.; Kawasaki, S.; Sogabe, H.; Kinoshita, S. Different expression of angiogenesis-related factors between human cultivated corneal and oral epithelial sheets. *Exp. Eye Res.* **2006**, *83*, 741–746.

36. Lim, P.; Fuchsluger, T.A.; Jurkunas, U.V. Limbal stem cell deficiency and corneal neovascularization. *Semin. Ophthalmol.* **2009**, *24*, 139–148.

37. Ang, L.P.; Nakamura, T.; Inatomi, T.; Sotozono, C.; Koizumi, N.; Yokoi, N.; Kinoshita, S. Autologous serum-derived cultivated oral epithelial transplants for severe ocular surface disease. *Arch. Ophthalmol.* **2006**, *124*, 1543–1551.

38. Burillon, C.; Huot, L.; Justin, V.; Nataf, S.; Chapuis, F.; Decullier, E.; Damour, O. Cultured autologous oral mucosal epithelial cell sheet (CAOMECS) transplantation for the treatment of corneal limbal epithelial stem cell deficiency. *Investig. Ophthalmol. Vis. Sci.* **2012**, *53*, 1325–1331.

39. Chen, H.C.; Chen, H.L.; Lai, J.Y.; Chen, C.C.; Tsai, Y.J.; Kuo, M.T.; Chu, P.H.; Sun, C.C.; Chen, J.K.; Ma, D.H. Persistence of transplanted oral mucosal epithelial cells in human cornea. *Investig. Ophthalmol. Vis. Sci.* **2009**, *50*, 4660–4668.

40. Gaddipati, S.; Muralidhar, R.; Sangwan, V.S.; Mariappan, I.; Vemuganti, G.K.; Balasubramanian, D. Oral epithelial cells transplanted on to corneal surface tend to adapt to the ocular phenotype. *Indian J. Ophthalmol.* **2014**, *62*, 644–648.

41. Hirayama, M.; Satake, Y.; Higa, K.; Yamaguchi, T.; Shimazaki, J. Transplantation of cultivated oral mucosal epithelium prepared in fibrin-coated culture dishes. *Investig. Ophthalmol. Vis. Sci.* **2012**, *53*, 1602–1609.

42. Inatomi, T.; Nakamura, T.; Kojyo, M.; Koizumi, N.; Sotozono, C.; Kinoshita, S. Ocular surface reconstruction with combination of cultivated autologous oral mucosal epithelial transplantation and penetrating keratoplasty. *Am. J. Ophthalmol.* **2006**, *142*, 757–764.

43. Kolli, S.; Ahmad, S.; Mudhar, H.S.; Meeny, A.; Lako, M.; Figueiredo, F.C. Successful application of *ex vivo* expanded human autologous oral mucosal epithelium for the treatment of total bilateral limbal stem cell deficiency. *Stem Cells* **2014**, *32*, 2135–2146.

44. Nakamura, T.; Inatomi, T.; Cooper, L.J.; Rigby, H.; Fullwood, N.J.; Kinoshita, S. Phenotypic investigation of human eyes with transplanted autologous cultivated oral mucosal epithelial sheets for severe ocular surface diseases. *Ophthalmology* **2007**, *114*, 1080–1088.

45. Nakamura, T.; Takeda, K.; Inatomi, T.; Sotozono, C.; Kinoshita, S. Long-term results of autologous cultivated oral mucosal epithelial transplantation in the scar phase of severe ocular surface disorders. *Br. J. Ophthalmol.* **2011**, *95*, 942–946.

46. Priya, C.G.; Arpitha, P.; Vaishali, S.; Prajna, N.V.; Usha, K.; Sheetal, K.; Muthukkaruppan, V. Adult human buccal epithelial stem cells: Identification, *ex vivo* expansion, and transplantation for corneal surface reconstruction. *Eye* **2011**, *25*, 1641–1649.

47. Satake, Y.; Dogru, M.; Yamane, G.Y.; Kinoshita, S.; Tsubota, K.; Shimazaki, J. Barrier function and cytologic features of the ocular surface epithelium after autologous cultivated oral mucosal epithelial transplantation. *Arch. Ophthalmol.* **2008**, *126*, 23–28.

48. Satake, Y.; Higa, K.; Tsubota, K.; Shimazaki, J. Long-term outcome of cultivated oral mucosal epithelial sheet transplantation in treatment of total limbal stem cell deficiency. *Ophthalmology* **2011**, *118*, 1524–1530.

49. Sotozono, C.; Inatomi, T.; Nakamura, T.; Koizumi, N.; Yokoi, N.; Ueta, M.; Matsuyama, K.; Kaneda, H.; Fukushima, M.; Kinoshita, S. Cultivated oral mucosal epithelial transplantation for persistent epithelial defect in severe ocular surface diseases with acute inflammatory activity. *Acta Ophthalmol.* **2014**, *92*, e447–e453.

50. Sotozono, C.; Inatomi, T.; Nakamura, T.; Koizumi, N.; Yokoi, N.; Ueta, M.; Matsuyama, K.; Miyakoda, K.; Kaneda, H.; Fukushima, M.; *et al.* Visual improvement after cultivated oral mucosal epithelial transplantation. *Ophthalmology* **2013**, *120*, 193–200.

51. Takeda, K.; Nakamura, T.; Inatomi, T.; Sotozono, C.; Watanabe, A.; Kinoshita, S. Ocular surface reconstruction using the combination of autologous cultivated oral mucosal epithelial transplantation and eyelid surgery for severe ocular surface disease. *Am. J. Ophthalmo.* **2011**, *152*, 195–201.

52. Rheinwald, J.G.; Green, H. Serial cultivation of strains of human epidermal keratinocytes: The formation of keratinizing colonies from single cells. *Cell* **1975**, *6*, 331–343.

53. Green, H.; Kehinde, O.; Thomas, J. Growth of cultured human epidermal cells into multiple epithelia suitable for grafting. *Proc. Natl. Acad. Sci. USA* **1979**, *76*, 5665–5668.

54. Mariappan, I.; Maddileti, S.; Savy, S.; Tiwari, S.; Gaddipati, S.; Fatima, A.; Sangwan, V.S.; Balasubramanian, D.; Vemuganti, G.K. *In vitro* culture and expansion of human limbal epithelial cells. *Nat. Protoc.* **2010**, *5*, 1470–1479.

55. Martin, M.J.; Muotri, A.; Gage, F.; Varki, A. Human embryonic stem cells express an immunogenic nonhuman sialic acid. *Nat. Med.* **2005**, *11*, 228–232.

56. Madhira, S.L.; Vemuganti, G.; Bhaduri, A.; Gaddipati, S.; Sangwan, V.S.; Ghanekar, Y. Culture and characterization of oral mucosal epithelial cells on human amniotic membrane for ocular surface reconstruction. *Mol. Vis.* **2008**, *14*, 189–196.

57. Nakamura, T.; Endo, K.; Cooper, L.J.; Fullwood, N.J.; Tanifuji, N.; Tsuzuki, M.; Koizumi, N.; Inatomi, T.; Sano, Y.; Kinoshita, S. The successful culture and autologous transplantation of rabbit oral mucosal epithelial cells on amniotic membrane. *Investig. Ophthalmol. Vis. Sci.* **2003**, *44*, 106–116.

58. Nakamura, T.; Kinoshita, S. Ocular surface reconstruction using cultivated mucosal epithelial stem cells. *Cornea* **2003**, *22*, S75–S80.

59. Promprasit, D.; Bumroongkit, K.; Tocharus, C.; Mevatee, U.; Tananuvat, N. Cultivation and phenotypic characterization of rabbit epithelial cells expanded *ex vivo* from fresh and cryopreserved limbal and oral mucosal explants. *Curr. Eye Res.* **2014**, 1–8.

60. Sen, S.; Sharma, S.; Gupta, A.; Gupta, N.; Singh, H.; Roychoudhury, A.; Mohanty, S.; Sen, S.; Nag, T.C.; Tandon, R. Molecular characterization of explant cultured human oral mucosal epithelial cells. *Investig. Ophthalmol. Vis. Sci.* **2011**, *52*, 9548–9554.

61. Sharma, S.M.; Fuchsluger, T.; Ahmad, S.; Katikireddy, K.R.; Armant, M.; Dana, R.; Jurkunas, U.V. Comparative analysis of human-derived feeder layers with 3T3 fibroblasts for the *ex vivo* expansion of human limbal and oral epithelium. *Stem cell Rev.* **2012**, *8*, 696–705.

62. Shimazaki, J.; Higa, K.; Kato, N.; Satake, Y. Barrier function of cultivated limbal and oral mucosal epithelial cell sheets. *Investig. Ophthalmol. Vis. Sci.* **2009**, *50*, 5672–5680.

63. Yokoo, S.; Yamagami, S.; Mimura, T.; Amano, S.; Saijo, H.; Mori, Y.; Takato, T. UV absorption in human oral mucosal epithelial sheets for ocular surface reconstruction. *Ophthalmic Res.* **2006**, *38*, 350–354.

64. Bardag-Gorce, F.; Oliva, J.; Wood, A.; Hoft, R.; Pan, D.; Thropay, J.; Makalinao, A.; French, S.W.; Niihara, Y. Carrier-free cultured autologous oral mucosa epithelial cell sheet (CAOMECS) for corneal epithelium reconstruction: A histological study. *Ocul. Surf.* **2015**, *13*, 150–163.

65. Hayashi, R.; Yamato, M.; Takayanagi, H.; Oie, Y.; Kubota, A.; Hori, Y.; Okano, T.; Nishida, K. Validation system of tissue-engineered epithelial cell sheets for corneal regenerative medicine. *Tissue Eng. Part C Methods* **2010**, *16*, 553–560.

66. Hayashida, Y.; Nishida, K.; Yamato, M.; Watanabe, K.; Maeda, N.; Watanabe, H.; Kikuchi, A.; Okano, T.; Tano, Y. Ocular surface reconstruction using autologous rabbit oral mucosal epithelial sheets fabricated *ex vivo* on a temperature-responsive culture surface. *Investig. Ophthalmol. Vis. Sci.* **2005**, *46*, 1632–1639.

67. Murakami, D.; Yamato, M.; Nishida, K.; Ohki, T.; Takagi, R.; Yang, J.; Namiki, H.; Okano, T. Fabrication of transplantable human oral mucosal epithelial cell sheets using temperature-responsive culture inserts without feeder layer cells. *J. Artif. Organs Off. J. Jpn. Soc. Artif. Organs* **2006**, *9*, 185–191.

68. Oie, Y.; Hayashi, R.; Takagi, R.; Yamato, M.; Takayanagi, H.; Tano, Y.; Nishida, K. A novel method of culturing human oral mucosal epithelial cell sheet using post-mitotic human dermal fibroblast feeder cells and modified keratinocyte culture medium for ocular surface reconstruction. *Br. J. Ophthalmol.* **2010**, *94*, 1244–1250.

69. Soma, T.; Hayashi, R.; Sugiyama, H.; Tsujikawa, M.; Kanayama, S.; Oie, Y.; Nishida, K. Maintenance and distribution of epithelial stem/progenitor cells after corneal reconstruction using oral mucosal epithelial cell sheets. *PloS One* **2014**, *9*.

70. Sugiyama, H.; Yamato, M.; Nishida, K.; Okano, T. Evidence of the survival of ectopically transplanted oral mucosal epithelial stem cells after repeated wounding of cornea. *Mol. Ther. J. Am. Soc. Gene Ther.* **2014**, *22*, 1544–1555.

71. Sheth, R.; Neale, M.H.; Shortt, A.J.; Massie, I.; Vernon, A.J.; Daniels, J.T. Culture and characterization of oral mucosal epithelial cells on a fibrin gel for ocular surface reconstruction. *Curr. Eye Res.* **2014**, 1–11.

72. Ilmarinen, T.; Laine, J.; Juuti-Uusitalo, K.; Numminen, J.; Seppanen-Suuronen, R.; Uusitalo, H.; Skottman, H. Towards a defined, serum- and feeder-free culture of stratified human oral mucosal epithelium for ocular surface reconstruction. *Acta Ophthalmol.* **2013**, *91*, 744–750.

73. Krishnan, S.; Iyer, G.K.; Krishnakumar, S. Culture & characterisation of limbal epithelial cells & oral mucosal cells. *Indian J. Med. Res.* **2010**, *131*, 422–428.

74. Hyun, D.W.; Kim, Y.H.; Koh, A.Y.; Lee, H.J.; Wee, W.R.; Jeon, S.; Kim, M.K. Characterization of biomaterial-free cell sheets cultured from human oral mucosal epithelial cells. *J. Tissue Eng. Regener. Med.* **2014**.

75. Schwab, I.R.; Reyes, M.; Isseroff, R.R. Successful transplantation of bioengineered tissue replacements in patients with ocular surface disease. *Cornea* **2000**, *19*, 421–426.

76. Schwab, I.R. Cultured corneal epithelia for ocular surface disease. *Trans. Am. Ophthalmol. Soc.* **1999**, *97*, 891–986.

77. Pellegrini, G.; Ranno, R.; Stracuzzi, G.; Bondanza, S.; Guerra, L.; Zambruno, G.; Micali, G.; de Luca, M. The control of epidermal stem cells (holoclones) in the treatment of massive full-thickness burns with autologous keratinocytes cultured on fibrin. *Transplantation* **1999**, *68*, 868–879.

78. Rama, P.; Bonini, S.; Lambiase, A.; Golisano, O.; Paterna, P.; de Luca, M.; Pellegrini, G. Autologous fibrin-cultured limbal stem cells permanently restore the corneal surface of patients with total limbal stem cell deficiency. *Transplantation* **2001**, *72*, 1478–1485.

79. Han, B.; Schwab, I.R.; Madsen, T.K.; Isseroff, R.R. A fibrin-based bioengineered ocular surface with human corneal epithelial stem cells. *Cornea* **2002**, *21*, 505–510.

80. De Rötth, A. Plastic repair of conjunctival defects with fetal membrane. *Arch. Ophthalmol.* **1940**, *23*, 522–525.

81. Sorsby, A.; Symons, H.M. Amniotic membrane grafts in caustic burns of the eye (burns of the second degree). *Br. J. Ophthalmol.* **1946**, *30*, 337–345.

82. Anderson, D.F.; Ellies, P.; Pires, R.T.; Tseng, S.C. Amniotic membrane transplantation for partial limbal stem cell deficiency. *Br. J. Ophthalmol.* **2001**, *85*, 567–575.

83. Tseng, S.C.; Prabhasawat, P.; Barton, K.; Gray, T.; Meller, D. Amniotic membrane transplantation with or without limbal allografts for corneal surface reconstruction in patients with limbal stem cell deficiency. *Arch. Ophthalmol.* **1998**, *116*, 431–441.

84. Shimazaki, J.; Shinozaki, N.; Tsubota, K. Transplantation of amniotic membrane and limbal autograft for patients with recurrent pterygium associated with symblepharon. *Br. J. Ophthalmol.* **1998**, *82*, 235–240.

85. Sato, H.; Shimazaki, J.; Shinozaki, N.; Tsubota, K. Role of growth factors for ocular surface reconstruction after amniotic membrane transplantation. *Investig. Ophthalmol. Vis. Sci.* **1998**, *39*, 428–430.

86. Tosi, G.M.; Massaro-Giordano, M.; Caporossi, A.; Toti, P. Amniotic membrane transplantation in ocular surface disorders. *J. Cell Physiol.* **2005**, *202*, 849–851.

87. Koizumi, N.J.; Inatomi, T.J.; Sotozono, C.J.; Fullwood, N.J.; Quantock, A.J.; Kinoshita, S. Growth factor mRNA and protein in preserved human amniotic membrane. *Curr. Eye Res.* **2000**, *20*, 173–177.

88. Nishida, K.; Yamato, M.; Hayashida, Y.; Watanabe, K.; Maeda, N.; Watanabe, H.; Yamamoto, K.; Nagai, S.; Kikuchi, A.; Tano, Y.; *et al.* Functional bioengineered corneal epithelial sheet grafts from corneal stem cells expanded *ex vivo* on a temperature-responsive cell culture surface. *Transplantation* **2004**, *77*, 379–385.

89. Yamato, M.; Utsumi, M.; Kushida, A.; Konno, C.; Kikuchi, A.; Okano, T. Thermo-responsive culture dishes allow the intact harvest of multilayered keratinocyte sheets without dispase by reducing temperature. *Tissue Eng.* **2001**, *7*, 473–480.

90. Kushida, A.; Yamato, M.; Konno, C.; Kikuchi, A.; Sakurai, Y.; Okano, T. Decrease in culture temperature releases monolayer endothelial cell sheets together with deposited fibronectin matrix from temperature-responsive culture surfaces. *J. Biomed. Mater. Res.* **1999**, *45*, 355–362.

91. Dunn, C.J.; Goa, K.L. Fibrin sealant: A review of its use in surgery and endoscopy. *Drugs* **1999**, *58*, 863–886.

92. Martinowitz, U.; Saltz, R. Fibrin sealant. *Curr. Opin. Hematol.* **1996**, *3*, 395–402.

93. Le Guehennec, L.; Goyenvalle, E.; Aguado, E.; Pilet, P.; Spaethe, R.; Daculsi, G. Influence of calcium chloride and aprotinin in the *in vivo* biological performance of a composite combining biphasic calcium phosphate granules and fibrin sealant. *J. Mater. Sci. Mater. Med.* **2007**, *18*, 1489–1495.

94. Radosevich, M.; Goubran, H.I.; Burnouf, T. Fibrin sealant: Scientific rationale, production methods, properties, and current clinical use. *Vox Sang.* **1997**, *72*, 133–143.

95. Tuan, T.L.; Song, A.; Chang, S.; Younai, S.; Nimni, M.E. *In vitro* fibroplasia: Matrix contraction, cell growth, and collagen production of fibroblasts cultured in fibrin gels. *Exp. Cell Res.* **1996**, *223*, 127–134.

96. Weisel, J.W. Fibrinogen and fibrin. *Adv. Protein. Chem.* **2005**, *70*, 247–299.

97. Dvorak, H.F.; Harvey, V.S.; Estrella, P.; Brown, L.F.; McDonagh, J.; Dvorak, A.M. Fibrin containing gels induce angiogenesis. Implications for tumor stroma generation and wound healing. *Lab. Invest.* **1987**, *57*, 673–686.

98. Pellegrini, G.; Dellambra, E.; Golisano, O.; Martinelli, E.; Fantozzi, I.; Bondanza, S.; Ponzin, D.; McKeon, F.; de Luca, M. P63 identifies keratinocyte stem cells. *Proc. Natl. Acad. Sci. USA* **2001**, *98*, 3156–3161.

99. Carpenter, G.; Cohen, S. Epidermal growth factor. *J. Biol. Chem.* **1990**, *265*, 7709–7712.

100. Herbst, R.S. Review of epidermal growth factor receptor biology. *Int. J. Radiat. Oncol. Biol. Phys.* **2004**, *59*, 21–26.

101. Dua, H.S.; Miri, A.; Alomar, T.; Yeung, A.M.; Said, D.G. The role of limbal stem cells in corneal epithelial maintenance: Testing the dogma. *Ophthalmology* **2009**, *116*, 856–863.

102. Kawakita, T.; Espana, E.M.; He, H.; Li, W.; Liu, C.Y.; Tseng, S.C. Intrastromal invasion by limbal epithelial cells is mediated by epithelial-mesenchymal transition activated by air exposure. *Am. J. Pathol.* **2005**, *167*, 381–393.

103. Ban, Y.; Cooper, L.J.; Fullwood, N.J.; Nakamura, T.; Tsuzuki, M.; Koizumi, N.; Dota, A.; Mochida, C.; Kinoshita, S. Comparison of ultrastructure, tight junction-related protein expression and barrier function of human corneal epithelial cells cultivated on amniotic membrane with and without air-lifting. *Exp. Eye Res.* **2003**, *76*, 735–743.

104. Henderson, H.W.; Collin, J.R. Mucous membrane grafting. *Dev. Ophthalmol.* **2008**, *41*, 230–242.

105. Meyer-Blazejewska, E.A.; Kruse, F.E.; Bitterer, K.; Meyer, C.; Hofmann-Rummelt, C.; Wunsch, P.H.; Schlotzer-Schrehardt, U. Preservation of the limbal stem cell phenotype by appropriate culture techniques. *Investig. Ophthalmol. Vis. Sci.* **2010**, *51*, 765–774.

Chapter 3:
Ocular Nanotechnology and Tissue Engineering

Lipid Nanoparticles for Ocular Gene Delivery

Yuhong Wang, Ammaji Rajala and Raju V. S. Rajala

Abstract: Lipids contain hydrocarbons and are the building blocks of cells. Lipids can naturally form themselves into nano-films and nano-structures, micelles, reverse micelles, and liposomes. Micelles or reverse micelles are monolayer structures, whereas liposomes are bilayer structures. Liposomes have been recognized as carriers for drug delivery. Solid lipid nanoparticles and lipoplex (liposome-polycation-DNA complex), also called lipid nanoparticles, are currently used to deliver drugs and genes to ocular tissues. A solid lipid nanoparticle (SLN) is typically spherical, and possesses a solid lipid core matrix that can solubilize lipophilic molecules. The lipid nanoparticle, called the liposome protamine/DNA lipoplex (LPD), is electrostatically assembled from cationic liposomes and an anionic protamine-DNA complex. The LPD nanoparticles contain a highly condensed DNA core surrounded by lipid bilayers. SLNs are extensively used to deliver drugs to the cornea. LPD nanoparticles are used to target the retina. Age-related macular degeneration, retinitis pigmentosa, and diabetic retinopathy are the most common retinal diseases in humans. There have also been promising results achieved recently with LPD nanoparticles to deliver functional genes and micro RNA to treat retinal diseases. Here, we review recent advances in ocular drug and gene delivery employing lipid nanoparticles.

Reprinted from *J. Funct. Biomater.* Cite as: Wang, Y.; Rajala, A.; Rajala, R.V.S. Lipid Nanoparticles for Ocular Gene Delivery. *J. Funct. Biomater.* **2015**, *6*, 379–394.

1. Introduction

The eye is made up of many components, and therapeutic agents could be easily applied to the anterior part of the eye. However, it is difficult to administer these agents to the posterior part of the eye. Intravitreal or subretinal routes are the only means of targeting agents to the posterior area of the eye. The eye is one of the sensory organs of the body, and frequent administration of drugs to the eye is undesirable. Therefore, gene therapy would be an ideal way to provide sustained gene expression that could overcome these limitations. The eyes have been early targets for gene therapy because they are small—that is, they require relatively little active dose—they are self-contained, and because the tools of eye surgery have advanced enough to make these treatments possible. The eye offers an excellent target for gene therapy studies, it is easily accessible and relatively immune privileged. If we inject any drug or gene systemically, the drug or gene must then cross the blood retinal barrier (BRB).

To our knowledge, most of the successful gene therapy trials use local administration of drug(s)/gene(s) into the eye.

2. Uses and Advantages of Nanoparticles in Medicine

Nanoparticles play important roles in the diagnosis of disease, delivery of drugs to target tissue, research into the organization of DNA, drug-mediated apoptosis of cancer cells, studies of the pharmacological efficiency of drugs, and tissue engineering. Their size and surface characteristics enable us to alter nanoparticle properties to allow for continuous discharge of drugs during transport and release at a defined location. Choosing the appropriate matrix is vital to drug delivery. Modifying the surface properties of nanoparticles will help to clear the drug from the patient's body with significantly fewer side effects.

These particles are currently conjugated with either drugs or genes and administered through several avenues, including the oral, nasal, intra-ocular, and arterial routes. Researchers are exploring the use of various polymers to conjugate drugs and genes to enhance therapeutic benefits while minimizing adverse effects.

Nanoparticles for gene therapy are broadly classified into three groups, metal-based nanoparticles, lipid based-nanoparticles, and polymer-based nanoparticles. These particles are different in size, charge, shape, and structure, and have their own modes of delivering cargo into cells and assimilating the cargo into the genetic machinery for gene expression [1–4]. Compacted DNA nanoparticles formulated with polyethylene glycol-substituted polylysine have been used for gene therapy in mouse models of eye diseases [5–8]. Solid lipid nanoparticles (SLNs) and nanostructured lipid carriers (NLCs) have been developed to improve the ocular delivery of acyclovir into excised corneal tissue [9].

A solid lipid nanoparticle is typically spherical, with an average diameter between 10 and 1000 nm, and possesses a solid lipid core matrix that can solubilize lipophilic molecules. The lipid core is stabilized by surfactants; the lipid component may be a triglyceride, diglyceride, monoglyceride, fatty acid, steroid, or wax. The lipid nanoparticle, called the liposome protamine/DNA lipoplex (LPD), is electrostatically assembled from cationic liposomes and an anionic complex of protamine and DNA. The LPD nanoparticles contain a highly condensed DNA core, surrounded by lipid bilayers with an average size of ~100 nm. Lipid nanoparticles have also been used to improve the efficiency of siRNA delivery in RPE cells and a laser-induced rat model for the treatment of choroidal neovascularization [10].

Methazolamide (MTA) is an anti-glaucoma drug; however, systemic administration produces side effects, while providing insufficient ocular therapeutic concentrations [11]. Solid lipid nanoparticles containing MTA have been shown to have higher therapeutic efficacy at low doses with more prolonged effects than those of the drug solution itself [11]. Lipid nanoparticles have also been shown to be feasible for the ocular delivery of anti-inflammatory drugs [12].

Since the 1990s, solid lipid nanoparticles have been examined as potential drug carrier systems. SLNs do not show bio-toxicity, as they are prepared from physiological lipids, and are especially useful in ocular drug delivery, as they enhance the corneal absorption of drugs and improve the ocular bioavailability of both hydrophilic and lipophilic drugs [13]. Cyclosporine is commonly prescribed for chronic dry eye, caused by inflammation, and cyclosporine A-loaded solid lipid nanoparticles have been shown to improve drug efficacy when administered to rabbit eyes [14,15].

Liquid lipid has also been incorporated into lipid nanoparticles to enhance ocular drug delivery [16]. These particles have been tested on human corneal epithelial cell lines and rabbit corneas [16]. The liquid lipid incorporation has been shown to improve the ocular retention and penetration of therapeutics [16]. Surface-modified solid lipid nanoparticles have been shown to provide an efficient way of improving the ocular bioavailability of drugs to bioengineered human corneas [17]. Solid lipid nanoparticles have also been used for retinal gene therapy and to study intracellular trafficking in RPE cells [18]. Solid lipid nanoparticles and lipid nanoparticles have been extensively reviewed and described in detailed in recently published articles [19,20].

The majority of solid lipid nanoparticles have been used to deliver drugs to the cornea [9,13,16,17,21,22]. We recently formulated a novel lipid nanoparticle, and examined its efficiency and delivery of genes and microRNA to the retina [23,24]. LPD has been used to successfully deliver the vascular endothelial growth factor gene into mesenchymal stem cells [25]. In this article, we review several important aspects of lipid nanoparticles, including their formulation, mechanism of internalization, cell-specific expression, and barriers that affect gene expression.

3. Gene Therapy and Viral Vectors

The success of gene therapy relies on the development of efficient, non-toxic gene carriers that can encapsulate and deliver foreign genetic materials into specific cell types [26]. Gene therapy carriers can be classified into two groups, viral and non-viral gene delivery systems. Although viral vectors, such as adeno-associated virus (AAV), have attractive features, particularly their high gene transduction capability, they face biosafety issues, especially innate and immune barriers [27], toxicity [28], and potential recombination of or complementation [29] to vector delivery. The size of viral vectors, which restricts the insertion of genes to <5 kb, is another limitation [30]. Table 1 lists various viral and non-viral carriers. All viral vectors have been used to deliver functional genes to the retina whereas non-viral vectors have been used to deliver both drugs (liposome nanoparticles and solid lipid nanoparticles) and genes (solid lipid nanoparticles, LPD/lipoplexes and CK30-PEG) to the retina.

Table 1. Viral and non-viral delivery systems for ocular gene delivery.

Vector	Carrier	Delivery	Ref.
Virus	AAV	Local/systemic	[31–35]
	Adenovirus	Local	[36]
	Baculovirus	Local	[37,38]
	Lentivirus	Local	[39]
Non-virus	Liposome nanoparticles	Local	[40–43]
	Solid lipid nanoparticles	Local	[9,13,16,17,21,22]
	LPD/lipoplexes	Local	[23,24,44]
	CK30-PEG	Local	[5–8]

Despite rapid advances in gene therapy during the last two decades, major obstacles to clinical applications for human diseases still exist. These impediments include immune response, vector toxicity, and the lack of sustained therapeutic gene expression. Therefore, new strategies are needed to achieve safe and effective gene therapy. The ideal vector should have low antigenic potential, high capacity to accommodate genetic material, high transduction efficiency, controlled and targeted transgene expression, and reasonable expense and safety for both the patients and the environment. These desired features led researchers to focus on non-viral vectors as an alternative to viral vectors.

4. Lipid-Based Nanoparticles

The main constituent of lipid nanoparticles is the liposome. A liposome is a spherical vesicle of a lamellar phase of the lipid bilayer. The liposome can be used as a transport vehicle to send nutrients and drugs into the body [45–47]. One can prepare these liposomes through disruption of biological membranes by sonication, a process of sending sound waves to disturb particles in a solution. Lipids can naturally form themselves into nano-films and nano-structures, called micelles, reverse micelles, and liposomes [20,48]. The monolayer structures are called micelles or reverse micelles, whereas the lipid bilayer structures are called liposomes (Figure 1A). In the lipid bilayer, phospholipids are principal lipids, which are amphiphilic molecules with hydrophilic (*water-loving*, polar) and lipophilic (*fat-loving*) properties, sometimes described as having hydrophobic tails and hydrophilic heads. Therefore, liposomes are artificial phospholipid bilayers; as a result, liposomes have biocompatible characteristics [49,50]. This biocompatibility accounts for their most important advantages as drug carriers, (1) liposomes have almost no toxicity and low antigenicity; (2) liposomes can be biodegraded and metabolized *in vivo*, and

(3) liposomal properties, such as membrane permeability, can be controlled to some extent [51–53]. Remarkably, liposomes can entrap and protect drug molecules or nucleic acids on the journey to the target site [54].

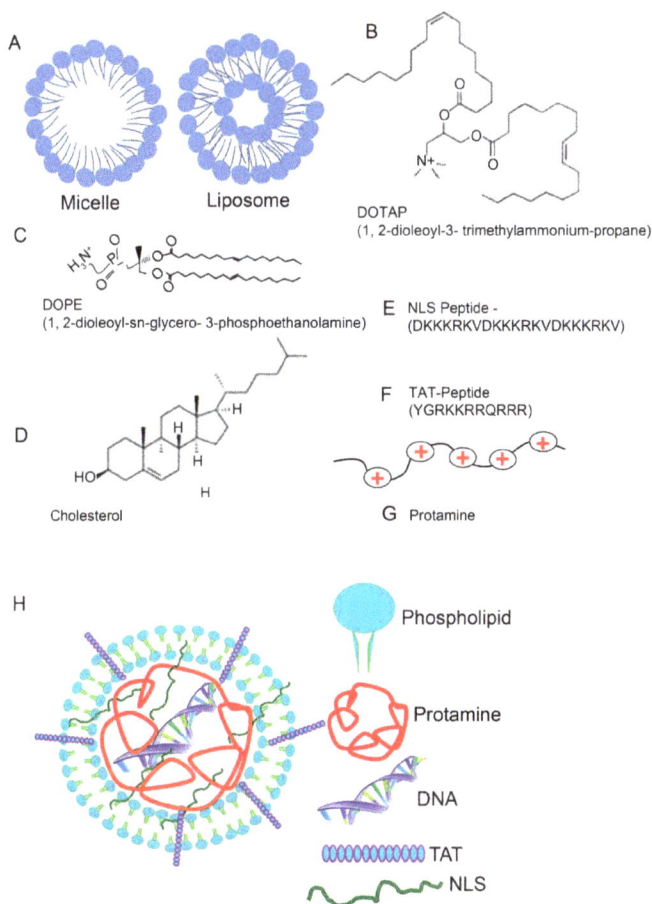

Figure 1. Lipid, peptide, and protein components of the lipid nanoparticle. The monolayer structures are called micelles, whereas the lipid bilayer structures are called liposomes (**A**). Chemical structures of DOTAP (**B**), DOPE (**C**), and Cholesterol (**D**). NLS (**E**) and TAT (**F**) peptide sequences and protamine (small, arginine-rich, nuclear protein) (**G**) are also presented. DOTAP, 1, 2-dioleoyl-3-trimethylammonium-propane, DOPE, 1, 2-dioleoyl-sn-glycero-3-phosphoethanolamine; NLS, nuclear localization signal; TAT, transactivator of transcription. Formulation of a peptide-based lipid nanoparticle (**H**). Peptide-based nanoparticles can be formulated by mixing liposome, protamine, DNA, TAT, and NLS. TAT, transactivator of transcription; NLS, nuclear localization signal.

191

When nucleic acids, molecules, or drugs are enclosed in a lipid-based coating, they have a lower degradation rates than do molecules without a lipid coating. Such enclosure also increases the likelihood of endocytosis and uptake of nucleic acids or drugs into cells [4,20,55]. These desired features led researchers to focus on non-viral vectors as an alternative to viral vectors. The non-viral vectors include polymers like polyethylenimine (PEI) [56] and poly L-lysine (PLL) [57], peptides, liposomes (tiny fat-like particles) [58], and liposomes-protamine-DNA (LPD) complexes [59,60]. However, the current non-viral vectors cannot achieve tissue-specific or cell-specific sustained gene expression, nor eliminate the unwanted and harmful effects on other cells.

The use of lipid nanoparticles as part of a system delivering drugs and genes to the retina has been attempted [44]. We recently developed an artificial virus, an LPD nanoparticle in combination with nuclear localization signaling (NLS) [61] peptide and transactivator of transcription (TAT) peptide [62], to produce efficient, cell-specific gene delivery to eye tissues, with sustained gene expression. The key to our success arises from three unique features, (1) the use of biocompatible lipid molecules to pack DNA and the biocompatible protamine molecules into the nanoparticles; (2) the integration of cell-penetrating and nuclei-targeting peptides into the nanoparticles, to improve the efficiency of gene transfer and the subsequent lasting gene expression; and (3) the use of a DNA that carries the target gene, and also bears a unique promoter to achieve cell-specific gene expression.

5. Composition of Lipid Nanoparticles

Liposomes were first identified in 1965 [63], and were successfully applied as cationic liposome complexes via intravenous DNA delivery into adult mice in 1993. Since then, liposomes have been successful and widely applied in nanotechnology [64] and in various medical fields [23,24,61,65,66]. One approach for more successful nanoparticle gene therapy is the liposome protamine/DNA lipoplex (LPD), which is applied as a two-step packaging technology employing a multilayering method [61,67]. First, the DNA is packaged into a condensed core via electrostatic interactions with protamine, and various peptides (NLS and TAT) and the plasmid DNA (pDNA) are mixed at various weight ratios (Figure 1). Then, the liposomes, consisting of a cationic lipid DOTAP (1,2-dioleoyl-3-trimethylammonium-propane), a neutral "helper" lipid DOPE (1,2-dioleoyl-sn-glycero-3-phosphoethanolamine) and neutral cholesterol, are added so that the positively charged DOTAP/DOPE/Chol liposome can form a complex with the negative protamine/DNA particles, leading to the formation of LPD nanoparticles (Figure 1B–G). The negatively charged DNA is complexed with protamine, an arginine-rich, positively charged nuclear protein that replaces histone late in the haploid phase of spermatogenesis, and is essential for sperm head condensation and DNA stabilization. The advantage of adding protamine to DNA is that protamine condenses the DNA and the subsequent mixing of the protamine/DNA complex to cationic liposomes, producing a small nanoparticle. Another advantage of protamine is that the encapsulated DNA is protected from nuclease degradation [23,59]. Inclusion of protamine in solid lipid nanoparticles (SLN) has previously been shown to yield a six-fold increase in the transfection of SLN in retinal cells, due to the presence of a nuclear localization signal [68].

6. Transfer Mechanism of LPD Nanoparticles into Cells

Successful gene delivery systems have their own transfer mechanisms into the cell. Viral vectors have the advantage in cellular entry, because they bind to the cellular receptors and co-receptors, which help them to internalize and traffic to the nucleus [69–73]. In contrast, cationic liposomes take advantage of biocompatible characteristics and are widely used to transfect DNA into cells in culture and *in vivo*, since the formation of cationic lipid-DNA complexes can facilitate the association with the cell membrane and allow the complex to enter the cell through the endocytotic pathway [4,58,74]. The complex is internalized into an endosome, which will destabilize the endosome membrane and result in a flip-flop of anionic lipids that are mainly on the cytoplasmic side of the membrane. The anionic lipids will then diffuse into the complex and form charge-neutralized ion pairs with cationic lipids. This displaces the DNA from the complex and allows DNA to enter into the cytoplasm [4,74,75]. Protamine in the solid lipid nanoparticles has been reported to shift the internalization mechanism from caveolae/raft-mediated to clathrin-mediated endocytosis [68]. Some researchers also proposed that LPD nanoparticles could use two different endocytosis pathways, macropinocytosis and clathrin-mediated endocytosis [58]. In the final analysis, liposomes depend on continually improving the formulation of the nanoparticles' coating and DNA design to increase the transfection efficiency [76,77]. The mechanisms by which peptide-modified liposome protamine/DNA lipoplex (LPD) nanoparticles improve transfer efficiency is charge-ratio-dependent and dose-dependent *in vivo*, and these mechanisms provide their own unique approaches to improve transfer efficiency [23,59,61].

7. Cellular Barriers in the Internalization of Lipid Nanoparticles

DNA packed into liposomes must overcome biological barriers before it can be integrated into the genome. These barriers are the cellular membrane, the nuclear membrane, and chromosomal integrity. Cell targeting and cell-internalization peptides have been extensively studied and used for efficient drug delivery and for image analysis [61]. Arginine-rich (RNA-binding, DNA-binding, and polyarginine) cell-permeable peptides have been shown to cross the cellular barrier [62]. Nuclear localization peptide of the SV40 T large antigen has been shown to promote high LPD-mediated transfection efficiency [23,24,61,78]. In designing our recently formulated lipid nanoparticle, we used a nuclear localization peptide derived from SV40 T antigen (DKKKRKVDKKKRKVDKKKRKV), and another peptide derived from human immunodeficiency virus transactivator of transcription (TAT; YGRKKRRQRRR) peptide [79–82]. The TAT-fusions have been shown to cross the blood–brain barrier [81]. A combination of these two peptides resulted in a high level of sustained gene expression *in vivo* (Figure 2) [23]. The TAT-peptide belongs to an arginine-rich family of peptides, which is an abundant source of membrane-permeable peptides that have potential as carriers for intracellular protein delivery [54,67]. Even with the omission of TAT-peptide, LPD nanoparticles were able to mediate gene delivery [24].

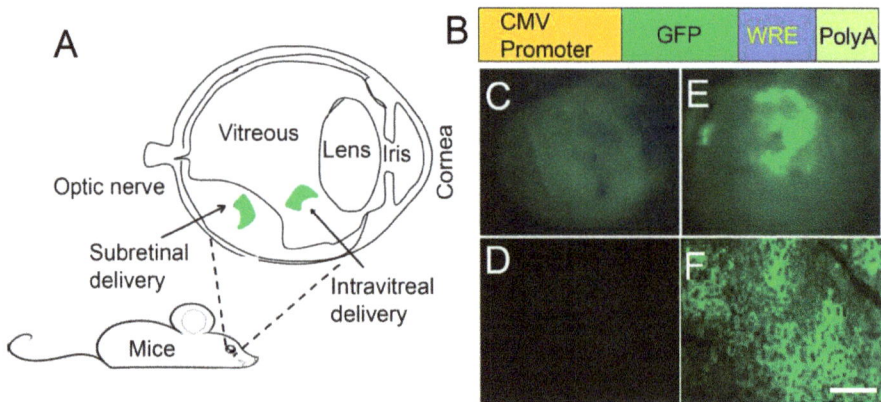

Figure 2. LPD-mediated gene delivery into the retina. Schematic illustration of the eye and route of administration. The most commonly used and preferred mode of administration to retinal layers is subretinal (**A**). Generation of green fluorescent protein construct under the control of CMV promoter (**B**). CMV, cytomegalovirus; GFP, green fluorescent protein; WRE, posttranscriptional regulatory element from the woodchuck hepatitis virus; PolyA, polyadenylation sequence; increases the stability of the molecule. Using BalbC mice, we injected the cDNA construct subretinally into one eye. LPD was complexed with CMV-GFP-WRE-PolyA construct. The other eye was injected with LPD, with a control vector without GFP. Seventy-two hours later, eyes were removed and examined for GFP expression under inverted fluorescence microscopy. GFP expression is clearly seen in the GFP-injected eye (**E**), but not in the control eye (**C**). Whole RPE flat mounts were prepared and examined for GFP expression under inverted fluorescence microscopy. GFP expression is seen in the GFP-injected eye (**F**), but not in the control eye (**D**). Scale bar, 20 μm.

The cell-penetrating peptides (CPPs) are short peptides that facilitate cellular uptake of various molecular cargo [63,64]. In 1988, the first CPP was sequenced from an HIV-1-encoded cell-penetrating transactivator of transcription (TAT) peptide, and delivered efficiently through cell membranes; TAT has been widely applied since then [83–85]. The TAT mechanism of action is still poorly understood, but we do know that this TAT may possess a common internalization mechanism that is ubiquitous to arginine-rich peptides. However, the mechanism is not explained by either adsorptive-mediated endocytosis or by receptor-mediated endocytosis [62,86,87].

8. LPD Nanoparticle-Mediated Delivery of Genes to Eye Tissues

In the eye, the photochemical 11-*cis*-retinal allows the visual pigment rhodopsin to absorb light in the visible range. Without the photochemical, we lose the ability to see light [88]. Pre-clinical studies with viral vectors demonstrated restoration of vision upon gene transfer into retinal cells in mice and dogs [31–34]. In clinical trials, three independent groups reported vision improvements upon the viral-mediated delivery of the Rpe65 gene in patients with Rpe65-associated Leber's congenital amaurosis (LCA) [89–91]. A mouse model lacking the Rpe65 gene has been commonly used for gene therapy studies [5,92–94].

Retinal pigment epithelium protein 65 (Rpe65) is the key enzyme in regulating the availability of photochemicals; a deficiency in this gene results in a blinding eye disease. We showed for the first time that LPD promotes efficient delivery in a cell specific-manner and long-term expression of the Rpe65 gene in mice lacking Rpe65 protein, leading to *in vivo* correction of blindness [23]. The efficacy of this method of restoring vision is comparable to AAV [93] and lentiviral [39] gene transfer of the Rpe65 gene to Rpe65 knockout mice. Our recently published data suggest that we successfully applied LPD to deliver miRNA-184 to the retina, to repress Wnt-mediated ischemia-induced neovascularization [24]. Thus, LPD nanoparticles could provide a promising, efficient, non-viral method of gene delivery with clinical applications in the treatment of eye disease.

9. Cell-Specific Delivery of LPD Nanoparticles

One disadvantage of nanoparticles could be cell specificity. Often, delivery and expression of genes in unwanted cells may lead to adverse or off target effects. We recently achieved specificity by cloning the genes under the control of cell-specific promoters [23]. VMD2-promoter specifically targets LPD to RPE cells, whereas rod opsin promoter specifically drives the expression into rod photoreceptor cells [23]. These studies suggest that other retinal cell specific promoters, such as cone opsin (cone), Thy1 (ganglion cell), and glial fibrillary acidic protein (Müller cells) could be used to achieve cell specificity in conjunction with LPD. The cytomegalovirus (CMV) promoter is widely used, due to its ability to induce protein expression in varied cell types [1,23]. Interestingly, our recent study suggests that CMV promoter exclusively drives expression in retinal pigment epithelial cells [23]. These features make lipid nanoparticles ideal for gene or drug delivery to ocular tissues.

10. Conclusions

Many unique genes have been associated with major retinal diseases, such as retinitis pigmentosa (RP), Leber's congenital amaurosis (LCA), and Stargardt disease [65,69,95]. Until now, Rpe65 defection-induced LCA has been most extensively researched retinal disease. LCA-Rpe65 gene therapy is an example of successful, innovative, translational research. Further studies are needed to determine how retinal gene therapy can be improved [96,97]. The LPD is modified with cell-penetrating peptide and NLS peptide, and carries DNA capable of cell-specific gene expression. Our recent studies suggest that LPD promotes efficient and lasting gene expression *in vivo* without any corresponding inflammatory response [23].

The LPD system could be a promising non-viral gene delivery vector yielding long-term expression and lasting gene transfer efficiency, making it a favorable gene carrier for future applications for eye cell-based therapies. The advantage is that this system allows us to simultaneously introduce multiple biomolecules to turn on the defective signaling pathway *in vivo*. Thus far, non-viral vectors have traditionally been acknowledged as safer. However, non-viral vectors present their own difficulties, with low gene expression efficiency and short transient expression. Recently, the peptide-modified liposome protamine/DNA lipoplex (LPD) nanoparticle has demonstrated the potential to overcome these barriers.

Based on the successful gene therapy of Rpe65 in peptide-modified LPD nanoparticles, the optimization of liposome nanoparticle formulations is safe and efficient. Improvements in gene expression are key to the further development of liposomal nanoparticle technology for retinal gene therapy. The development of modified and safe delivery systems to optimize transfection efficiency will be a critical step toward clinical trials for human gene therapy. Thus, these new peptide-modified LPD nanoparticles open avenues to investigate and develop highly efficient liposome nanoparticles that can overcome the shortcomings of other viral vectors in the treatment of ocular diseases.

These peptide-modified LPD have many advantages for future clinical applications. First, liposome nanoparticles are able to deliver large molecular cargo. Second, the optimization of peptide-modified LPD nanoparticles allows multiple mutant genes to be simultaneously co-delivered to one vector. Third, peptide-modified LPD formulations are more biocompatible and safe. On the whole, a successful delivery formulation for gene therapy should encapsulate and protect the nucleic acid materials, escape endosomal degradation, and reach the specific target site. These new peptide-modified LPD nanoparticles offer new hope for gene therapy for ocular and other related diseases.

Acknowledgments: This study was supported by grants from the National Institutes of Health (EY016507, EY00871, and EY021725), bridge funding from the Presbyterian Health Foundation, and unrestricted departmental grants from Research to Prevent Blindness, Inc. The authors acknowledge Kathy J. Kyler, Staff Editor, University of Oklahoma Health Sciences Center, for editing this manuscript.

Conflicts of Interest: The authors declare no conflicts of interest.

References

1. Adijanto, J.; Naash, M.I. Nanoparticle-based technologies for retinal gene therapy. *Eur. J. Pharm. Biopharm.* **2015**.
2. Kuznetsova, N.R.; Vodovozova, E.L. Differential binding of plasma proteins by liposomes loaded with lipophilic prodrugs of methotrexate and melphalan in the bilayer. *Biochemistry (Mosc.)* **2014**, *79*, 797–804.
3. Wang, F.; Liu, J. Liposome supported metal oxide nanoparticles, interaction mechanism, light controlled content release, and intracellular delivery. *Small* **2014**, *10*, 3927–3931.
4. Xu, Y.; Szoka, F.C., Jr. Mechanism of DNA release from cationic liposome/DNA complexes used in cell transfection. *Biochemistry* **1996**, *35*, 5616–5623.
5. Koirala, A.; Makkia, R.S.; Conley, S.M.; Cooper, M.J.; Naash, M.I. S/MAR-containing DNA nanoparticles promote persistent RPE gene expression and improvement in RPE65-associated LCA. *Hum. Mol. Genet.* **2013**, *22*, 1632–1642.
6. Han, Z.; Conley, S.M.; Makkia, R.S.; Cooper, M.J.; Naash, M.I. DNA nanoparticle-mediated ABCA4 delivery rescues Stargardt dystrophy in mice. *J. Clin. Invest.* **2012**, *122*, 3221–3226.
7. Rowe-Rendleman, C.L.; Durazo, S.A.; Kompella, U.B.; Rittenhouse, K.D.; Di, P.A.; Weiner, A.L.; Grossniklaus, H.E.; Naash, M.I.; Lewin, A.S.; Horsager, A.; *et al.* Drug and gene delivery to the back of the eye, from bench to bedside. *Invest. Ophthalmol. Vis. Sci.* **2014**, *55*, 2714–2730.
8. Koirala, A.; Conley, S.M.; Makkia, R.; Liu, Z.; Cooper, M.J.; Sparrow, J.R.; Naash, M.I. Persistence of non-viral vector mediated RPE65 expression, case for viability as a gene transfer therapy for RPE-based diseases. *J. Control. Release* **2013**, *172*, 745–752.
9. Seyfoddin, A.; Al-Kassas, R. Development of solid lipid nanoparticles and nanostructured lipid carriers for improving ocular delivery of acyclovir. *Drug Dev. Ind. Pharm.* **2013**, *39*, 508–519.
10. Liu, H.A.; Liu, Y.L.; Ma, Z.Z.; Wang, J.C.; Zhang, Q. A lipid nanoparticle system improves siRNA efficacy in RPE cells and a laser-induced murine CNV model. *Invest. Ophthalmol. Vis. Sci.* **2011**, *52*, 4789–4794.

11. Li, R.; Jiang, S.; Liu, D.; Bi, X.; Wang, F.; Zhang, Q.; Xu, Q. A potential new therapeutic system for glaucoma, solid lipid nanoparticles containing methazolamide. *J. Microencapsul.* **2011**, *28*, 134–141.

12. Souto, E.B.; Doktorovova, S.; Gonzalez-Mira, E.; Egea, M.A.; Garcia, M.L. Feasibility of lipid nanoparticles for ocular delivery of anti-inflammatory drugs. *Curr. Eye Res.* **2010**, *35*, 537–552.

13. Seyfoddin, A.; Shaw, J.; Al-Kassas, R. Solid lipid nanoparticles for ocular drug delivery. *Drug Deliv.* **2010**, *17*, 467–489.

14. Gokce, E.H.; Sandri, G.; Bonferoni, M.C.; Rossi, S.; Ferrari, F.; Guneri, T.; Caramella, C. Cyclosporine A loaded SLNs: Evaluation of cellular uptake and corneal cytotoxicity. *Int. J. Pharm.* **2008**, *364*, 76–86.

15. Gokce, E.H.; Sandri, G.; Egrilmez, S.; Bonferoni, M.C.; Guneri, T.; Caramella, C. Cyclosporine A-loaded solid lipid nanoparticles, ocular tolerance and *in vivo* drug release in rabbit eyes. *Curr. Eye Res.* **2009**, *34*, 996–1003.

16. Shen, J.; Sun, M.; Ping, Q.; Ying, Z.; Liu, W. Incorporation of liquid lipid in lipid nanoparticles for ocular drug delivery enhancement. *Nanotechnology* **2010**, *21*.

17. Attama, A.A.; Reichl, S.; Muller-Goymann, C.C. Sustained release and permeation of timolol from surface-modified solid lipid nanoparticles through bioengineered human cornea. *Curr. Eye Res.* **2009**, *34*, 698–705.

18. Del Pozo-Rodriguez, A.; Delgado, D.; Solinis, M.A.; Gascon, A.R.; Pedraz, J.L. Solid lipid nanoparticles for retinal gene therapy, transfection and intracellular trafficking in RPE cells. *Int. J. Pharm.* **2008**, *360*, 177–183.

19. Montasser, I.; Shahgaldian, P.; Perret, F.; Coleman, A.W. Solid lipid nanoparticle-based calix[n]arenes and calix-resorcinarenes as building blocks, synthesis, formulation and characterization. *Int. J. Mol. Sci.* **2013**, *14*, 21899–21942.

20. Mashaghi, S.; Jadidi, T.; Koenderink, G.; Mashaghi, A. Lipid nanotechnology. *Int. J. Mol. Sci.* **2013**, *14*, 4242–4282.

21. Attama, A.A.; Reichl, S.; Muller-Goymann, C.C. Diclofenac sodium delivery to the eye, *in vitro* evaluation of novel solid lipid nanoparticle formulation using human cornea construct. *Int. J. Pharm.* **2008**, *355*, 307–313.

22. Cavalli, R.; Gasco, M.R.; Chetoni, P.; Burgalassi, S.; Saettone, M.F. Solid lipid nanoparticles (SLN) as ocular delivery system for tobramycin. *Int. J. Pharm.* **2002**, *238*, 241–245.

23. Rajala, A.; Wang, Y.; Zhu, Y.; Ranjo-Bishop, M.; Ma, J.X.; Mao, C.; Rajala, R.V. Nanoparticle-assisted targeted delivery of eye-specific genes to eyes significantly improves the vision of blind mice *in vivo*. *Nano. Lett.* **2014**, *14*, 5257–5263.

24. Takahashi, Y.; Chen, Q.; Rajala, R.V.; Ma, J. Micro RNA-184 modulates cannocial Wnt signaling through regulation of frizzled-7 expression in the retina with ischemia-induced neovascularization. *FEBS Lett.* **2015**, *589*, 1143–1149.

25. Cao, B.; Qiu, P.; Mao, C. Mesoporous iron oxide nanoparticles prepared by polyacrylic acid etching and their application in gene delivery to mesenchymal stem cells. *Microsc. Res. Tech.* **2013**, *76*, 936–941.

26. Mulligan, R.C. The basic science of gene therapy. *Science* **1993**, *260*, 926–932.

27. Herz, J.; Gerard, R.D. Adenovirus-mediated transfer of low density lipoprotein receptor gene acutely accelerates cholesterol clearance in normal mice. *Proc. Natl. Acad. Sci USA* **1993**, *90*, 2812–2816.

28. Simon, R.H.; Engelhardt, J.F.; Yang, Y.; Zepeda, M.; Weber-Pendleton, S.; Grossman, M.; Wilson, J.M. Adenovirus-mediated transfer of the CFTR gene to lung of nonhuman primates, toxicity study. *Hum. Gene Ther.* **1993**, *4*, 771–780.

29. Ali, M.; Lemoine, N.R.; Ring, C.J. The use of DNA viruses as vectors for gene therapy. *Gene Ther.* **1994**, *1*, 367–384.

30. Wu, Z.; Yang, H.; Colosi, P. Effect of genome size on AAV vector packaging. *Mol. Ther.* **2010**, *18*, 80–86.

31. Acland, G.M.; Aguirre, G.D.; Ray, J.; Zhang, Q.; Aleman, T.S.; Cideciyan, A.V.; Pearce-Kelling, S.E.; Anand, V.; Zeng, Y.; Maguire, A.M.; *et al.* Gene therapy restores vision in a canine model of childhood blindness. *Nat. Genet.* **2001**, *28*, 92–95.

32. Li, X.; Li, W.; Dai, X.; Kong, F.; Zheng, Q.; Zhou, X.; Lu, F.; Chang, B.; Rohrer, B.; Hauswirth, W.W.; *et al.* Gene therapy rescues cone structure and function in the 3-month-old rd12 mouse, a model for midcourse RPE65 leber congenital amaurosis. *Invest. Ophthalmol. Vis. Sci.* **2011**, *52*, 7–15.

33. Beltran, W.A.; Cideciyan, A.V.; Lewin, A.S.; Iwabe, S.; Khanna, H.; Sumaroka, A.; Chiodo, V.A.; Fajardo, D.S.; Roman, A.J.; Deng, W.T.; *et al.* Gene therapy rescues photoreceptor blindness in dogs and paves the way for treating human X-linked retinitis pigmentosa. *Proc. Natl. Acad. Sci USA* **2012**, *109*, 2132–2137.

34. Cepko, C.L. Emerging gene therapies for retinal degenerations. *J. Neurosci.* **2012**, *32*, 6415–6420.

35. Byrne, L.C.; Dalkara, D.; Luna, G.; Fisher, S.K.; Clerin, E.; Sahel, J.A.; Leveillard, T.; Flannery, J.G. Viral-mediated RdCVF and RdCVFL expression protects cone and rod photoreceptors in retinal degeneration. *J. Clin. Invest.* **2015**, *125*, 105–116.

36. Park, K. Cornea-targeted gene therapy using adenovirus vector. *J. Control. Release* **2014**, *181*.

37. Kaikkonen, M.U.; Yla-Herttuala, S.; Airenne, K.J. How to avoid complement attack in baculovirus-mediated gene delivery. *J. Invertebr. Pathol.* **2011**, *107* Suppl, S71–S79.

38. Luz-Madrigal, A.; Clapp, C.; Aranda, J.; Vaca, L. *In vivo* transcriptional targeting into the retinal vasculature using recombinant baculovirus carrying the human flt-1 promoter. *Virol. J.* **2007**, *4*.

39. Bemelmans, A.P.; Kostic, C.; Crippa, S.V.; Hauswirth, W.W.; Lem, J.; Munier, F.L.; Seeliger, M.W.; Wenzel, A.; Arsenijevic, Y. Lentiviral gene transfer of RPE65 rescues survival and function of cones in a mouse model of Leber congenital amaurosis. *PLoS. Med.* **2006**, *3*, e347.

40. Vadlapudi, A.D.; Mitra, A.K. Nanomicelles, an emerging platform for drug delivery to the eye. *Ther. Deliv.* **2013**, *4*, 1–3.

41. Vadlapudi, A.D.; Vadlapatla, R.K.; Earla, R.; Sirimulla, S.; Bailey, J.B.; Pal, D.; Mitra, A.K. Novel biotinylated lipid prodrugs of acyclovir for the treatment of herpetic keratitis (HK), transporter recognition, tissue stability and antiviral activity. *Pharm. Res.* **2013**, *30*, 2063–2076.

42. Vaishya, R.D.; Gokulgandhi, M.; Patel, S.; Minocha, M.; Mitra, A.K. Novel dexamethasone-loaded nanomicelles for the intermediate and posterior segment uveitis. *AAPS PharmSciTech.* **2014**, *15*, 1238–1251.

43. Gaudana, R.; Ananthula, H.K.; Parenky, A.; Mitra, A.K. Ocular drug delivery. *AAPS. J.* **2010**, *12*, 348–360.

44. Del Pozo-Rodriguez, A.; Delgado, D.; Gascon, A.R.; Solinis, M.A. Lipid nanoparticles as drug/gene delivery systems to the retina. *J. Ocul. Pharmacol. Ther.* **2013**, *29*, 173–188.

45. Paphadjopoulos, D.; Wilson, T.; Taber, R. Liposomes as vehicles for cellular incorporation of biologically active macromolecules. *In Vitro* **1980**, *16*, 49–54.

46. Sessa, G.; Weissmann, G. Phospholipid spherules (liposomes) as a model for biological membranes. *J. Lipid Res.* **1968**, *9*, 310–318.

47. Torchilin, V.P. Multifunctional nanocarriers. *Adv. Drug Deliv. Rev.* **2006**, *58*, 1532–1555.

48. Hamley, I.W. Nanotechnology with soft materials. *Angew. Chem Int. Ed. Engl.* **2003**, *42*, 1692–1712.

49. Jain, P.K.; Huang, X.; El-Sayed, I.H.; El-Sayed, M.A. Noble metals on the nanoscale, optical and photothermal properties and some applications in imaging, sensing, biology, and medicine. *Acc. Chem Res.* **2008**, *41*, 1578–1586.

50. Immordino, M.L.; Dosio, F.; Cattel, L. Stealth liposomes, review of the basic science, rationale, and clinical applications, existing and potential. *Int. J. Nanomedicine.* **2006**, *1*, 297–315.

51. Honda, M.; Asai, T.; Oku, N.; Araki, Y.; Tanaka, M.; Ebihara, N. Liposomes and nanotechnology in drug development, focus on ocular targets. *Int. J. Nanomed.* **2013**, *8*, 495–503.

52. Lopez-Berestein, G.; Mehta, R.; Hopfer, R.; Mehta, K.; Hersh, E.M.; Juliano, R. Effects of sterols on the therapeutic efficacy of liposomal amphotericin B in murine candidiasis. *Cancer Drug Deliv.* **1983**, *1*, 37–42.

53. Van Rooijen, N.; van Nieuwmegen, R. Liposomes in immunology, multilamellar phosphatidylcholine liposomes as a simple, biodegradable and harmless adjuvant without any immunogenic activity of its own. *Immunol. Commun.* **1980**, *9*, 243–256.

54. Schnyder, A.; Huwyler, J. Drug transport to brain with targeted liposomes. *NeuroRx.* **2005**, *2*, 99–107.

55. Oberle, V.; Bakowsky, U.; Zuhorn, I.S.; Hoekstra, D. Lipoplex formation under equilibrium conditions reveals a three-step mechanism. *Biophys. J.* **2000**, *79*, 1447–1454.

56. Bragonzi, A.; Dina, G.; Villa, A.; Calori, G.; Biffi, A.; Bordignon, C.; Assael, B.M.; Conese, M. Biodistribution and transgene expression with nonviral cationic vector/DNA complexes in the lungs. *Gene Ther.* **2000**, *7*, 1753–1760.

57. Kollen, W.J.; Mulberg, A.E.; Wei, X.; Sugita, M.; Raghuram, V.; Wang, J.; Foskett, J.K.; Glick, M.C.; Scanlin, T.F. High-efficiency transfer of cystic fibrosis transmembrane conductance regulator cDNA into cystic fibrosis airway cells in culture using lactosylated polylysine as a vector. *Hum. Gene Ther.* **1999**, *10*, 615–622.

58. El-Aneed, A. An overview of current delivery systems in cancer gene therapy. *J. Control. Release* **2004**, *94*, 1–14.

59. Li, S.; Huang, L. *In vivo* gene transfer via intravenous administration of cationic lipid-protamine-DNA (LPD) complexes. *Gene Ther.* **1997**, *4*, 891–900.

60. Li, S.; Rizzo, M.A.; Bhattacharya, S.; Huang, L. Characterization of cationic lipid-protamine-DNA (LPD) complexes for intravenous gene delivery. *Gene Ther.* **1998**, *5*, 930–937.

61. Ma, K.; Wang, D.D.; Lin, Y.; Wang, J.; Petrenko, V.; Mao, C. Synergetic Targeted Delivery of Sleeping-Beauty Transposon System to Mesenchymal Stem Cells Using LPD Nanoparticles Modified with a Phage-Displayed Targeting Peptide. *Adv. Funct. Mater.* **2013**, *23*, 1172–1181.

62. Futaki, S.; Suzuki, T.; Ohashi, W.; Yagami, T.; Tanaka, S.; Ueda, K.; Sugiura, Y. Arginine-rich peptides. An abundant source of membrane-permeable peptides having potential as carriers for intracellular protein delivery. *J. Biol. Chem.* **2001**, *276*, 5836–5840.

63. Bangham, A.D.; Standish, M.M.; Watkins, J.C. Diffusion of univalent ions across the lamellae of swollen phospholipids. *J. Mol. Biol.* **1965**, *13*, 238–252.

64. Zhu, N.; Liggitt, D.; Liu, Y.; Debs, R. Systemic gene expression after intravenous DNA delivery into adult mice. *Science* **1993**, *261*, 209–211.

65. Sundaram, V.; Moore, A.T.; Ali, R.R.; Bainbridge, J.W. Retinal dystrophies and gene therapy. *Eur. J. Pediatr.* **2012**, *171*, 757–765.

66. Mao, Y.; Triantafillou, G.; Hertlein, E.; Towns, W.; Stefanovski, M.; Mo, X.; Jarjoura, D.; Phelps, M.; Marcucci, G.; Lee, L.J.; *et al.* Milatuzumab-conjugated liposomes as targeted dexamethasone carriers for therapeutic delivery in CD74+ B-cell malignancies. *Clin. Cancer Res.* **2013**, *19*, 347–356.

67. Gandra, N.; Wang, D.D.; Zhu, Y.; Mao, C. Virus-mimetic cytoplasm-cleavable magnetic/silica nanoclusters for enhanced gene delivery to mesenchymal stem cells. *Angew. Chem Int. Ed. Engl.* **2013**, *52*, 11278–11281.

68. Delgado, D.; del Pozo-Rodriguez, A.; Solinis, M.A.; Rodriguez-Gascon, A. Understanding the mechanism of protamine in solid lipid nanoparticle-based lipofection, the importance of the entry pathway. *Eur. J. Pharm. Biopharm.* **2011**, *79*, 495–502.

69. Han, Z.; Conley, S.M.; Naash, M.I. AAV and compacted DNA nanoparticles for the treatment of retinal disorders, challenges and future prospects. *Invest. Ophthalmol. Vis. Sci* **2011**, *52*, 3051–3059.

70. Day, T.P.; Byrne, L.C.; Schaffer, D.V.; Flannery, J.G. Advances in AAV vector development for gene therapy in the retina. *Adv. Exp. Med. Biol.* **2014**, *801*, 687–693.

71. Koirala, A.; Conley, S.M.; Naash, M.I. A review of therapeutic prospects of non-viral gene therapy in the retinal pigment epithelium. *Biomaterials* **2013**, *34*, 7158–7167.

72. Han, Z.; Conley, S.M.; Makkia, R.; Guo, J.; Cooper, M.J.; Naash, M.I. Comparative analysis of DNA nanoparticles and AAVs for ocular gene delivery. *PLoS. On.* **2012**, *7*, e52189.

73. Powell, S.K.; Rivera-Soto, R.; Gray, S.J. Viral expression cassette elements to enhance transgene target specificity and expression in gene therapy. *Discov. Med.* **2015**, *19*, 49–57.

74. Pozzi, D.; Marchini, C.; Cardarelli, F.; Salomone, F.; Coppola, S.; Montani, M.; Zabaleta, M.E.; Digman, M.A.; Gratton, E.; Colapicchioni, V.; *et al.* Mechanistic evaluation of the transfection barriers involved in lipid-mediated gene delivery, interplay between nanostructure and composition. *Biochim. Biophys. Acta* **2014**, *1838*, 957–967.

75. Conley, S.M.; Naash, M.I. Nanoparticles for retinal gene therapy. *Prog. Retin. Eye Res.* **2010**, *29*, 376–397.

76. Torchilin, V.P.; Rammohan, R.; Weissig, V.; Levchenko, T.S. TAT peptide on the surface of liposomes affords their efficient intracellular delivery even at low temperature and in the presence of metabolic inhibitors. *Proc. Natl. Acad. Sci. USA* **2001**, *98*, 8786–8791.

77. Felgner, P.L.; Gadek, T.R.; Holm, M.; Roman, R.; Chan, H.W.; Wenz, M.; Northrop, J.P.; Ringold, G.M.; Danielsen, M. Lipofection, a highly efficient, lipid-mediated DNA-transfection procedure. *Proc. Natl. Acad. Sci. USA* **1987**, *84*, 7413–7417.

78. Hoare, M.; Greiser, U.; Schu, S.; Mashayekhi, K.; Aydogan, E.; Murphy, M.; Barry, F.; Ritter, T.; O'Brien, T. Enhanced lipoplex-mediated gene expression in mesenchymal stem cells using reiterated nuclear localization sequence peptides. *J. Gene Med.* **2010**, *12*, 207–218.

79. Becker-Hapak, M.; McAllister, S.S.; Dowdy, S.F. TAT-mediated protein transduction into mammalian cells. *Methods* **2001**, *24*, 247–256.

80. Gump, J.M.; Dowdy, S.F. TAT transduction, the molecular mechanism and therapeutic prospects. *Trends Mol. Med.* **2007**, *13*, 443–448.

81. Schwarze, S.R.; Ho, A.; Vocero-Akbani, A.; Dowdy, S.F. *In vivo* protein transduction, delivery of a biologically active protein into the mouse. *Science* **1999**, *285*, 1569–1572.

82. Schwarze, S.R.; Dowdy, S.F. *In vivo* protein transduction, intracellular delivery of biologically active proteins, compounds and DNA. *Trends Pharmacol. Sci.* **2000**, *21*, 45–48.

83. Farkhani, S.M.; Valizadeh, A.; Karami, H.; Mohammadi, S.; Sohrabi, N.; Badrzadeh, F. Cell penetrating peptides, efficient vectors for delivery of nanoparticles, nanocarriers, therapeutic and diagnostic molecules. *Peptides* **2014**, *57*, 78–94.

84. Green, M.; Loewenstein, P.M. Autonomous functional domains of chemically synthesized human immunodeficiency virus tat trans-activator protein. *Cell* **1988**, *55*, 1179–1188.

85. Green, M.; Ishino, M.; Loewenstein, P.M. Mutational analysis of HIV-1 Tat minimal domain peptides, identification of trans-dominant mutants that suppress HIV-LTR-driven gene expression. *Cell* **1989**, *58*, 215–223.

86. Suzuki, T.; Futaki, S.; Niwa, M.; Tanaka, S.; Ueda, K.; Sugiura, Y. Possible existence of common internalization mechanisms among arginine-rich peptides. *J. Biol. Chem.* **2002**, *277*, 2437–2443.

87. Peng, L.H.; Niu, J.; Zhang, C.Z.; Yu, W.; Wu, J.H.; Shan, Y.H.; Wang, X.R.; Shen, Y.Q.; Mao, Z.W.; Liang, W.Q.; *et al.* TAT conjugated cationic noble metal nanoparticles for gene delivery to epidermal stem cells. *Biomaterials* **2014**, *35*, 5605–5618.

88. Redmond, T.M.; Yu, S.; Lee, E.; Bok, D.; Hamasaki, D.; Chen, N.; Goletz, P.; Ma, J.X.; Crouch, R.K.; Pfeifer, K. Rpe65 is necessary for production of 11-cis-vitamin A in the retinal visual cycle. *Nat. Genet.* **1998**, *20*, 344–351.

89. Maguire, A.M.; Simonelli, F.; Pierce, E.A.; Pugh, E.N., Jr.; Mingozzi, F.; Bennicelli, J.; Banfi, S.; Marshall, K.A.; Testa, F.; Surace, E.M.; *et al.* Safety and efficacy of gene transfer for Leber's congenital amaurosis. *N. Engl. J. Med.* **2008**, *358*, 2240–2248.

90. Cideciyan, A.V.; Aleman, T.S.; Boye, S.L.; Schwartz, S.B.; Kaushal, S.; Roman, A.J.; Pang, J.J.; Sumaroka, A.; Windsor, E.A.; Wilson, J.M.; *et al.* Human gene therapy for RPE65 isomerase deficiency activates the retinoid cycle of vision but with slow rod kinetics. *Proc. Natl. Acad. Sci USA* **2008**, *105*, 15112–15117.

91. Bainbridge, J.W.; Smith, A.J.; Barker, S.S.; Robbie, S.; Henderson, R.; Balaggan, K.; Viswanathan, A.; Holder, G.E.; Stockman, A.; *et al.* Effect of gene therapy on visual function in Leber's congenital amaurosis. *N. Engl. J. Med.* **2008**, *358*, 2231–2239.

92. Bemelmans, A.P.; Kostic, C.; Hornfeld, D.; Jaquet, M.; Crippa, S.V.; Hauswirth, W.W.; Lem, J.; Wang, Z.; Schorderet, D.E.; Munier, F.L.; *et al.* Lentiviral vectors containing a retinal pigment epithelium specific promoter for leber congenital amaurosis gene therapy. Lentiviral gene therapy for LCA. *Adv. Exp. Med. Biol* **2006**, *572*, 247–253.

93. Lai, C.M.; Yu, M.J.; Brankov, M.; Barnett, N.L.; Zhou, X.; Redmond, T.M.; Narfstrom, K.; Rakoczy, P.E. Recombinant adeno-associated virus type 2-mediated gene delivery into the Rpe65$^{-/-}$ knockout mouse eye results in limited rescue. *Genet. Vaccines. Ther.* **2004**, *2*.

94. Chen, Y.; Moiseyev, G.; Takahashi, Y.; Ma, J.X. RPE65 gene delivery restores isomerohydrolase activity and prevents early cone loss in Rpe65$^{-/-}$ mice. *Invest. Ophthalmol. Vis. Sci.* **2006**, *47*, 1177–1184.

95. Ali, R.R. Gene therapy for retinal dystrophies, twenty years in the making. *Hum. Gene Ther.* **2012**, *23*, 337–339.

96. Stein, L.; Roy, K.; Lei, L.; Kaushal, S. Clinical gene therapy for the treatment of RPE65-associated Leber congenital amaurosis. *Expert. Opin. Biol. Ther.* **2011**, *11*, 429–439.

97. Mowat, F.M.; Breuwer, A.R.; Bartoe, J.T.; Annear, M.J.; Zhang, Z.; Smith, A.J.; Bainbridge, J.W.; Petersen-Jones, S.M.; Ali, R.R. RPE65 gene therapy slows cone loss in Rpe65-deficient dogs. *Gene Ther.* **2013**, *20*, 545–555.

Bioengineered Lacrimal Gland Organ Regeneration *in Vivo*

Masatoshi Hirayama, Kazuo Tsubota and Takashi Tsuji

Abstract: The lacrimal gland plays an important role in maintaining a homeostatic environment for healthy ocular surfaces via tear secretion. Dry eye disease, which is caused by lacrimal gland dysfunction, is one of the most prevalent eye disorders and causes ocular discomfort, significant visual disturbances, and a reduced quality of life. Current therapies for dry eye disease, including artificial tear eye drops, are transient and palliative. The lacrimal gland, which consists of acini, ducts, and myoepithelial cells, develops from its organ germ via reciprocal epithelial-mesenchymal interactions during embryogenesis. Lacrimal tissue stem cells have been identified for use in regenerative therapeutic approaches aimed at restoring lacrimal gland functions. Fully functional organ replacement, such as for tooth and hair follicles, has also been developed via a novel three-dimensional stem cell manipulation, designated the Organ Germ Method, as a next-generation regenerative medicine. Recently, we successfully developed fully functional bioengineered lacrimal gland replacements after transplanting a bioengineered organ germ using this method. This study represented a significant advance in potential lacrimal gland organ replacement as a novel regenerative therapy for dry eye disease. In this review, we will summarize recent progress in lacrimal regeneration research and the development of bioengineered lacrimal gland organ replacement therapy.

Reprinted from *J. Funct. Biomater.* Cite as: Hirayama, M.; Tsubota, K.; Tsuji, T. Bioengineered Lacrimal Gland Organ Regeneration *in Vivo*. *J. Funct. Biomater.* **2015**, *6*, 634–649.

1. Introduction

Advances in regenerative medicine, influenced by our understanding of developmental biology, stem cell biology, and tissue engineering, are expected to underlie next-generation medical therapies [1–3]. Regenerative medicine for various organs, such as stem cell transplants of enriched or purified tissue-derived stem cells and cytokine therapies that activate tissue stem cell differentiation, have been clinically developed and applied [4,5]. These therapies represent attractive concepts with the potential to partially restore lost organ functionality in damaged tissues, malignant diseases, myocardial infarction, neurological diseases, and hepatic dysfunction [6–9]. Current tissue engineering technologies have established two-dimensional tissue regeneration approaches, including the cell sheet transplant technique [10]. The concept of regenerative medicine in ophthalmology includes corneal limbal stem cell transplants, which are based on the understanding of stem cell biology, and regenerative cell sheets, such as cultivated corneal epithelial cell sheets and cultivated oral mucosal epithelial cell sheets, and this has contributed to effective ocular surface reconstruction in clinics for severe ocular surface disorders [11–13]. Regenerative therapies in ophthalmology have steadily advanced to overcome vision-threatening eye diseases, including those of the cornea and retina [14].

Clinically transplanting donor organs is an important therapeutic approach for severe organ dysfunctions; however, there are related medical issues, including allogenic immunological rejection and critical donor shortage [15]. The use of fully functional substitute organs, including artificial organs made from mechanical devices and bio-artificial organs, which consist of living cells and artificial polymers, has been demonstrated to reproduce physiological functions for various organs [16–19]. Organ replacement regenerative therapy for tissue repair, via reconstruction of a fully functional, bioengineered organ from stem cells using *in vitro* three-dimensional cell manipulation, is one of the ultimate goals for regenerative medicine: the replacement of dysfunctional organs arising from disease, injury, or aging [20]. Developing cell manipulation techniques *in vitro*, through the precise arrangement of several different cell species and organ culture methods, is required to realize the next generation of three-dimensional, functional, bioengineered organ replacement regenerative therapy [21].

This review details the physiological functions, diseases, and development of the lacrimal gland obtained from published stem cell research. We illustrate that there is potential for novel, fully functional lacrimal gland regeneration as a next-generation regenerative medicine [22,23].

2. Physiological Function of the Lacrimal Glands

The lacrimal glands are essential for maintaining the physiological function and homeostasis of the ocular surface microenvironment via tear secretion [24,25]. The lacrimal gland consists of the main lacrimal gland, which primarily secretes aqueous tears, and small accessory lacrimal glands [25]. Mature lacrimal glands are organized into a tubuloalveolar system, which includes the acini, the ducts that carry fluid from the acini to a mucosal surface, and the myoepithelial cells that envelop the acini and early duct elements [25]. For physiological tear secretion, establishing the secretagogue stimulus-secretion coupling mechanisms and innervation is required. A tear film consisting of lipid, aqueous, and mucin layers contributes to the microenvironment homeostasis and optical properties of the ocular surface [26–30]. The aqueous layer of the tear film is secreted by the lacrimal glands and contains water and various tear proteins, such as lactoferrin, with biological functions including moisturizing capacity and antimicrobial activity [31–36]. The lacrimal gland and tear functions are indispensable in protecting the epithelial surface and visual function.

3. Dry Eye Disease

Dry-eye disease (DED) is caused by a tear shortage due to lacrimal gland dysfunction that results from systemic diseases and environmental exposures, such as Sjogren's syndrome and ocular cicatricial pemphigoid, or other causes, including aging, long-term work with visual displays, the use of contact lenses, low-humidity environments, and refractive surgery [37–49]. DED is one of the most common eye diseases, and it causes ocular surface epithelial damage, which leads to ocular discomfort, significant loss of vision, and a reduced quality of life [12,50,51]. Current therapies for DED, such as artificial tear solutions, are palliative and do not completely substitute normal tear complexes that contain water, salts, hydrocarbons, proteins, and lipids [52–54]. A therapeutic approach using regenerative medicine is expected to restore lacrimal gland function as a cure for DED [55].

4. Organogenesis of the Lacrimal Glands

Organs, including the lacrimal glands, are functional units composed of various cells with the appropriate three-dimensiona histological architecture, which is achieved through developmental processes in the embryo, to work efficiently. Almost all ectodermal organs, such as teeth, hair follicles, and lacrimal glands, exhibit similar embryonic development from their organ germs that involves reciprocal epithelial and mesenchymal interactions [56]. Branching morphogenesis, which is a fundamental process for developing lacrimal glands, leads to the specification of the ocular surface epithelium and the induction of the lacrimal gland germ (Figure 1a,b) [57,58]. The development of the murine lacrimal gland occurs on embryonic day (ED) 13.5 via a tubular invagination of the conjunctival epithelium at the temporal region of the eye [59]. After the epithelium invaginates and elongates, the lacrimal gland germ invades the mesenchymal sac on ED 16.5 and begins to rapidly proliferate and branch to form a lobular structure [59–62]. The development of lacrimal gland structures is essentially completed by ED 19. By the time the eyes open, seven days after birth, secretory tear components including proteins and lipids are produced [63,64]. Mouse harderian glands, which secrete lipids, also play an important role in protecting the ocular surface [65]. The harderian glands originate from the nasal region of the conjunctival epithelium at ED 16 via a developmental branching process similar to that of the lacrimal glands, and they are located behind the eye [65,66]. The harderian glands are either degenerated or do not exist in primates, including humans [65]. This comprehensive developmental mechanism, involving branching morphogenesis, modulates lacrimal gland maturation.

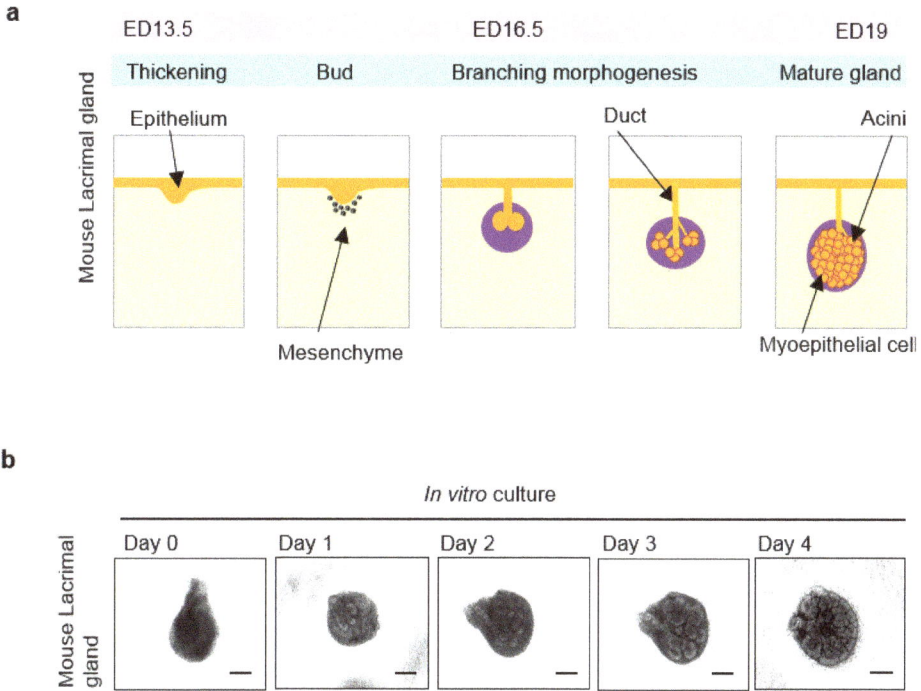

Figure 1. Lacrimal gland organogenesis via epithelial-mesenchymal interactions: (**a**) Schematic representation of the lacrimal gland development during embryogenesis; (**b**) Phase-contrast images of the *in vitro* lacrimal gland germ organ culture development. Scale bars, 100 μm. Modified and reprinted from Hirayama *et al.* [23].

5. Tissue Stem Cells in the Lacrimal Gland

To restore lacrimal gland function, several therapeutic approaches have been reported, such as ectopic salivary gland transplantation *in vivo* [67,68] and regenerative medicine [69]. Secretory glands, including salivary glands, the pancreas [70,71], and mammary glands [72], can self-renew after tissue injury, and this process is mediated by tissue stem cells. Many studies aimed at restoring secretory gland function have attempted to use various stem cells derived from adult tissues [73,74]. For salivary glands, long-term abnormal ligation of the salivary excretory duct leads to inflammation and cell death, which results in gland atrophy; however, some repair processes, including the proliferation of the tubuloalveolar structure, do occur when the ligation is released [75–81]. The salivary gland can potentially regenerate using various stem cells, such as intercalated duct cells from the salivary gland [76], c-kit-positive duct cells in human salivary glands [75], salivary gland-derived progenitor cells isolated from duct-ligated animals, and bone marrow-derived Sca-1- and c-kit-positive cells [73]. For stem cell therapy of the lacrimal glands, the potential existence of stem cells or progenitor cells has been previously described [69,82]. Tissue stem/progenitor cells, which express nestin and Ki67, and mesenchymal cells both contribute to tissue repair after interleukin-1-induced inflammation in murine-lacrimal glands [83–86]. Stem cell candidates expressing stem cell markers such as c-kit, ABCG2, and ALDH1 have been identified in human lacrimal gland cells [87,88]. Tissue regeneration using transplanted stem cells in adult tissues to restore lacrimal gland function is an area of intense research because of its potential clinical benefits [89,90].

6. A Novel Three-Dimensional Cell Manipulation Method Termed the Organ Germ Method

To further these biological technologies, the development of methods for the manipulation of multiple cells is required to realize three-dimensional organ regeneration for functional bioengineered organ replacement therapy [20]. A novel strategy for developing bioengineered organs by reproducing the developmental process during organogenesis has been proposed for the functional replacement and complete restoration of lost organs [21]. This bioengineered organ germ method, which manipulates epithelial and mesenchymal cells via cell compartmentalization at a high cell density in a type I collagen gel matrix, was developed to reconstruct bioengineered organ germs *in vitro* as an organ engineering technology (Figure 2a,b) [91,92]. This method successfully developed bioengineered ectodermal organs, such as teeth and hair follicle germs, through multicellular assembly and epithelial and mesenchymal interactions similar to those in natural organ germs (Figure 2c,d) [91–95]. Importantly, the bioengineered tooth and hair follicle germ transplants could restore physiological functions via cooperation with peripheral tissues at the lost tooth or hair follicle [93–96]. Developing this method was a substantial advance towards potentially regenerating other ectodermal secretory organs, including the salivary glands [97,98] and lacrimal glands [23].

7. Fully Functional Lacrimal Gland Organ Regeneration

We investigated whether our organ germ method could regenerate a bioengineered lacrimal gland and restore its physiological function. The bioengineered lacrimal gland germ, which was reconstituted using the epithelial and mesenchymal cells from the lacrimal gland germ of an ED 16.5 mouse, successfully developed branching morphogenesis followed by stalk elongation and cleft formation in organ culture *in vitro*. Bioengineered harderian gland germs were also regenerated via the organ germ method (Figure 3a).

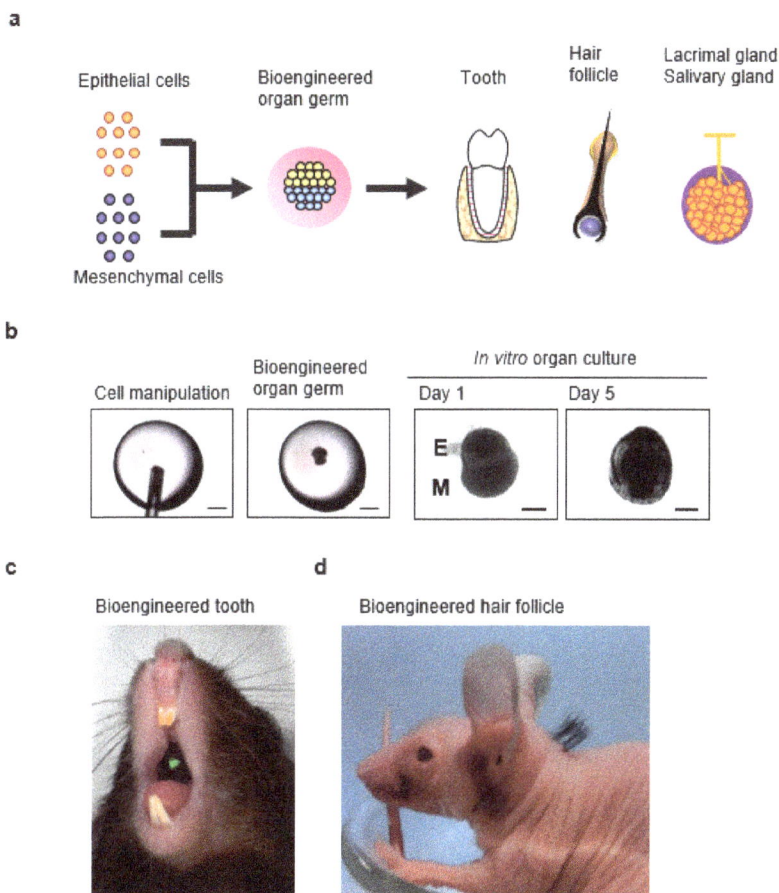

Figure 2. Strategy for bioengineered organ regeneration using the organ germ method: (**a**) Functional organs, such as teeth, hair follicles, salivary glands, and lacrimal glands, can now be regenerated *in vivo* by transplanting bioengineered organ germs reconstituted from epithelial and mesenchymal cells via the organ germ method; (**b**) Representative image of our developed three-dimensional cell processing system, the organ germ method. A high density of dissociated mesenchymal cells is injected into the center of a collagen drop (left panel). Dissociated epithelial cells are subsequently injected into the drop adjacent to the mesenchymal cell aggregate (center-left panel). The bioengineered tooth regenerated via the organ germ method could develop into an appropriate tooth germ via organ culturing (center-right and right panels); (**c**) Photograph showing the green fluorescence protein (GFP)-labeled bioengineered tooth engrafted in an adult mouse (green); (**d**) Photograph of the developed bioengineered hair follicles, which were successfully engrafted into a nude mouse. Modified and reprinted from Nakao *et al.* [21].

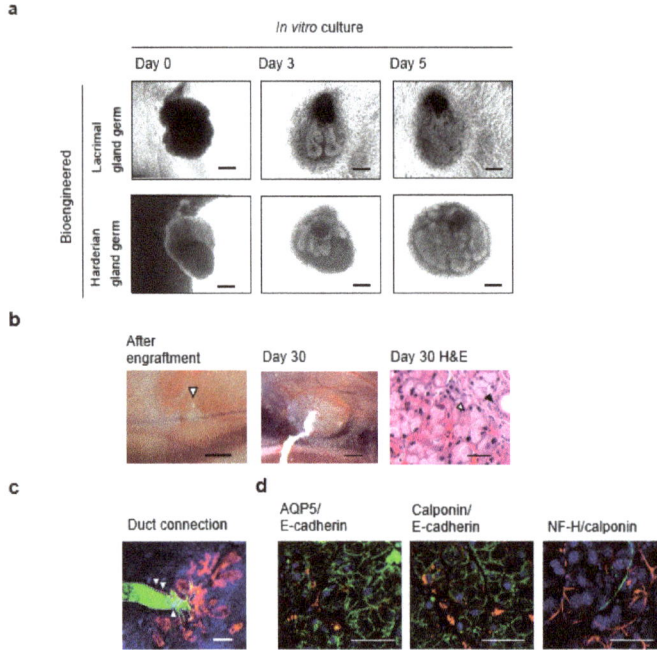

Figure 3. Functional lacrimal gland regeneration via bioengineered lacrimal gland germ transplant: (**a**) Phase-contrast images of the bioengineered lacrimal gland germ (upper line) and bioengineered harderian gland germ (lower line) development. Scale bar, 100 μm; (**b**) Photographs of the bioengineered lacrimal gland germ after transplanting into a mouse with the extra-orbital lacrimal gland removed (arrowhead) (left panel; Scale bar, 1 mm). At 30 days after transplantation, the bioengineered lacrimal gland was successfully engrafted and developed (center; Scale bar, 500 μm). The hematoxylin-eosin(H.E.) staining revealed that the bioengineered lacrimal gland achieved a mature secretory gland structure including acini (white arrowhead) and duct (black arrowhead) (right; Scale bar, 50 μm); (**c**) Histological analysis of the duct connection between the bioengineered lacrimal gland and recipient lacrimal excretory duct. Bioengineered lacrimal glands regenerated using DsRed transgenic mouse-derived epithelial cells (red) and normal mouse-derived mesenchymal cells developed with the correct duct association in the recipient mouse (arrowhead). Fluorescein isothiocyanate (FITC) -gelatin (green), which was injected from the recipient lacrimal excretory duct, could successfully reach the bioengineered lacrimal gland. 4′,6-diamidino-2-phenylindole (DAPI; blue) and the excretory duct (dotted line) are shown. Scale bars, 100 μm; (**d**) Immunohistochemical analysis of the bioengineered lacrimal gland after transplantation. Aquaporin-5 is red and E-cadherin is green in the left panel. Calponin is red and E-cadherin is green in the center panel. Calponin is red, neurofilament-H (NF-H) is green, and DAPI is blue in the right panel. Scale bars, 50 μm. Modified and reprinted from Hirayama *et al.* [23].

7.1. Engraftment of Bioengineered Lacrimal Gland Germ with Duct Association

A duct association between the bioengineered lacrimal gland and the mouse receiving the ocular surface discharge is required for tear secretion from the bioengineered lacrimal gland. The bioengineered lacrimal gland germ and the bioengineered harderian gland germ were successfully engrafted to a mouse from which an extra-orbital lacrimal gland had been removed, and the bioengineered lacrimal gland duct was connected to the recipient lacrimal excretory duct using our thread-guided transplant method (Figure 3b,c). After the transplant, the bioengineered lacrimal and harderian glands formed the appropriate histo-architecture, including acini-expressing aquaporin 5 and myoepithelial cells, duct, and nerve fibers, by reproducing the developmental process that occurs during organogenesis (Figure 3c,d). Thus, the bioengineered lacrimal gland and harderian gland can develop after *in vivo* orthotopic or ectopic transplantation.

7.2. Tear Secretion Ability of the Bioengineered Lacrimal Gland

Reconstituting neural pathways between the bioengineered lacrimal gland and the recipient's neural system is important to protect the ocular surface via restored tear secretion [99–101]. Tearing resulting from a cooling stimulation at the ocular surface that is activated via corneal thermoreceptors and is a representative neural pathway for lacrimal gland function (Figure 4a) [102,103]. We demonstrated that the bioengineered lacrimal gland could secret tears in cooperation with peripheral tissues, including neural systems, because the tear secretion volume from the bioengineered lacrimal gland increased after ocular surface cooling stimulation. Tear components secreted from acini in the lacrimal and harderian glands, such as lactoferrin and lipids, respectively, are essential for physiological tear functions such as increased stability, wound healing, and anti-bacterial effects [104–109]. Current therapies for severe lacrimal gland dysfunction include medical treatments such as albumin eye drops and autologous serum eye drops that attempt to substitute tear protein function [54,110–114]. We have shown that tears from the bioengineered lacrimal gland contained major tear proteins, including lactoferrin. In addition, the lipid concentration increased significantly in tears from the bioengineered harderian gland. These results indicated that these bioengineered glands can produce appropriate tear components. Functional replacements of the bioengineered lacrimal gland would be an attractive strategy for treating severe dry eye disease.

7.3. Ocular Surface Protection Effect by the Bioengineered Lacrimal Gland

Protecting the ocular surface is the main purpose of using the bioengineered lacrimal gland to restore lacrimal gland function. Punctate staining of the impaired area on the ocular surface [115,116] and corneal epithelial changes including thinning and stromal fibroblast activation [117,118] were observed in a mouse with an extra-orbital lacrimal gland defect, which mimics the corneal epithelial damage caused by lacrimal gland dysfunction. However, these changes were prevented using a bioengineered lacrimal gland (Figure 4b,c). Our results indicate that the bioengineered lacrimal gland can develop and provide sufficient function to maintain a healthy ocular surface.

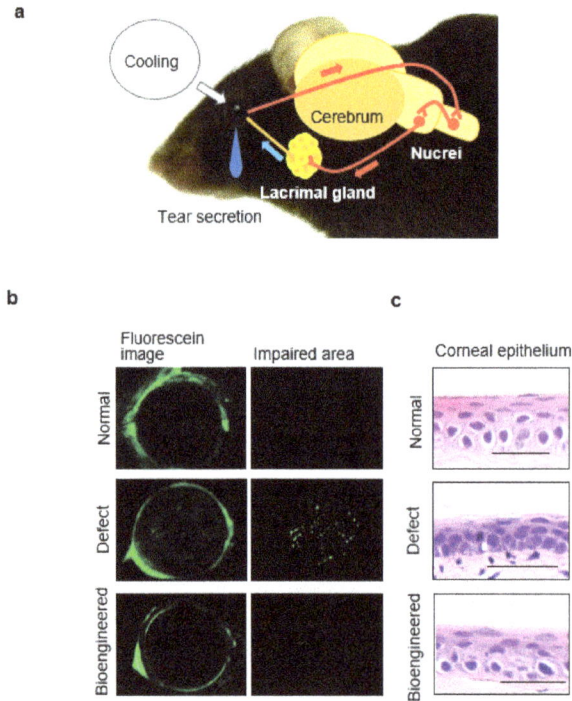

Figure 4. Tear secretion and ocular surface protection for the bioengineered lacrimal gland: (**a**) Schematic representation of the neural reflex loop for tear secretion. Cooling on the ocular surface stimulates tear secretion from the lacrimal gland via the central nervous system; (**b**) Representative images of the corneal surface of a normal lacrimal gland (upper), a lacrimal gland-defect mouse (center), and a bioengineered lacrimal gland–engrafted mouse (lower). The punctate staining area by fluorescein showed impaired area on corneal surface. Scale bar, 1 mm. Modified and reprinted from Hirayama *et al.* [23]; (**c**) Representative microscopic images of the corneal epithelium, including a normal mouse (upper), lacrimal gland–defective mouse (center), and bioengineered lacrimal gland–transplanted mouse (lower) are shown. Chronic dry eye status in lacrimal gland–defective mouse induced corneal thickening as shown in the center panel, whereas these changes were not observed in the bioengineered lacrimal gland-transplanted mouse. Scale bars, 25 μm. Modified and reprinted from Hirayama *et al.* [23].

8. Conclusions and Future Directions

Bioengineered lacrimal gland germs exhibit appropriate physiological functions, such as tear secretion, in response to nervous stimulation and ocular surface protection. These studies are a proof-of-concept for bioengineered organs that can functionally restore the lacrimal gland. Our bioengineered organ regeneration concept, which has also been applied to salivary gland regeneration [98], provides substantial advances for regenerative therapies for dry eye disease and xerostomia. Epithelial and mesenchymal stem cells, which have organ-inductive potential for bioengineered organs, have not been reported in adult tissues. To realize the future practical clinical applications of organ replacement regenerative therapy, studies to develop technologies for organ regeneration, such as investigations of available cell sources (e.g., pluripotent stem cells represented as embryonic stem cells and induced pluripotent stem cells) and the efficacy of disease models (e.g., Sjogren syndrome and Stevens-Johnson syndrome) for these methods, technical procedures for culture methods to create bioengineered organs, and appropriate transplantation methods for human patients, are required. Bioengineered organ regenerative therapy is expected to be an essential therapeutic strategy for the next generation of regenerative medicine.

Acknowledgments: This work was partially supported by a Grant-in-Aid for Scientific Research A (to T.T.) from the Ministry of Education, Culture, Sports, Science, and Technology, Japan. T.T. has already acquired patents related to this article's subject of organ regeneration, and they have been applied for in other countries.

Conflicts of Interest: The authors declare no conflict of interest.

References

1. Okano, H.; Yamanaka, S. iPS cell technologies: Significance and applications to CNS regeneration and disease. *Mol. Brain.* **2014**, 7.
2. Miyahara, Y.; Nagaya, N.; Kataoka, M.; Yanagawa, B.; Tanaka, K.; Hao, H.; Ishino, K.; Ishida, H.; Shimizu, T.; Kangawa, K.; *et al.* Monolayered mesenchymal stem cells repair scarred myocardium after myocardial infarction. *Nat. Med.* **2006**, *12*, 459–465.
3. Sekine, H.; Shimizu, T.; Yang, J.; Kobayashi, E.; Okano, T. Pulsatile myocardial tubes fabricated with cell sheet engineering. *Circulation* **2006**, *114*, I87–I93.
4. Appelbaum, F.R. The current status of hematopoietic cell transplantation. *Annu. Rev. Med.* **2003**, *54*, 491–512.
5. Kim, J.H.; Auerbach, J.M.; Rodriguez-Gomez, J.A.; Velasco, I.; Gavin, D.; Lumelsky, N.; Lee, S.H.; Nguyen, J.; Sanchez-Pernaute, R.; Bankiewicz, K.; *et al.* Dopamine neurons derived from embryonic stem cells function in an animal model of parkinson's disease. *Nature* **2002**, *418*, 50–56.
6. Lindvall, O.; Kokaia, Z. Stem cells for the treatment of neurological disorders. *Nature* **2006**, *441*, 1094–1096.

7.   Copelan, E.A. Hematopoietic stem-cell transplantation. *New Engl. J. Med.* **2006**, *354*, 1813–1826.

8.   Segers, V.F.; Lee, R.T. Stem-cell therapy for cardiac disease. *Nature* **2008**, *451*, 937–942.

9.   Wang, X.; Foster, M.; Al-Dhalimy, M.; Lagasse, E.; Finegold, M.; Grompe, M. The origin and liver repopulating capacity of murine oval cells. *Proc. Natl. Acad. Sci. USA* **2003**, *100*, 11881–11888.

10.   Haraguchi, Y.; Shimizu, T.; Matsuura, K.; Sekine, H.; Tanaka, N.; Tadakuma, K.; Yamato, M.; Kaneko, M.; Okano, T. Cell sheet technology for cardiac tissue engineering. *Methods Mol. Biol.* **2014**, *1181*, 139–155.

11.   Tsubota, K.; Satake, Y.; Kaido, M.; Shinozaki, N.; Shimmura, S.; Bissen-Miyajima, H.; Shimazaki, J. Treatment of severe ocular-surface disorders with corneal epithelial stem-cell transplantation. *N. Engl. J. Med.* **1999**, *340*, 1697–1703.

12.   Tsubota, K.; Satake, Y.; Shimazaki, J. Treatment of severe dry eye. *Lancet* **1996**, *348*, 123.

13.   Nishida, K.; Yamato, M.; Hayashida, Y.; Watanabe, K.; Yamamoto, K.; Adachi, E.; Nagai, S.; Kikuchi, A.; Maeda, N.; Watanabe, H.; *et al.* Corneal reconstruction with tissue-engineered cell sheets composed of autologous oral mucosal epithelium. *New Engl. J. Med.* **2004**, *351*, 1187–1196.

14.   Osakada, F.; Hirami, Y.; Takahashi, M. Stem cell biology and cell transplantation therapy in the retina. *Biotechnol. Genet. Eng. Rev.* **2010**, *26*, 297–334.

15.   Abouna, G.M. Organ shortage crisis: Problems and possible solutions. *Transplant. Proc.* **2008**, *40*, 34–38.

16.   Oshima, M.; Inoue, K.; Nakajima, K.; Tachikawa, T.; Yamazaki, H.; Isobe, T.; Sugawara, A.; Ogawa, M.; Tanaka, C.; Saito, M.; *et al.* Functional tooth restoration by next-generation bio-hybrid implant as a bio-hybrid artificial organ replacement therapy. *Sci. Rep.* **2014**, *4*.

17.   Wolf, A.V. The artificial kidney. *Science* **1952**, *115*, 193–199.

18.   Copeland, J.G.; Smith, R.G.; Arabia, F.A.; Nolan, P.E.; Sethi, G.K.; Tsau, P.H.; McClellan, D.; Slepian, M.J. Cardiac replacement with a total artificial heart as a bridge to transplantation. *New Engl. J. Med.* **2004**, *351*, 859–867.

19.   Fort, A.; Fort, N.; Ricordi, C.; Stabler, C.L. Biohybrid devices and encapsulation technologies for engineering a bioartificial pancreas. *Cell Transplant.* **2008**, *17*, 997–1003.

20.   Ikeda, E.; Tsuji, T. Growing bioengineered teeth from single cells: Potential for dental regenerative medicine. *Expert Opin. Boil. Ther.* **2008**, *8*, 735–744.

21.   Nakao, K.; Morita, R.; Saji, Y.; Ishida, K.; Tomita, Y.; Ogawa, M.; Saitoh, M.; Tomooka, Y.; Tsuji, T. The development of a bioengineered organ germ method. *Nat. Methods* **2007**, *4*, 227–230.

22.   Hirayama, M.; Oshima, M.; Tsuji, T. Development and prospects of organ replacement regenerative therapy. *Cornea* **2013**, *32*, S13–S21.

23.   Hirayama, M.; Ogawa, M.; Oshima, M.; Sekine, Y.; Ishida, K.; Yamashita, K.; Ikeda, K.; Shimmura, S.; Kawakita, T.; Tsubota, K.; *et al.* Functional lacrimal gland regeneration by transplantation of a bioengineered organ germ. *Nat. Commun.* **2013**, *4*.

24.   Dilly, P.N. Structure and function of the tear film. *Adv. Exp. Med. Boil.* **1994**, *350*, 239–247.

25. Schechter, J.E.; Warren, D.W.; Mircheff, A.K. A lacrimal gland is a lacrimal gland, but rodent's and rabbit's are not human. *Ocul. Surf.* **2010**, *8*, 111–134.

26. Holly, F.J. Formation and stability of the tear film. *Int. Ophthalmol. Clin.* **1973**, *13*, 73–96.

27. Schoenwald, R.D.; Vidvauns, S.; Wurster, D.E.; Barfknecht, C.F. The role of tear proteins in tear film stability in the dry eye patient and in the rabbit. *Adv. Exp. Med. Boil.* **1998**, *438*, 391–400.

28. Sweeney, D.F.; Millar, T.J.; Raju, S.R. Tear film stability: A review. *Exp. Eye. Res.* **2013**, *117*, 28–38.

29. Mishima, S. Some physiological aspects of the precorneal tear film. *Arch. Ophthalmol.* **1965**, *73*, 233–241.

30. Tiffany, J.M.; Winter, N.; Bliss, G. Tear film stability and tear surface tension. *Curr. Eye Res.* **1989**, *8*, 507–515.

31. Balasubramanian, S.A.; Pye, D.C.; Willcox, M.D. Levels of lactoferrin, secretory IgA and serum albumin in the tear film of people with keratoconus. *Exp. Eye Res.* **2012**, *96*, 132–137.

32. Broekhuyse, R.M. Tear lactoferrin: A bacteriostatic and complexing protein. *Invest. Ophthalmol.* **1974**, *13*, 550–554.

33. Danjo, Y.; Lee, M.; Horimoto, K.; Hamano, T. Ocular surface damage and tear lactoferrin in dry eye syndrome. *Acta Ophthalmol.* **1994**, *72*, 433–437.

34. Ohashi, Y.; Dogru, M.; Tsubota, K. Laboratory findings in tear fluid analysis. *Clin. Chim. Acta Int. J. Clin. Chem.* **2006**, *369*, 17–28.

35. Seal, D.V.; McGill, J.I.; Mackie, I.A.; Liakos, G.M.; Jacobs, P.; Goulding, N.J. Bacteriology and tear protein profiles of the dry eye. *Br. J. Ophthalmol.* **1986**, *70*, 122–125.

36. Delaire, A.; Lassagne, H.; Gachon, A.M. New members of the lipocalin family in human tear fluid. *Exp. Eye Res.* **1992**, *55*, 645–647.

37. Ahn, J.M.; Lee, S.H.; Rim, T.H.; Park, R.J.; Yang, H.S.; Kim, T.I.; Yoon, K.C.; Seo, K.Y. Prevalence of and risk factors associated with dry eye: The Korea national health and nutrition examination survey 2010–2011. *Am. J. Ophthalmol.* **2014**, *158*, 1205–1214.

38. Paulsen, A.J.; Cruickshanks, K.J.; Fischer, M.E.; Huang, G.H.; Klein, B.E.; Klein, R.; Dalton, D.S. Dry eye in the beaver dam offspring study: Prevalence, risk factors, and health-related quality of life. *Am. J. Ophthalmol.* **2014**, *157*, 799–806.

39. Galor, A.; Feuer, W.; Lee, D.J.; Florez, H.; Carter, D.; Pouyeh, B.; Prunty, W.J.; Perez, V.L. Prevalence and risk factors of dry eye syndrome in a united states veterans affairs population. *Am. J. Ophthalmol.* **2011**, *152*, 377–384.

40. Shoja, M.R.; Besharati, M.R. Dry eye after lasik for myopia: Incidence and risk factors. *Eur. J. Ophthalmol.* **2007**, *17*, 1–6.

41. De Paiva, C.S.; Chen, Z.; Koch, D.D.; Hamill, M.B.; Manuel, F.K.; Hassan, S.S.; Wilhelmus, K.R.; Pflugfelder, S.C. The incidence and risk factors for developing dry eye after myopic lasik. *Am. J. Ophthalmol.* **2006**, *141*, 438–445.

42. Lee, A.J.; Lee, J.; Saw, S.M.; Gazzard, G.; Koh, D.; Widjaja, D.; Tan, D.T. Prevalence and risk factors associated with dry eye symptoms: A population based study in indonesia. *Br. J. Ophthalmol.* **2002**, *86*, 1347–1351.

43. Moss, S.E.; Klein, R.; Klein, B.E. Prevalence of and risk factors for dry eye syndrome. *Arch. Ophthalmol.* **2000**, *118*, 1264–1268.

44. Alves, M.; Fonseca, E.C.; Alves, M.F.; Malki, L.T.; Arruda, G.V.; Reinach, P.S.; Rocha, E.M. Dry eye disease treatment: A systematic review of published trials and a critical appraisal of therapeutic strategies. *Ocul. Surf.* **2013**, *11*, 181–192.

45. Mantelli, F.; Massaro-Giordano, M.; Macchi, I.; Lambiase, A.; Bonini, S. The cellular mechanisms of dry eye: From pathogenesis to treatment. *J. Cell. Physiol.* **2013**, *228*, 2253–2256.

46. Toda, I.; Asano-Kato, N.; Hori-Komai, Y.; Tsubota, K. Ocular surface treatment before laser *in situ* keratomileusis in patients with severe dry eye. *J. Refract. Surg.* **2004**, *20*, 270–275.

47. Uchino, M.; Nishiwaki, Y.; Michikawa, T.; Shirakawa, K.; Kuwahara, E.; Yamada, M.; Dogru, M.; Schaumberg, D.A.; Kawakita, T.; Takebayashi, T.; *et al.* Prevalence and risk factors of dry eye disease in japan: Koumi study. *Ophthalmology* **2011**, *118*, 2361–2367.

48. Tsubota, K.; Nakamori, K. Dry eyes and video display terminals. *N. Engl. J. Med.* **1993**, *328*, 584.

49. Uchino, M.; Yokoi, N.; Uchino, Y.; Dogru, M.; Kawashima, M.; Komuro, A.; Sonomura, Y.; Kato, H.; Kinoshita, S.; Schaumberg, D.A.; *et al.* Prevalence of dry eye disease and its risk factors in visual display terminal users: The osaka study. *Am. J. Ophthalmol.* **2013**, *156*, 759–766.

50. Kaido, M.; Ishida, R.; Dogru, M.; Tsubota, K. Visual function changes after punctal occlusion with the treatment of short but type of dry eye. *Cornea* **2012**, *31*, 1009–1013.

51. Gadaria-Rathod, N.; Lee, K.I.; Asbell, P.A. Emerging drugs for the treatment of dry eye disease. *Expert Opin. Emerging Drugs* **2013**, *18*, 121–136.

52. Kojima, T.; Ishida, R.; Dogru, M.; Goto, E.; Matsumoto, Y.; Kaido, M.; Tsubota, K. The effect of autologous serum eyedrops in the treatment of severe dry eye disease: A prospective randomized case-control study. *Am. J. Ophthalmol.* **2005**, *139*, 242–246.

53. Messmer, E.M. The pathophysiology, diagnosis, and treatment of dry eye disease. *Deutsch. Arztebl. Int.* **2015**, *112*, 71–81.

54. Kojima, T.; Higuchi, A.; Goto, E.; Matsumoto, Y.; Dogru, M.; Tsubota, K. Autologous serum eye drops for the treatment of dry eye diseases. *Cornea* **2008**, *27*, S25–S30.

55. Brockes, J.P.; Kumar, A. Appendage regeneration in adult vertebrates and implications for regenerative medicine. *Science* **2005**, *310*, 1919–1923.

56. Pispa, J.; Thesleff, I. Mechanisms of ectodermal organogenesis. *Dev. Biol.* **2003**, *262*, 195–205.

57. Sakai, T. Epithelial branching morphogenesis of salivary gland: Exploration of new functional regulators. *J. Med. Invest. JMI* **2009**, *56*, 234–238.

58. Pan, Y.; Carbe, C.; Powers, A.; Zhang, E.E.; Esko, J.D.; Grobe, K.; Feng, G.S.; Zhang, X. Bud specific N-sulfation of heparan sulfate regulates Shp2-dependent FGF signaling during lacrimal gland induction. *Development* **2008**, *135*, 301–310.

59. Makarenkova, H.P.; Ito, M.; Govindarajan, V.; Faber, S.C.; Sun, L.; McMahon, G.; Overbeek, P.A.; Lang, R.A. FGF10 is an inducer and Pax6 a competence factor for lacrimal gland development. *Development* **2000**, *127*, 2563–2572.

60. Dean, C.; Ito, M.; Makarenkova, H.P.; Faber, S.C.; Lang, R.A. Bmp7 regulates branching morphogenesis of the lacrimal gland by promoting mesenchymal proliferation and condensation. *Development* **2004**, *131*, 4155–4165.

61. Tsau, C.; Ito, M.; Gromova, A.; Hoffman, M.P.; Meech, R.; Makarenkova, H.P. Barx2 and Fgf10 regulate ocular glands branching morphogenesis by controlling extracellular matrix remodeling. *Development* **2011**, *138*, 3307–3317.

62. Govindarajan, V.; Ito, M.; Makarenkova, H.P.; Lang, R.A.; Overbeek, P.A. Endogenous and ectopic gland induction by Fgf-10. *Dev. Biol.* **2000**, *225*, 188–200.

63. Franklin, R.M. The ocular secretory immune system: A review. *Curr. Eye Res.* **1989**, *8*, 599–606.

64. Wang, Y.L.; Tan, Y.; Satoh, Y.; Ono, K. Morphological changes of myoepithelial cells of mouse lacrimal glands during postnatal development. *Histol. Histopathol.* **1995**, *10*, 821–827.

65. Payne, A.P. The harderian gland: A tercentennial review. *J. Anat.* **1994**, *185*, 1–49.

66. Satoh, Y.; Gesase, A.P.; Habara, Y.; Ono, K.; Kanno, T. Lipid secretory mechanisms in the mammalian harderian gland. *Microsc. Res. Tech.* **1996**, *34*, 104–110.

67. Sant' Anna, A.E.; Hazarbassanov, R.M.; de Freitas, D.; Gomes, J.A. Minor salivary glands and labial mucous membrane graft in the treatment of severe symblepharon and dry eye in patients with stevens-johnson syndrome. *Br. J. Ophthalmol.* **2012**, *96*, 234–239.

68. Soares, E.J.; Franca, V.P. Transplantation of labial salivary glands for severe dry eye treatment. *Arq. Bras. Oftalmol.* **2005**, *68*, 481–489.

69. Zoukhri, D. Mechanisms involved in injury and repair of the murine lacrimal gland: Role of programmed cell death and mesenchymal stem cells. *Ocul. Surf.* **2010**, *8*, 60–69.

70. Walker, N.I. Ultrastructure of the rat pancreas after experimental duct ligation. I. The role of apoptosis and intraepithelial macrophages in acinar cell deletion. *Am J. Pathol.* **1987**, *126*, 439–451.

71. Scoggins, C.R.; Meszoely, I.M.; Wada, M.; Means, A.L.; Yang, L.; Leach, S.D. P53-dependent acinar cell apoptosis triggers epithelial proliferation in duct-ligated murine pancreas. *Am. J. Physiol. Gastrointest. Liver Physiol.* **2000**, *279*, G827–G836.

72. Walker, N.I.; Bennett, R.E.; Kerr, J.F. Cell death by apoptosis during involution of the lactating breast in mice and rats. *Am. J. Anat.* **1989**, *185*, 19–32.

73. Sumita, Y.; Liu, Y.; Khalili, S.; Maria, O.M.; Xia, D.; Key, S.; Cotrim, A.P.; Mezey, E.; Tran, S.D. Bone marrow-derived cells rescue salivary gland function in mice with head and neck irradiation. *Int. J. Biochem. Cell Boil.* **2011**, *43*, 80–87.

74. Lombaert, I.M.; Brunsting, J.F.; Wierenga, P.K.; Faber, H.; Stokman, M.A.; Kok, T.; Visser, W.H.; Kampinga, H.H.; de Haan, G.; Coppes, R.P. Rescue of salivary gland function after stem cell transplantation in irradiated glands. *PloS One* **2008**, *3*.

217

75. Feng, J.; van der Zwaag, M.; Stokman, M.A.; van Os, R.; Coppes, R.P. Isolation and characterization of human salivary gland cells for stem cell transplantation to reduce radiation-induced hyposalivation. *Radiother. Oncol. J. Eur. Soc. Ther. Radiol. Oncol.* **2009**, *92*, 466–471.

76. Takahashi, S.; Schoch, E.; Walker, N.I. Origin of acinar cell regeneration after atrophy of the rat parotid induced by duct obstruction. *Int. J. Exp. Pathol.* **1998**, *79*, 293–301.

77. Cummins, M.; Dardick, I.; Brown, D.; Burford-Mason, A. Obstructive sialadenitis: A rat model. *J. Otolaryngol.* **1994**, *23*, 50–56.

78. Burgess, K.L.; Dardick, I.; Cummins, M.M.; Burford-Mason, A.P.; Bassett, R.; Brown, D.H. Myoepithelial cells actively proliferate during atrophy of rat parotid gland. *Oral Surg. Oral Med. Oral Pathol. Oral Radiol. Endod.* **1996**, *82*, 674–680.

79. Takahashi, S.; Shinzato, K.; Domon, T.; Yamamoto, T.; Wakita, M. Mitotic proliferation of myoepithelial cells during regeneration of atrophied rat submandibular glands after duct ligation. *J. Oral Pathol. Med. Off. Publ. Int. Assoc. Oral Pathol. Am. Acad. Oral Pathol.* **2004**, *33*, 430–434.

80. Takahashi, S.; Shinzato, K.; Nakamura, S.; Domon, T.; Yamamoto, T.; Wakita, M. The roles of apoptosis and mitosis in atrophy of the rat sublingual gland. *Tissue Cell* **2002**, *34*, 297–304.

81. Takahashi, S.; Nakamura, S.; Suzuki, R.; Islam, N.; Domon, T.; Yamamoto, T.; Wakita, M. Apoptosis and mitosis of parenchymal cells in the duct-ligated rat submandibular gland. *Tissue Cell* **2000**, *32*, 457–463.

82. Kobayashi, S.; Kawakita, T.; Kawashima, M.; Okada, N.; Mishima, K.; Saito, I.; Ito, M.; Shimmura, S.; Tsubota, K. Characterization of cultivated murine lacrimal gland epithelial cells. *Mol. Vision* **2012**, *18*, 1271–1277.

83. You, S.; Avidan, O.; Tariq, A.; Ahluwalia, I.; Stark, P.C.; Kublin, C.L.; Zoukhri, D. Role of epithelial-mesenchymal transition in repair of the lacrimal gland after experimentally induced injury. *Inv. Ophthalmol. Vis. Sci.* **2012**, *53*, 126–135.

84. Zoukhri, D.; Fix, A.; Alroy, J.; Kublin, C.L. Mechanisms of murine lacrimal gland repair after experimentally induced inflammation. *Invest. Ophthalmol. Vis. Sci.* **2008**, *49*, 4399–4406.

85. Zoukhri, D.; Macari, E.; Kublin, C.L. A single injection of interleukin-1 induces reversible aqueous-tear deficiency, lacrimal gland inflammation, and acinar and ductal cell proliferation. *Exp. Eye Res.* **2007**, *84*, 894–904.

86. You, S.; Kublin, C.L.; Avidan, O.; Miyasaki, D.; Zoukhri, D. Isolation and propagation of mesenchymal stem cells from the lacrimal gland. *Invest. Ophthalmol. Vis. Sci.* **2011**, *52*, 2087–2094.

87. Tiwari, S.; Ali, M.J.; Vemuganti, G.K. Human lacrimal gland regeneration: Perspectives and review of literature. *Saudi J. Ophthalmol. Off. J. Saudi Ophthalmol. Soc.* **2014**, *28*, 12–18.

88. Tiwari, S.; Ali, M.J.; Balla, M.M.; Naik, M.N.; Honavar, S.G.; Reddy, V.A.; Vemuganti, G.K. Establishing human lacrimal gland cultures with secretory function. *PLoS One* **2012**, *7*.

89. Purnell, B. New release: The complete guide to organ repair. *Science* **2008**, *322*.

90. Kagami, H.; Wang, S.; Hai, B. Restoring the function of salivary glands. *Oral Dis.* **2008**, *14*, 15–24.

91. Oshima, M.; Mizuno, M.; Imamura, A.; Ogawa, M.; Yasukawa, M.; Yamazaki, H.; Morita, R.; Ikeda, E.; Nakao, K.; Takano-Yamamoto, T.; *et al.* Functional tooth regeneration using a bioengineered tooth unit as a mature organ replacement regenerative therapy. *PLoS One* **2011**, *6*.

92. Ikeda, E.; Morita, R.; Nakao, K.; Ishida, K.; Nakamura, T.; Takano-Yamamoto, T.; Ogawa, M.; Mizuno, M.; Kasugai, S.; Tsuji, T. Fully functional bioengineered tooth replacement as an organ replacement therapy. *Proc. Natl. Acad. Sci. USA* **2009**, *106*, 13475–13480.

93. Sato, A.; Toyoshima, K.E.; Toki, H.; Ishibashi, N.; Asakawa, K.; Iwadate, A.; Kanayama, T.; Tobe, H.; Takeda, A.; Tsuji, T. Single follicular unit transplantation reconstructs arrector pili muscle and nerve connections and restores functional hair follicle piloerection. *J. Dermatol.* **2012**, *39*, 682–687.

94. Asakawa, K.; Toyoshima, K.E.; Ishibashi, N.; Tobe, H.; Iwadate, A.; Kanayama, T.; Hasegawa, T.; Nakao, K.; Toki, H.; Noguchi, S.; *et al.* Hair organ regeneration via the bioengineered hair follicular unit transplantation. *Sci. Rep.* **2012**, *2*, 424.

95. Toyoshima, K.E.; Asakawa, K.; Ishibashi, N.; Toki, H.; Ogawa, M.; Hasegawa, T.; Irie, T.; Tachikawa, T.; Sato, A.; Takeda, A.; *et al.* Fully functional hair follicle regeneration through the rearrangement of stem cells and their niches. *Nat. Commun.* **2012**, *3*.

96. Oshima, M.; Tsuji, T. Functional tooth regenerative therapy: Tooth tissue regeneration and whole-tooth replacement. *Odontol. Soc. Nippon. Dent. Univ.* **2014**, *102*, 123–136.

97. Ogawa, M.; Tsuji, T. Fully functional salivary gland regeneration as a next-generation regenerative therapy. *Jpn J. Clin. Immunol.* **2015**, *38*, 93–100.

98. Ogawa, M.; Oshima, M.; Imamura, A.; Sekine, Y.; Ishida, K.; Yamashita, K.; Nakajima, K.; Hirayama, M.; Tachikawa, T.; Tsuji, T. Functional salivary gland regeneration by transplantation of a bioengineered organ germ. *Nat. Commun.* **2013**, *4*.

99. Rock, J.R.; Hogan, B.L. Developmental biology. Branching takes nerve. *Science* **2010**, *329*, 1610–1611.

100. Kumar, A.; Brockes, J.P. Nerve dependence in tissue, organ, and appendage regeneration. *Trends Neurosci.* **2012**, *35*, 691–699.

101. Rios, J.D.; Horikawa, Y.; Chen, L.L.; Kublin, C.L.; Hodges, R.R.; Dartt, D.A.; Zoukhri, D. Age-dependent alterations in mouse exorbital lacrimal gland structure, innervation and secretory response. *Exp. Eye Res.* **2005**, *80*, 477–491.

102. Robbins, A.; Kurose, M.; Winterson, B.J.; Meng, I.D. Menthol activation of corneal cool cells induces TRPM8-mediated lacrimation but not nociceptive responses in rodents. *Invest. Ophthalmol. Vis. Sci.* **2012**, *53*, 7034–7042.

103. Parra, A.; Madrid, R.; Echevarria, D.; del Olmo, S.; Morenilla-Palao, C.; Acosta, M.C.; Gallar, J.; Dhaka, A.; Viana, F.; Belmonte, C. Ocular surface wetness is regulated by TRPM8-dependent cold thermoreceptors of the cornea. *Nat. Med.* **2010**, *16*, 1396–1399.

104. Shimazaki, J.; Sakata, M.; Tsubota, K. Ocular surface changes and discomfort in patients with meibomian gland dysfunction. *Arch. Ophthalmol.* **1995**, *113*, 1266–1270.

105. Shimazaki, J.; Goto, E.; Ono, M.; Shimmura, S.; Tsubota, K. Meibomian gland dysfunction in patients with Sjögren syndrome. *Ophthalmology* **1998**, *105*, 1485–1488.

106. Driver, P.J.; Lemp, M.A. Meibomian gland dysfunction. *Surv. Ophthalmol.* **1996**, *40*, 343–367.

107. Mathers, W.D.; Lane, J.A. Meibomian gland lipids, evaporation, and tear film stability. *Adv. Exp. Med. Boil.* **1998**, *438*, 349–360.

108. Mathers, W.D.; Billborough, M. Meibomian gland function and giant papillary conjunctivitis. *Am. J. Ophthalmol.* **1992**, *114*, 188–192.

109. Nelson, J.D.; Shimazaki, J.; Benitez-del-Castillo, J.M.; Craig, J.P.; McCulley, J.P.; Den, S.; Foulks, G.N. The international workshop on meibomian gland dysfunction: Report of the definition and classification subcommittee. *Invest. Ophthalmol. Vis. Sci.* **2011**, *52*, 1930–1937.

110. Shimmura, S.; Ueno, R.; Matsumoto, Y.; Goto, E.; Higuchi, A.; Shimazaki, J.; Tsubota, K. Albumin as a tear supplement in the treatment of severe dry eye. *Br. J. Ophthalmol.* **2003**, *87*, 1279–1283.

111. Tsubota, K.; Goto, E.; Fujita, H.; Ono, M.; Inoue, H.; Saito, I.; Shimmura, S. Treatment of dry eye by autologous serum application in Sjögren's syndrome. *Br. J. Ophthalmol.* **1999**, *83*, 390–395.

112. Ogawa, Y.; Okamoto, S.; Mori, T.; Yamada, M.; Mashima, Y.; Watanabe, R.; Kuwana, M.; Tsubota, K.; Ikeda, Y.; Oguchi, Y. Autologous serum eye drops for the treatment of severe dry eye in patients with chronic graft-versus-host disease. *Bone Marrow Transplant.* **2003**, *31*, 579–583.

113. Lee, G.A.; Chen, S.X. Autologous serum in the management of recalcitrant dry eye syndrome. *Clin. Exp. Ophthalmol.* **2008**, *36*, 119–122.

114. Pan, Q.; Angelina, A.; Zambrano, A.; Marrone, M.; Stark, W.J.; Heflin, T.; Tang, L.; Akpek, E.K. Autologous serum eye drops for dry eye. *Cochrane Database Syst. Rev.* **2013**, *8*.

115. Ward, K.W. Superficial punctate fluorescein staining of the ocular surface. *Optom. Vis. Sci.* **2008**, *85*, 8–16.

116. Korb, D.R.; Herman, J.P.; Finnemore, V.M.; Exford, J.M.; Blackie, C.A. An evaluation of the efficacy of fluorescein, rose bengal, lissamine green, and a new dye mixture for ocular surface staining. *Eye Contact Lens* **2008**, *34*, 61–64.

117. Xiong, C.; Chen, D.; Liu, J.; Liu, B.; Li, N.; Zhou, Y.; Liang, X.; Ma, P.; Ye, C.; Ge, J.; *et al.* A rabbit dry eye model induced by topical medication of a preservative benzalkonium chloride. *Invest. Ophthalmol. Vis. Sci.* **2008**, *49*, 1850–1856.

118. Nyunt, A.K.; Ishida, Y.; Yu, Y.; Shimada, S. Topical apolipoprotein A-1 may have a beneficial effect on the corneal epithelium in a mouse model of dry eye: A pilot study. *Eye Contact Lens* **2008**, *34*, 287–292.

Peptide Amphiphiles in Corneal Tissue Engineering

Martina Miotto, Ricardo M. Gouveia and Che J. Connon

Abstract: The increasing interest in effort towards creating alternative therapies have led to exciting breakthroughs in the attempt to bio-fabricate and engineer live tissues. This has been particularly evident in the development of new approaches applied to reconstruct corneal tissue. The need for tissue-engineered corneas is largely a response to the shortage of donor tissue and the lack of suitable alternative biological scaffolds preventing the treatment of millions of blind people worldwide. This review is focused on recent developments in corneal tissue engineering, specifically on the use of self-assembling peptide amphiphiles for this purpose. Recently, peptide amphiphiles have generated great interest as therapeutic molecules, both *in vitro* and *in vivo*. Here we introduce this rapidly developing field, and examine innovative applications of peptide amphiphiles to create natural bio-prosthetic corneal tissue *in vitro*. The advantages of peptide amphiphiles over other biomaterials, namely their wide range of functions and applications, versatility, and transferability are also discussed to better understand how these fascinating molecules can help solve current challenges in corneal regeneration.

Reprinted from *J. Funct. Biomater.* Cite as: Miotto, M.; Gouveia, R.M.; Connon, C.J. Peptide Amphiphiles in Corneal Tissue Engineering. *J. Funct. Biomater.* **2015**, *6*, 687–707.

1. Introduction

The cornea is the transparent, outermost part of the eye that serves as the primary refractive organ in the visual system [1]. Diseases, traumas, or injuries are the leading causes of corneal blindness and its prevalence varies from country to country, and even from one population to another, depending on many factors such as availability and general standards of eye care. It is estimated that 180 million people worldwide have severely impaired vision in both eyes, resulting in a considerable social and economic impact [2]. Corneal disease remains a major cause of blindness, second only to cataracts. Although multi-factorial, the vast majority of corneal clinical cases would benefit from a suitable corneal replacement. However, there is currently a lack of donor cornea availability. The main factor behind this donor shortage is that, in many parts of the world, there are limitations in the storage and distribution of corneal tissue, as well as cultural and/or religious barriers [3]. Moreover, the supply of human corneal tissue is expected to diminish even further due to the increasing popularity of refractive surgery (such as LASIK), a technique that renders these corneas unsuitable for donation. However, even considering the best conditions, donor grafts are typically variable in quality and usually fail due to immunological rejection or endothelial decompensation resulting in an 18% failure rate for initial grafts [4].

In the context of these limitations, the field of corneal tissue engineering has made considerable advances in the last 10 years, focusing on alternative means of replacing damaged corneal tissue. Approaches have included development of fully artificial keratoprostheses, use of decellularized tissue scaffolding from animal or human sources, and use of acellular, cross-linked collagen constructs as corneal replacements [5,6]. Presently, bioengineered corneal substitutes are already available for experimental clinical purposes such as corneal grafting [7,8]. In addition to their clinical applications for transplantation and wound healing enhancement, these engineered tissues also represent attractive *in vitro* models of human tissues for various biological purposes. However, whilst much work is going on in this research area [9,10], this review will instead focus on ongoing studies using different biomaterials to create new corneal tissues, and more specifically, work involving peptide amphiphiles (PAs) in corneal tissue engineering. The advantages of using these biomaterials and the significant challenges involved will also be discussed, along with the many future perspectives in the field.

2. Challenges in Corneal Tissue Engineering

The final purpose of corneal tissue engineering is the fabrication of corneal tissue equivalents able to improve the function of their injured or diseased natural counterparts. However, constructing a cornea presents several challenges to the field of tissue engineering due to the very specific structural and cellular properties of the organ. Strength, shape, transparency, biocompatibility, and molecular and cellular compositions are important properties of the cornea that remain difficult to replicate *in vitro*. Moreover, assembly and recovery of the engineered corneal tissues whilst maintaining minimal manipulation before and during grafting remains an important part of the bio-fabrication process and still requires intense study and optimization.

At a macroscopic level, from an anterior to a posterior location, the human cornea is composed of a non-keratinized multi-layered epithelium, the Bowman's membrane, a 0.5 mm-thick stroma, the Descemet's membrane, and an endothelium [11] (Figure 1). The stroma accounts for 90% of the volume of the cornea, and is essential to support the mechanical and refractive properties of the organ. These properties are based on the ultrastructural organization of the stroma's extracellular matrix, comprised by a pseudocrystalline lattice of highly-ordered collagen fibers and proteoglycans, and sparsely populated by quiescent stromal cells, the keratocytes. This arrangement plays a fundamental role in the structure and function of the cornea. Specifically, the orderly array of collagen fibers and the refractive index matching of these fibrils by interstitial proteoglycans play a significant role in the transparency of the tissue [12]. Stromal collagen type-I fibers have a 20–35 nm diameter, and are aligned parallel to each other with regular 30-nm spacing between fibrils. This regular spacing is thought to be regulated by stromal-specific proteoglycans, which have been observed to form ring-like structures around collagen fibrils in the normal cornea [13]. The aligned fibers are grouped into layers called lamellae, which are stacked in an alternating lattice [14]. The thickness of the stroma and arrangement of the collagen fibers are optimal for light transmission through the cornea, with minimal light scatter.

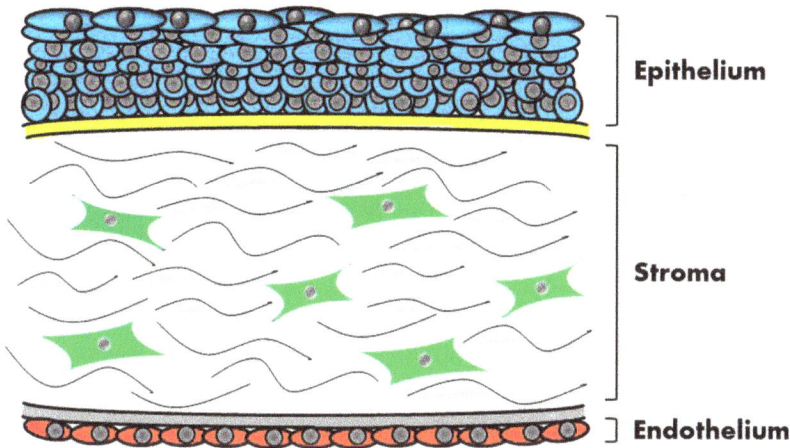

Figure 1. The human central cornea in cross-section. The outer-, anterior-most surface of the cornea comprises a non-keratinized, multi-layered epithelium (*blue*) supported by a basement membrane and above the Bowman's layer (*yellow*). The middle stromal tissue comprises 90% of the cornea's thickness, and is sparsely populated with keratocytes (*green*) interspersed within approximately 200 lamellae of dense, collagen- and proteoglycan-rich extracellular matrix (*lines*). The innermost posterior tissue consists of a single layer of endothelial cells (*red*) supported by the Descemet's membrane (*grey*).

Corneal transparency also has a cellular contribution. Keratocytes express certain proteins, known as corneal crystallins, which are thought to match the refractive index of the cell cytoplasm with the surrounding matrix material [15]. In addition, the cornea is avascular, a property maintained by its anti-angiogenic milieu [16] which, if eliminated, can lead to blood vessel ingrowth and loss of transparency [17]. Moreover, the overall structure of the stroma is dependent on its hydration state, a feature regulated by the corneal epithelial [18] and endothelial layers [19]. Furthermore, the correct development and function of these tissue layers are dependent on the presence of a well-developed network of nociceptive neurons [20]. From this perspective, any attempts to engineer corneal tissues must take into consideration the multi-cellular nature of the cornea, and the intricate direct and indirect interaction maintained between the various tissues.

3. Previous Approaches to Engineer Corneal Tissue: Top-Down or Bottom-Up?

In recent years, the development of tissue-engineered corneal substitutes emerged as an alternative method to overcome several issues related to corneal transplantation, namely the relatively high host-rejection rate of keratoprosthesis. In this context, different corneal stromal equivalents have been developed using various biomaterials such as decellularized corneas [21], amniotic membrane [22] or scaffolds produced from collagen type I [23], fibrin agarose [8,24–26], fish scales [27], chitosan [28], caprolactone [29], or poly(ester urethane) urea [30]. For instance, Du *et al.* [22] used amniotic membrane as a biomaterial upon which human corneal epithelium stem cells were tested as therapy for limbal stem cell deficiency. Among all the materials used for the production of biocompatible scaffolds, collagen-based constructs seem to be the most interesting. A number of examples can be seen in the work of Griffith and co-workers, where considerable effort was made to create and optimize scaffolds produced from cross-linked collagen [31], recombinant human collagen [32], and bio-functionalized collagen [33] for corneal tissue engineering purposes. Although many of these approaches are currently being tested in the clinic, none has had the broad success and acceptance of fresh tissue transplantation. The reason for this discrepancy might be due to the different strategies used to produce bioengineered tissues in general, and corneal tissue equivalents in particular.

Traditional tissue engineering typically employs what is called a top-down approach. This is based on the use of scaffolds, necessarily biocompatible and optionally biodegradable, to recreate the appropriate microarchitecture of the natural tissue and serve as support for cell attachment and growth. Cells seeded on such materials are expected to populate them while maintaining their native phenotype, and use these scaffolds as support while creating a suitable growth environment (e.g., by depositing their own extracellular matrix). Theoretically, a 3D scaffold with a precise shape, composition and internal organization can provide a perfect microenvironment allowing the organization of individual cells into a functional tissue [34]. However, the design *a priori* of scaffolds with mechanical, physiochemical and biological properties ideal for a specific tissue has not been yet realized, and is probably beyond current knowledge and technology. On the other hand, bottom-up approaches are emerging as an alternative for creating highly organized tissues, and using these modular units as building blocks to engineer biological tissues. These modular units can be fabricated using different methods such as self-assembled aggregation [35], microfabrication of cell-laden hydrogels [36] and extracellular matrix [37], overlapping of cell sheets [38], or direct printing of tissues [39]. Once bio-fabricated, these blocks can be stacked, assembled, or combined to form larger tissues or whole organs [40]. Commonly, the bottom-up approaches aim at providing cells with a guiding template to direct cell-driven organization and tissue formation. In other words, cells are instructed to recapitulate natural tissue differentiation, growth and morphogenesis *in vitro*. These strategies have allowed the creation of modular tissues with native-type composition and micro-architecture, without the need to introduce scaffolds and with better perceived outcomes in downstream applications (e.g., grafting). The difference between top-down and bottom-up strategies constitutes an important topic for future approaches to corneal tissue engineering, as discussed in a recent review article focused on the subject [41].

4. Peptide Amphiphiles in Tissue Engineering

Recently, small bioactive molecules capable of self-assembly have attracted considerable interest as new functional materials with broad applications in tissue engineering and regenerative medicine [42–44]. Specifically, these are self-assembling molecules used to produce biocompatible materials for three dimensional cell culture [45,46], drug delivery [47,48], inhibition of bacterial growth [49,50], delivery of therapeutic molecules [51], or as scaffolds for cell therapy [52–55]. Concerning their use as delivery systems, it is important to understand if and how supramolecular nanostructures can cross the diffusion barriers present in the human body such as the blood-brain barrier or, relevant to corneal applications, the corneal epithelial or endothelial layers. One of the most promising types of such molecules comprises small synthetic peptides. These molecules incorporate small bioactive or bio-inspired peptide sequences, with several advantages over the use of whole-protein matrixes, including sourcing (*i.e.*, easier isolation/production and purification), reduced immunogenicity [56], presentation (*i.e.*, more effective and controlled density and orientation of the bioactive motives [57]), and stability [58]. In addition, they can be rationally designed to have amphiphilic characteristics, *i.e.*, to contain both hydrophilic and hydrophobic domains that help them self-assemble into a variety of supramolecular 3D nanostructures, such as tubes, tapes, fibres, vesicles and micelles, among other architectures [52,59–61] (Figure 2). However, despite this variability, or maybe because of it, there are currently no set rules for this rational design. In other words, there is still much work to be done regarding the development of a supramolecular code that will allow us to predict the self-assembly of hierarchical architectures and bio-function based solely on the primary structure of amphiphilic peptides [62].

Amphiphilic peptides can be classified as peptide sequences with amphiphilic properties arising from hydrophobic and hydrophilic residues, whereas peptide amphiphiles (PAs) constitute a subset of the former comprising a peptide sequence linked to a hydrophobic tail [63]. PAs can be easily synthesized by standard peptide synthesis protocols by standard solid phase chemistry that ends with the alkylation of the NH$_2$ terminus of the peptide; their structural folding and stability have been extensively characterized [64,65]. However, and although small and medium-sized peptides are easily obtained in high yields, large peptide sequences (*i.e.*, longer than 50 amino acids) are still difficult to produce and purify by direct chemical synthesis [66]. An example of a representative PA contains three segments: a hydrophobic sequence, commonly a lipid chain that guides aggregation through hydrophobic collapse, a β-sheet-forming peptide that promotes nanofiber formation through the formation of hydrogen bonds, and a peptide segment, usually less than 15 amino acids long, with ionisable side chains and a bio-functional amino acid sequence [67]. In this review we will give specific attention to the characteristics and applications of such PAs.

The self-assembly mechanism involved in these single-tailed PAs usually occurs after changes in pH [68], mixing of oppositely charged PAs [69], or addition of multivalent cations [54] to generate electrostatic repulsion between the PA molecules. The supramolecular self-assembly of PAs in aqueous environments is governed by at least three major forces: the interactions between the hydrophobic tails, the electrostatic repulsions between charged groups, and the hydrogen bonding among the middle peptide segments [52]. The derived ultrastructure of self-assembled PAs reflects a balance of each force contribution, and has dimensions similar to those of fibrils from natural extracellular matrix. Specifically, taking advantage of their amphiphilic properties, the hydrophobic alkyl tails of PAs solubilized in aqueous solutions are packed in the center of the fiber while the hydrophilic peptide segments are exposed to the aqueous environment, forming an external corona. As such, these molecules can be designed to display bioactive epitopes at the surface of the self-assembled nanostructure, while keeping intermolecular hydrogen bonds parallel to the long axis of the fiber [52]. To date, a considerable number of PAs have been reported in the literature [70], including molecules with different hydrophobic tails [71–73]. For example, PAs comprised of a similar peptide sequence but with either saturated or diene-containing hexadecyl lipid chains self-assemble into polydisperse nanotapes or spherical micelles, respectively [44,74]. These examples illustrate the versatility of PAs, where increasing unsaturation and length of the lipid chains, or changing from alkyl to aromatic tails dramatically alters the final architecture of the self-assembled nanostructures [75]. However, this feature might compromise the stability, physical properties, and function of the PA, namely when exposed to UV [76] and γ-irradiation [77], two common sterilization methods used in materials for biological applications. Moreover, the concentration in which these molecules are used may constitute an important factor defining their self-assembled architecture [78] and bio-compatibility [42].

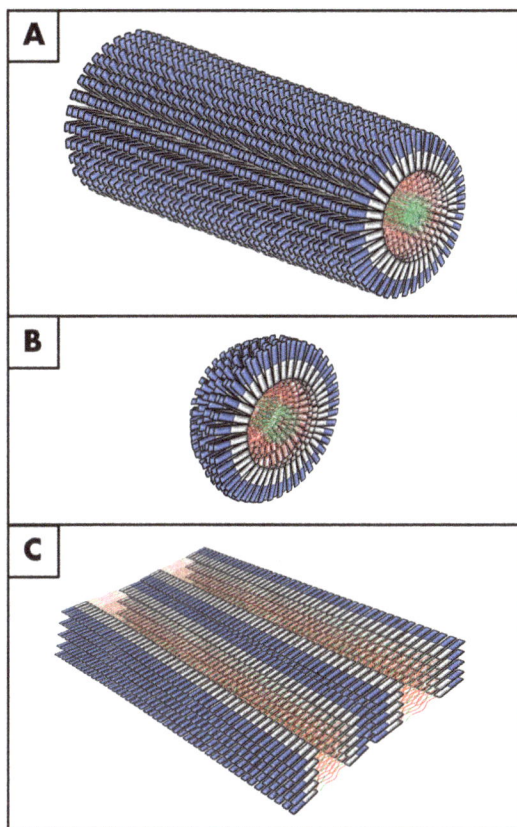

Figure 2. Examples of supramolecular PA nanostructures. Graphical representation of some representative structures obtained by PAs self-assembly: (**A**) nanofibers; (**B**) micelles; and (**C**) multi-layered nanotapes. All three structures have a hydrophilic outer corona comprised of bioactive peptide (*blue*) and self-assembly-inducing/spacer sequence (*white*), and a hydrophobic inner core with organized and/or non-organized PA tails (*red* and *green*, respectively) (*adapted from* [63,79]).

Some of the main applications of PAs in regenerative medicine are summarized in Table 1. As reported in Table 1, PAs can be used in different forms, such as in coatings, in solution, or as hydrogels. All of these forms have been quite extensively tested, however, in spite of the advantages of using PAs in hydrogel form, it has to be considered that they exhibit poor mechanical characteristics [80]. In this context, the use of PAs as hydrogels is less suited to the production of scaffolds for engineering tissues requiring high mechanical strength and integrity. Considering the application of PAs in the field of tissue engineering, these molecules have been used by several groups towards the development of engineered constructs, particularly for the regeneration of connective tissues with a collagen-rich extracellular matrix. In order to achieve this objective using scaffold-based approaches, an artificial PA scaffold should mimic the structure and biological function of the native extracellular matrix as much as possible, both in terms of chemical cues and physical and mechanical properties. The native extracellular matrix provides structural support to body tissues, acting not only as a physical framework for arranging cells within the connective tissues, but also as a dynamic and flexible substance defining cellular behaviour and tissue function [81]. Indeed, it has been shown that the supramolecular network formed by self-assembled PA mimics, from a structural point of view, the natural extracellular matrix, albeit in a simplified way. In this context, the main applications of PAs molecules, so far, have been in the repair of bone [82,83], cartilage and tendon [84], blood vessels [85,86], cavernous nerves [87], skin [88,89], and importantly for this review, the cornea [44,74].

5. Peptide Amphiphiles as Versatile Templates to Recreate Human Corneas *in vitro*

Various biomaterials have been explored for use in tissue engineered corneal substitutes, including, but by no means limited to, collagen [23,90,91], fibrin-agarose [8], decellularised cornea [92], and amniotic membrane [93,94]. Considering the high transparency of the corneal tissue, it is fundamental to design a scaffold preserving this characteristic whilst maintaining a high biocompatibility and a low immunogenicity [95]. In addition, there is a strong need to develop novel bioactive materials able to support cell adhesion and proliferation. In this context, PAs are well placed in that they can be designed to support a range of cell types important to corneal function, *i.e.* keratocytes, epithelial and endothelial cells. In this case the use of PAs would not only support the formation of extracellular matrix-inspired nanofibers following the self-assembly, but also enhance adhesion, proliferation, and alignment of human corneal stromal fibroblasts due to the insertion of specific bioactive motives. Notable research involving PAs in cornea tissue engineering are reported in Table 2.

Table 1. Main works involving the use of peptide amphiphiles in regenerative medicine. The table reports the PA used, its chemical structure, the aim of the studies, the source of the bioactive sequence, the concentration of PA used, and its form.

PA	Chemical Structure	Aim	Source	[PA] wt %	PA Form	Reference
C_{16}-C_4-G_3-S-RGD		Bone regeneration	Fibronectin	0.1	coating	[83]
C_{12}-HSNGLPLGGGS EEEAAAVVV(K)		Cartilage regeneration	De novo synthetized	1	hydrogel	[84]
C_{16}-V_2-A_2-E_2-NH_2		Cavernous nerve regeneration	De novo synthetized	0.85	hydrogel	[87]
C_{16}-C_4-G_3- LRKKLGKA		Blood vessels regeneration	Heparin binding consensus sequence	3	hydrogel	[85]
C_{16}-KTTKS		Skin regeneration	Procollagen I	0.0003	solution	[88]

229

Table 2. Main works involving the use of peptide amphiphiles in corneal tissue engineering. The table reports the PA used, the source of the bioactive sequence, the PA biological effect, the concentration of PA used, and its form.

PA	Source	Biological Effect	[PA] wt %	PA Form	Reference
C_{16}-G_3-RGD/RGDS + C_{16}-ETTES	Fibronectin	Enhanced adhesion and proliferation of hCSFs	1 to 0.005	coating	[42,96]
A_6-RGDS	Fibronectin	Enhanced adhesion and proliferation of hCSFs	1 to 0.1	coating	[97]
C16-TPGPQGIAGQ-RGDS	MMP cleavage sequence + Fibronectin	Promoted adhesion and growth of hCSFs. Stimulated collagen production. Governed tissues lift-up	2	coating	[44,74]
Fmoc-RGDS	Fibronectin	Enhanced cell attachment, proliferation and viability	1	solution	[43]
C_{16}-KTTKS	Procollagen I	Stimulated collagen production from hCSFs	0.002, 0.004, 0.008	solution	[98,99]
C_{12}-VVAGKYIGSR	Laminin	Enhanced keratocyte proliferation and migration, and stimulated collagen I synthesis	0.2	coating	[100]
C_{16}-YEALRVANEVTLN	Lumican	Stimulated collagen I production	0.01, 0.005, 0.0025 0.00125, 0.000625	solution	[101]

For corneal tissue engineering one of the most used bioactive sequences is represented by the fibronectin derived RGD motive. As extensively reported, this sequence is frequently used for creating adhesive biomaterials able to promote cell interaction and adhesion through binding to integrin subgroups like those composed of subunits αV, $\alpha 3\beta 1$, and $\alpha 5\beta 1$ [102,103]. However, even if the tripeptide motif RGD has been identified as a minimal essential cell adhesion sequence [104], the definition of the *minimo optimo* bioactive epitope required in order to have the same function as the whole protein, is still not known. In this respect, Castelletto *et al.* [96] studied the self-assembly of two PAs designed *ad hoc* to enhance the potential for cell attachment. These peptides contained RGD or RGDS bioactive motives as well as a functional spacer in the β-sheet domain, consisting of a sequence of three consecutive glycine (G) residues. In addition, they optimized the potential for cell attachment by co-assembling the bioactive PA (RGD or RGDS) with a non-bioactive PA acting as a diluent molecule able to vary the RGD density within its supramolecular form following self-assembly above critical aggregation concentration. In particular, they used a negatively charged PA composed of C_{16}-Glu-Thr-Thr-Glu-Ser (C_{16}-ETTES) as this diluent. Castelletto *et al.* posited that by tuning the diluent concentration, so the distance between neighbouring RGD groups would be greater than that in the undiluted RGD(S)-PA following self-assembly, they could optimize the PA for maximal cell attachment. When such a PA mixture was subsequently used as a coating for 2D human keratocyte attachment and growth, the optimal molar ratio of C_{16}-RGD(S):C_{16}-ETTES was found to be 15:85. Subsequently, this approach was employed by Gouveia *et al.* [42] to produce stable biocompatible film coatings which enhanced adhesion, proliferation and alignment of human corneal stromal fibroblasts whilst inducing the formation of 3D lamellar-like stromal tissue. These early results suggested that such mixtures of PAs constitute a promising new material capable of directing corneal stromal cells to produce appropriate amounts and type of extracellular matrix that are likely to be important to both corneal wound healing and tissue engineering. However, the biophysical, mechanical, and biological properties of these functional coatings require further extensive research. Castelletto *et al.* [97] also demonstrated the ability to produce bioactive film coatings for corneal stromal cell growth using an alanine-rich amphiphilic peptide containing the RGD motive (A_6-RGD). This PA was designed to simultaneously ensure solubility in water and specific binding to cells. They found that the self-assembly motive depended on the concentration of surfactant-like peptide (SLP) and demonstrated the co-existence of vesicles and fibres with an increase in vesicle population relative to fibres when the SLP concentration increases. Moreover, at low concentrations (0.1 wt % –1.0 wt %), this SLP promoted adhesion and enhanced proliferation of human corneal stromal cells. Thus, A_6-RGD PA represents another promising bioactive peptide for the manufacture of dry coatings for cornea tissue engineering.

More recently, a study by Gouveia *et al.* [74] described an inventive way to take advantage of PA film coatings for corneal tissue engineering. In this work, they described a biologically interactive PA coating that integrated both the abilities to induce tissue bio-fabrication and the subsequent tissue self-release *in vitro*. This was made achievable by employing a custom designed PA comprised a matrix metalloprotease (MMP)-cleavable sequence (Thr-Pro-Gly-Pro-Gln-Gly-Ile-Ala-Gly-Gln) followed by the RGDS bioactive motive which, as before, was mixed with C_{16}-ETTES at a molar ratio of 15:85. The subsequent self-assembly studies revealed that the PA self-assembled into nanotapes that were also capable of forming film coatings providing a stable surface for the attachment and growth of human corneal stromal cells in a quiescent phenotype. Furthermore the supplementation of all-*trans* retinoic acid (RA) to the culture media facilitated both retained cellular attachment and ultimately tissue formation via cell stratification. Previously Gouveia and Connon [105] had shown that the addition of RA inhibits the expression of several MMPs from the cells while enhancing their native extracellular matrix production. Thus, in the presence of the MMP-sensitive PA coating, the cells increased MMP expression and endogenous proteolytic activity following RA removal. The resulting increase in proteolytic activity in the culture supernatant cleaved the RGD peptide from the self-assembled PA (*i.e.*, the nanotape structures underpinning the tissue growth) facilitating the complete detachment of the tissue from the bioactive surface, and creating a free-floating construct (Figure 3). As such, this smart PA material represents a new and fascinating method for the bio-fabrication of certain structural tissues including corneal stromal equivalents.

PAs have not only been shown to augment the amounts of extracellular matrix produced by corneal stromal cells. For instance, PAs were also used to successfully control the form and shape of corneal tissue-engineered constructs. It is known that during embryonic development, corneal fibroblasts play a central role in exerting physical forces that organize the extracellular matrix into a unique pattern providing structural support whilst maintaining tissue transparency [106–110]. Furthermore, corneal fibroblasts guide wound contraction and matrix remodeling after injury or refractive surgery [14,111,112]. In the physiological process of corneal wound healing, quiescent corneal keratocytes switch to the fibroblast phenotype and migrate through the tridimensional matrix to restore it. In order to investigate cell-matrix mechanical interactions during fibroblast migration, Petroll *et al.* [113,114] developed a model in which cell-seeded compressed collagen matrices are nested within acellular uncompressed matrices. They found that matrices cultured in medium containing exogenous serum become significantly deformed due to keratocytes' activation and migration into the outer matrix [115]. Similar levels of matrix contraction can disrupt or damage the unique and functional architecture of the corneal stroma leading to the formation of scars or fibrosis. Therefore, a method able to limit it or to stimulate a low contractility migration is highly required. In this respect, a new approach for the containment of the collagen contraction by corneal fibroblasts has been developed through the novel use of PAs. In this example, again by Gouveia *et al.* [43], hybrid materials were constructed comprising fibrillar collagen type I and a PA formed from Fmoc-RGDS. Once encapsulated within this mixture, and exposed to serum, human keratocytes were unable to contract the collagen gel as they would normally (Figure 4). It is believed that this hybrid collagen PA system forms an interpenetrating "gel within a gel" and that the cells preferential bind to the Fmoc-RGDS motif inhibiting the cells from binding to and contracting the collagen gel in the presence of serum. Thus the degree of tissue contraction and shape was controlled by the concentration of the Fmoc-RGDS PA within the system. This interesting study opens up the possibility of using PAs to control localized effects on cells in a three-dimensional environment, thereby lifting the veil on much more complicated tissue-engineered constructs rather than the homogenous forms that are prevalent today.

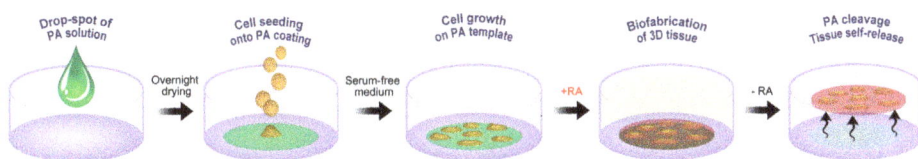

Figure 3. Bio-fabrication and controlled self-release of live tissues using PA coating templates. Schematic representation of the method used for the *in vitro* bio-fabrication and lift-off of human corneal stromal tissues, previously reported in [74]. Cells isolated from human donors were seeded and grown on low-attachment plates previously coated with a PA carrying both the MMP1-sensitive sequence and the RGDS cell adhesion motive. Cells were cultured in serum-free medium containing retinoic acid (RA) for 90 days and accumulated large quantities of corneal-specific stromal extracellular matrix. Subsequently, the bio-fabricated tissues were induced to express MMPs due to RA removal from the medium. In three days, the tissues expressed enough endogenous MMPs into the culture supernatant to provide the cue to degrade the adhesive PA coating, and induce their own release. The resulting free-floating corneal stromal equivalents were scaffold-free, easy to handle, and retained their shape and structural integrity for more than 18 months in storage.

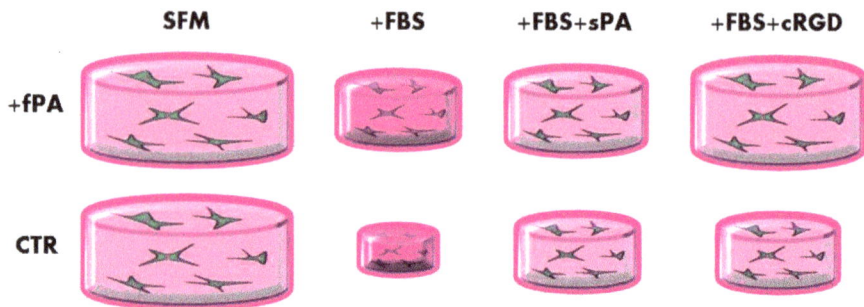

Figure 4. Schematic representation of the effect of Fmoc-RGDS PA on collagen gel contraction under different culture conditions. Human corneal stromal fibroblasts were encapsulated within uncompressed collagen gels that have been functionalized with the fibril-forming Fmoc-RGDS PA (+fPA), or produced without it (CTR). The relative contraction of collagen gels after seven days in serum-free medium (SFM) was negligible, but significantly minimized by the presence of the structural PA in serum-containing media alone (+FBS) or supplemented with 50 μM of soluble PA (+sPA) or cyclic RGD peptide (+cRGD) (adapted from [43]).

Another PA showing promising application in corneal wound repair is represented by C_{16}-KTTKS. This particular peptide is used in anti-wrinkle cosmeceutical applications under the trade name of Matrixyl. In the first published study on the effects of this commercially available PA, Castelletto *et al.* [98] investigated the self-assembly of this PA in aqueous solution, observing tape-like nanostructures with a broad distribution of widths and reporting that the internal structure of these nanotapes comprised PA bilayers. Subsequently, the work of Jones *et al.* [99] showed that this peptide was able to stimulate collagen production from human corneal stromal cells (as well as dermal fibroblasts) in a concentration-dependent manner. This finding paved the way for a more recent investigation using another potent extracellular matrix stimulatory peptide sequence in the form of a PA. In this case the bioactivity of a PA presenting a lumican-derived bioactive sequence (C_{16}-YEALRVANEVTLN) was studied [101]. Lumican is a proteoglycan playing a structural role through binding to fibrillar collagens and modulating fibril formation whilst regulating interfibrillar spacing. The choice of this specific sequence is due to the fact that previous studies have shown it to have matrikine properties [116] such as creating chemokine gradients [117] and promoting the healing of corneal epithelial wounds [118]. Interestingly, this PA self-assembled in twisting and curving tape-like structures, on the borderline between nanotape and fibril structures. Regarding its functionality, this PA has been shown to stimulate collagen production in a concentration-dependent manner, indicating its potential use in tissue engineering and bio-fabrication. A further highly-represented protein family in the extracellular matrix is that of laminins. It is well known that laminins trigger and control many cellular functions [119], rendering them suitable candidates for the design of new bioactive PAs. In this respect, Uzunalli *et al.* [100] reported an interesting study regarding the application of a PA carrying a laminin-derived sequence for corneal stroma regeneration. In particular, this peptide contained the sequence YIGSR derived from laminin β-chain, which regulates cell adhesion through binding to the laminin binding protein (LBP). Their results demonstrated that the PA containing the laminin-derived sequence enhanced cell proliferation, keratocyte migration, and collagen I synthesis, both from *in vitro* and *in vivo*, and to a greater extent compared to the more commonly used RGD-PA. For this reason, this specific bioactive sequence should be considered (alongside the others previously mentioned) in future studies involving corneal stroma bio-fabrication and regeneration.

6. Future Perspectives

It can be argued that PAs are "eclectic" molecules, a notion related to the fact that: (1) these molecules can be used in different forms (e.g., as colloidal solutions or hydrogels, as well as thin coating films); (2) PAs can self-assemble in distinct supramolecular structures depending on the molecule's chemical structure; (3) PAs can have a wide range of biological activities depending on the type and number of bio-functional motives incorporated; (4) PAs can be easily designed and used by researchers with limited expertise in synthetic and organic chemistry; and (5) PAs can be used for various and specific applications in multiple fields, including biochemistry, stem cell biology, biotechnology, and regenerative medicine. Upon evaluating the state-of-the-art on the application of PAs in corneal tissue engineering, it is evident that intriguing and promising results have been achieved recently. In this regard, the potential next step would involve studies concerning the response and the mechanism of integration of these constructs in animal models, eventually followed by clinical trials. From this review, the possibility of abandoning scaffolds (*i.e.*, the top-down approach) also emerges . Indeed, it has been shown that the bio-fabrication of corneal stromal tissue can be achieved entirely by human keratocytes instructed by a smart, multi-functional PA coating template. Despite the challenges involved in this work, current research constitutes the foundation and inspiration for the bio-fabrication of other corneal tissues. With such a premise, it is reasonable to conduct further studies in order to further produce multi-cellular, multi-layered, whole-thickness corneas *in vitro*. Moreover, the production of such artificial constructs paves the way for a wider range of applications. For instance, bio-fabricated corneal constructs could be used in pharmacological studies as models to test the effect of new drugs rather than using animals. Indeed, the development of alternative methods that avoid or significantly reduce the use of animals for scientific purposes is still an important issue. In addition, PA templates could be designed to bio-fabricate both corneal and other connective tissues using other cell sources, such as mesenchymal or induced pluripotent stem cells, instead of adult cells. This strategy would be useful for creating multi-cellular constructs in one step, possibly through the use of patterned, multi-component PA systems with numerous specific bioactive peptide motives designed to affect individual cell types. Moreover, since the native ECM proteins are exceptionally multifunctional while PAs typically only possess one bioactive sequence, it would be extremely interesting to design and extensively test the self-assembling and function of PAs carrying more than one epitope, or combine different bioactive PAs molecules in the same system.

7. Conclusions

The range of applications for PA-based nanomaterials is steadily expanding, with emerging uses as modulators of extracellular matrix production, and as templates of increasingly complex structure and composition. This wide range of purposes is rooted in the fact that PAs can be rationally designed, and where the primary, secondary, and tertiary structures of PAs depend on the *a priori* aim of the study. Moreover, the use of PAs is continuously evolving, with new discoveries on the function of specific peptide motifs and advances in understanding supramolecular chemistry facilitating the development of novel molecules. Although there are several studies involving the use of PAs in biological systems, few are focusing on their application in corneal tissue engineering and repair. This review has highlighted this nascent but potentially useful specialized field, and demonstrates what can be achieved from interdisciplinary collaboration among researchers from the physical, chemical, biological, and clinical sciences. In this context, the current PA studies represent an important step in the change of paradigm for tissue engineering and regenerative medicine.

Acknowledgments: This work was supported by BBSRC grant BB/I008187/1 and EPSRC grant EP/L020599/1.

Author Contributions: The manuscript was written through the contributions of all the authors. All authors have given approval to the final version of the manuscript.

Conflicts of Interest: The authors declare no conflict of interest.

References

1. Ruberti, J.W.; Zieske, J.D. Prelude to corneal tissue engineering—Gaining control of collagen organization. *Prog. Retin. Eye Res.* **2008**, *27*, 549–577.
2. Whitcher, J.P.; Srinivasan, I.M.; Upadhyay, M.P. Corneal blindness: A global perspective. *Bull. World Health Organ.* **2001**, *79*, 214–221.
3. Trinkaus-Randall, V. Principles of tissue engineering. *Cornea* **2000**, 471–491.

4. Thompson, R.W.J.; Price, M.O.; Bowers, P.J.; Price, F.W.J. Long-term graft survival after penetrating keratoplasty. *Ophthalmology* **2003**, *110*, 1396–1402.

5. McLaughlin, C.R.; Tsai, R.J.; Latorre, M.A.; Griffith, M. Bioengineered corneas for transplantation and *in vitro* toxicology. *Front. Biosci.* **2009**, *14*, 3326–3337.

6. Fagerholm, P.; Lagali, N.S.; Merrett, K.; Jackson, W.B.; Munger, R.; Liu, Y.; Polarek, J.W.; Söderqvist, M.; Griffith, M. A biosynthetic alternative to human donor tissue for inducing corneal regeneration: 24-month follow-up of a phase 1 clinical study. *Sci. Transl. Med.* **2010**, *2*, 46–61.

7. Liu, W.G.; Deng, C.; McLaughlin, C.R.; Fagerholm, P.; Lagali, N.S.; Heyne, B.; Scaiano, J.C.; Watsky, M.A.; Kato, Y.; Munger, R.; *et al.* Collagen-phosphorylcholine interpenetrating network hydrogels as corneal substitutes. *Biomaterials* **2009**, *30*, 1551–1559.

8. Alaminos, M.; Del Carmen Sanchez-Quevedo, M.; Munoz-Avila, J.I.; Serrano, D.; Medialdea, S.; Carreras, I.; Campos, A. Construction of a complete rabbit cornea substitute using a fibrin-agarose scaffold. *Investig. Ophthalmol. Vis. Sci.* **2006**, *47*, 3311–3317.

9. Levis, H.J.; Kureshi, A.K.; Massie, I.; Morgan, L.; Vernon, A.J.; Daniels, J.T. Tissue engineering the cornea: The evolution of raft. *J. Funct. Biomater.* **2015**, *6*, 50–65.

10. Chaurasia, S.S.; Lim, R.R.; Lakshminarayanan, R.; Mohan, R.R. Nanomedicine approaches for corneal diseases. *J. Funct. Biomater.* **2015**, *6*, 50–65.

11. Utheim, T.P. Limbal epithelial cell theraphy: Past, present, and future. In *Corneal Regenerative Medicine: Methods and Protocols*; Wright, B.C., Connon, C.J., Eds.; Humana Press: New York, NY, USA, 2013; Volume 1014, pp. 3–43.

12. Goldman, J.N.; Benedek, G.B.; Dohlman, C.H.; Kravitt, B. Structural alterations affecting transparency in swollen human corneas. *Investig. Ophthalmol.* **1968**, *7*, 501–519.

13. Müller, L.J.; Pels, E.; Schurmans, L.R.H.M.; Vrensen, G.F.J.M. A new three-dimensional model of the organization of proteoglycans and collagen fibrils in the human corneal stroma. *Exp. Eye Res.* **2004**, *78*, 493–501.

14. Jester, J.V.; Petroll, W.M.; Cavanagh, H.D. Corneal stromal wound healing in refractive surgery: The role of the myofibroblast. *Prog. Retin. Eye Res.* **1999**, *18*, 311–356.

15. Jester, J.V.; Moller-Pedersen, T.; Huang, J.; Sax, C.M.; Kays, W.T.; Cavangh, H.D.; Petroll, W.M.; Piatigorsky, J. The cellular basis of corneal transparency: Evidence for "corneal crystallins". *J. Cell Sci.* **1999**, *112*, 613–622.

16. Azar, D.T. Corneal angiogenic privilege: Angiogenic and antiangiogenic factors in corneal avascularity, vasculogenesis, and wound healing. *Trans. Am. Ophthalmol. Soc.* **2006**, *104*, 264–302.

17. Qazi, Y.; Wong, G.; Monson, B.; Stringham, J.; Ambati, B.K. Corneal transparency: Genesis, maintenance and dysfunction. *Brain Res. Bull.* **2010**, *81*, 198–210.

18. Nakamura, K. Interaction between injured corneal epithelial cells and stromal cells. *Cornea* **2003**, *22*, S35–S47.

19. Joice, N.C. Proliferative capacity of the corneal endothelium. *Prog. Retin. Eye Res.* **2003**, *22*, 359–389.

20. De Castro, F.; Silos-Santiago, I.; Lopez de Armentia, M.; Barbacid, M.; Belmonte, C. Corneal innervation and sensitivity to noxious stimuli intrka knockout mice. *Eur. J. Neurosci.* **2010**, *10*, 146–152.

21. Gonzalez-Andrades, M.; De la Cruz Cardona, J.; Ionescu, A.M.; Campos, A.; Del Mar Perez, M.; Alaminos, M. Generation of bioengineered corneas with decellularized xenografts and human keratocytes. *Investig. Ophthalmol. Vis. Sci.* **2011**, *52*, 215–222.

22. Du, Y.; Chen, J.; Funderburgh, J.L.; Zhu, X.; Li, L. Functional reconstruction of rabbit corneal epithelium by human limbal cells cultured on amniotic membrane. *Mol. Vis.* **2003**, *9*, 635–643.

23. Mi, S.; Chen, B.; Wright, B.; Connon, C.J. Ex vivo construction of an artificial ocular surface by combination of corneal limbal epithelial cells and a compressed collagen scaffold containing keratocytes. *Tissue Eng. Part A* **2010**, *1*, 2091–2100.

24. Ionescu, A.M.; de la Cruz Cardona, J.; Gonzalez-Andrades, M.; Alaminos, M.; Campos, A.; Hita, E.; del Mar Perez, M. UV absorbance of a bioengineered corneal stroma substitute in the 240–400 nm range. *Cornea* **2010**, *29*, 895–898.

25. Ionescu, A.M.; Alaminos, M.; de la Cruz Cardona, J.; de Dios García-López Durán, J.; Gonzalez-Andrades, M.; Ghinea, R.; Campos, A.; Hita, E.; del Mar Pérez, M. Investigating a novel nanostructured fibrin-agarose biomaterial for human cornea tissue engineering: Rheological properties. *J. Mech. Behav. Biomed. Mater.* **2011**, *4*, 1963–1973.

26. De la Cruz Cardona, J.; Ionescu, A.M.; Gomez-Sotomayor, R.; Gonzalez-Andrades, M.; Campos, A.; Alaminos, M.; del Mar Pérez, M. Transparency in a fibrin and fibrin-agarose corneal stroma substitute generated by tissue engineering. *Cornea* **2001**, *30*, 1428–1435.

27. Lin, C.C.; Ritch, R.; Lin, S.M.; Ni, M.H.; Chang, Y.C.; Lu, Y.L.; Lai, H.J.; Lin, F.H. A new fish scale derived scaffold for corneal regeneration. *Eur. Cell. Mater.* **2010**, *19*, 50–57.

28. Liang, Y.; Liu, W.; Han, B.; Yang, C.; Ma, Q.; Zhao, W.; Rong, M.; Li, M. Fabrication and characters of a corneal endothelial cells scaffold based on chitosan. *J. Mater. Sci. Mater. Med.* **2011**, *22*, 175–183.

29. Sharma, S.; Mohanty, S.; Gupta, D.; Jassal, M.; Agrawal, A.K.; Tandon, R. Cellular response of limbal epithelial cells on electrospun poly-epsilon-caprolactone nanofibrous scaffolds for ocular surface bioengineering: A preliminary *in vitro* study. *Mol. Vis.* **2011**, *17*, 2898–2910.

30. Wu, J.; Du, Y.; Watkins, S.C.; Funderburgh, J.L.; Wagner, W.R. The engineering of organized human corneal tissue through the spatial guidance of corneal stromal stem cells. *Biomaterials* **2012**, *33*, 1343–1352.

31. Liu, W.; Gan, L.; Carlsson, D.J.; Fagerholm, P.; Lagali, N.; Watsky, M.A.; Munger, R.; Hodge, W.G.; Priest, D.; Griffith, M. A simple, cross-linked collagen tissue substitute for corneal implantation. *Investig. Ophthalmol. Vis. Sci.* **2006**, *47*, 1869–1875.

32. Liu, W.; Merrett, K.; Griffith, M.; Fagerholm, P.; Dravida, S.; Heyne, B.; Scaiano, J.C.; Watsky, M.A.; Shinozaki, N.; Lagali, N.; *et al.* Recombinant human collagen for tissue engineered corneal substitutes. *Biomaterials* **2008**, *29*, 1147–1158.

33. Duan, X.; McLaughlin, C.; Griffith, M.; Sheardown, H. Biofunctionalization of collagen for improved biological response: Scaffolds for corneal tissue engineering. *Biomaterials* **2007**, *28*, 78–88.

34. Ott, H.C.; Matthiesen, T.S.; Goh, S.K.; Black, L.D.; Kren, S.M.; Netoff, T.I.; Taylor, D.A. Perfusion-decellularized matrix: Using nature's platform to engineer a bioartificial heart. *Nat. Med.* **2008**, *14*, 213–221.

35. Dean, D.M.; Napolitano, A.P.; Youssef, J.; Morgan, J.R. Rods, tori, and honeycombs: The directed self-assembly of microtissues with prescribed microscale geometries. *FASEB J.* **2007**, *21*, 4005–4012.

36. Yeh, J.; Ling, Y.; Karp, J.M.; Gantze, J.; Chandawarkard, A.; Eng, G.; Blumling, J.; Langer, R.; Khademhosseini, A. Micromolding of shape-controlled, harvestable cell-laden hydrogels. *Biomaterials* **2006**, *27*, 5391–5398.

37. Karamichos, D.; Funderburgh, M.L.; Hutcheon, A.E.; Zieske, J.D.; Du, Y.; Wu, J.; Funderburgh, J.L. A role for topographic cues in the organization of collagenous matrix by corneal fibroblasts and stem cells. *PLoS ONE* **2014**, *9*.

38. L'Heureux, N.; Paquet, S.; Labbe, R.; Germain, L.; Auger, F.A. A completely biological tissue-engineered human blood vessel. *FASEB J.* **1998**, *12*, 47–56.

39. Mironov, V.; Boland, T.; Trusk, T.; Forgacs, G.; Markwald, R.R. Organ printing: Computer-aided jet-based 3d tissue engineering. *Trends Biotechnol.* **2003**, *21*, 157–161.

40. Tsang, V.L.; Chen, A.A.; Cho, L.M.; Jadin, K.D.; Sah, R.L.; DeLong, S.; West, J.L.; Bhatia, S.N. Fabrication of 3D hepatic tissues by additive photopatterning of cellular hydrogels. *FASEB J.* **2007**, *21*, 790–801.

41. Connon, C.J. Approaches to corneal tissue engineering: Top-down or bottom-up? *Procedia Eng.* **2015**, *110*, 15–20.

42. Gouveia, R.M.; Castelletto, V.; Alcock, S.G.; Hamley, I.W.; Connon, C.J. Bioactive films produced from self-assembling peptide amphiphiles as versatile substrates for tuning cell adhesion and tissue architecture in serum-free conditions. *J. Mater. Chem. B* **2013**, *1*, 6157–6169.

43. Gouveia, R.M.; Jones, R.R.; Hamley, I.W.; Connon, C.J. The bioactivity of composite fmoc-rgds-collagen gels. *Biomater. Sci.* **2014**, *2*, 1222–1229.

44. Dehsorkhi, A.; Gouveia, R.M.; Smith, A.M.; Hamley, I.W.; Castelletto, V.; Connon, C.J.; Reza, M.; Ruokolainen, J. Self-assembly of a dual functional bioactive peptide amphiphile incorporating both matrix metalloprotease substrate and cell adhesion motifs. *Soft Matter* **2015**, *11*, 3115–3124.

45. Storrie, H.; Guler, M.O.; Abu-Amara, S.N.; Volberg, T.; Rao, M.; Geiger, B.; Stupp, S.I. Supramolecular crafting of cell adhesion. *Biomaterials* **2007**, *28*, 4608–4618.

46. Zhou, M.; Smith, A.M.; Das, A.K.; Hodson, N.W.; Collins, R.F.; Ulijin, R.V.; Gough, J.E. Self-assembled peptide-based hydrogels as scaffolds for anchorage-dependent cells. *Biomaterials* **2009**, *30*, 2523–2530.

47. Nagai, Y.; Unsworth, L.D.; Koutsopoulos, S.; Zhang, S.G. Slow release of molecules in self-assembling peptide nanofiber scaffold. *J. Control. Release* **2006**, *115*, 18–25.

48. Branco, M.C.; Pochan, D.J.; Wagner, N.J.; Schneider, J.P. Macromolecular diffusion and release from self-assembled beta-hairpin peptide hydrogels. *Biomaterials* **2009**, *30*, 1339–1347.

49. Salick, D.A.; Kretsinger, J.K.; Pochan, D.J.; Schneider, J.P. Inherent antibacterial activity of a peptide-based beta-hairpin hydrogel. *J. Am. Chem. Soc.* **2007**, *129*, 14793–14799.

50. Yang, Z.M.; Liang, G.L.; Guo, Z.F.; Guo, Z.H.; Xu, B. Intracellular hydrogelation of small molecules inhibits bacterial growth. *Angew. Chem. Intl. Ed.* **2007**, *46*, 8216–8219.

51. Sundar, S.; Chen, Y.; Tong, Y.W. Delivery of therapeutics and molecules using self-assembled peptides. *Curr. Med. Chem.* **2014**, *21*, 2469–2479.

52. Cui, H.G.; Webber, M.J.; Stupp, S.I. Self-assembly of peptide amphiphiles: From molecules to nanostructures to biomaterials. *Biopolymers* **2010**, *94*, 1–18.

53. Webber, M.J.; Tongers, J.; Renault, M.A.; Roncalli, J.G.; Losordo, D.W.; Stupp, S.I. Development of bioactive peptide amphiphiles for therapeutic cell delivery. *Acta Biomater.* **2010**, *6*, 3–11.

54. Beniash, E.; Hartgerink, J.D.; Storrie, H.; Stendahl, J.C.; Stupp, S.I. Self-assembling peptide amphiphile nanofiber matrices for cell entrapment. *Acta Biomater.* **2005**, *1*, 387–397.

55. Galler, K.M.; Cavender, A.; Yuwono, V.; Dong, H.; Shi, S.T.; Schmalz, G.; Hartgerink, J.D.; D'Souza, R.N. Self-assembling peptide amphiphile nanofibers as a scaffold for dental stem cells. *Tissue Eng. Part A* **2008**, *14*, 2051–2058.

56. Weib, N.; Klee, D.; Hocker, H. Konzept zur bioaktiven ausrüstung. *Biomaterialien* **2001**, *2*, 81–86.

57. Shroff, K.; Rexeisen, E.L.; Arunagirinathan, M.A.; Kokkoli, E. Fibronectin-mimetic peptide-amphiphile nanofiber gels support increased cell adhesion and promote ecm production. *Soft Matter* **2010**, *6*, 5064–5072.

58. Rexeisen, E.L.; Fan, W.; Pangburn, T.O.; Taribagil, R.R.; Bates, F.S.; Lodge, T.P.; Tsapatsis, M.; Kokkoli, E. Self-assembly of fibronectin mimetic peptide-amphiphile nanofibers. *Langmuir* **2010**, *26*, 1953–1959.

59. Stupp, S.I.; Pralle, M.U.; Tew, G.N.; Li, L.M.; Sayar, M.; Zubarev, E.R. Self-assembly of organic nano-objects into functional materials. *MRS Bull.* **2000**, *25*, 42–48.

60. Ghadiri, M.R.; Granja, J.R.; Milligan, R.A.; McRee, D.E.; Khazanovich, N. Self-assembling organic nanotubes based on a cyclic peptide architecture. *Nature* **1993**, *366*, 324–327.

61. Cui, H.; Muraoka, T.; Cheetham, A.G.; Stupp, S.I. Self-assembly of giant peptide nanobelts. *Nano Lett.* **2009**, *9*, 945–951.

62. Stupp, S.I.; Zha, R.H.; Palmer, L.C.; Cui, H.G.; Bitton, R. Self-assembly of biomolecular soft matter. *Faraday Discuss.* **2013**, *166*, 9–30.

63. Hamley, I.W. Self-assembly of amphiphilic peptides. *Soft Matter* **2011**, *7*, 4122–4138.

64. Maude, S.; Tai, L.R.; Davies, R.P.W.; Liu, B.; Harris, S.A.; Kocienski, P.J.; Aggeli, A. Peptide synthesis and self-assembly. *Top. Curr. Chem.* **2012**, *310*, 27–69.

65. Nagarkar, R.P.; Schneider, J.P. Synthesis and primary characterization of self-assembled peptide-based hydrogels. *Methods Mol. Biol.* **2008**, *474*, 61–77.

66. Cavalli, S.; Albericio, F.; Kros, A. Amphiphilic peptides and their cross-disciplinary role as building blocks for nanoscience. *Chem. Soc. Rev.* **2010**, *39*, 241–263.

67. Pashuck, E.T.; Cui, H.; Stupp, S.I. Tuning supramolecular rigidity of peptide fibers through molecular structure. *J. Am. Chem. Soc.* **2010**, *132*, 6041–6046.

68. Matson, J.B.; Stupp, S.I. Self-assembling peptide scaffolds for regenerative medicine. *Chem. Commun.* **2012**, *48*, 26–33.

69. Behanna, H.A.; Donners, J.J.J.M.; Gordon, A.C.; Stupp, S.I. Coassembly of amphiphiles with opposite peptide polarities into nanofibers. *J. Am. Chem. Soc.* **2005**, *127*, 1193–1200.

70. Kokkoli, E.; Mardilovich, A.; Wedekind, A.; Rexeisen, E.L.; Garg, A.; Craig, J.A. Self-assembly and applications of biomimetic and bioactive peptide amphiphiles. *Soft Matter* **2006**, *2*, 1015–1024.

71. Kunitake, T. Synthetic bilayer membranes: Molecular design, self-organization, and application. *Angew. Chem. Intl. Ed.* **1992**, *31*, 709–726.

72. Gore, T.; Dori, Y.; Talmon, Y.; Tirrell, M.; Bianco-Peled, H. Self-assembly of model collagen peptide amphiphiles. *Langmuir* **2001**, *17*, 5352–5360.

73. Yu, Y.C.; Tirrell, M.; Fields, G.B. Minimal lipidation stabilizes protein-like molecular architecture. *J. Am. Chem. Soc.* **1998**, *120*, 9979–9987.

74. Gouveia, R.M.; Castelletto, V.; Hamley, I.W.; Connon, C.J. New self-assembling multi-functional templates for the biofabrication and controlled self-release of cultured tissues. *Tissue Eng. Part A* **2015**, *21*, 1772–1784.

75. Xu, X.D.; Jin, Y.; Liu, Y.; Zhang, X.Z.; Zhuo, R.X. Self-assembly behaviour of peptide amphiphiles (PAs) with different length of hydrophobic alkyl tails. *Colloid Surf. B* **2010**, *81*, 329–335.

76. Morandat, S.; Bortolato, M.; Anker, G.; Doutheau, A.; Lagarde, M.; Chauvet, J.P.; Roux, B. Plasmalogens protect unsaturated lipids against uv-induced oxidation in monolayer. *BBA Biomembr.* **2003**, *1616*, 137–146.

77. Reinitz, S.D.; Currier, B.H.; Levine, R.A.; Van Citters, D.W. Crosslink density, oxidation and chain scission in retrieved, highly cross-linked uhmwpe tibial bearings. *Biomaterials* **2014**, *35*, 4436–4440.

78. Aulisa, L.; Forraz, N.; McGuckin, C.; Hartgerink, J.D. Inhibition of cancer cell proliferation by designed peptide amphiphiles. *Acta Biomater.* **2009**, *5*, 842–853.

79. Hashidzume, A.; Harada, A. Micelles and vesicles. In *Encyclopedia of polymeric Nanomaterials*; Kobayashi, S., Müllen, K., Eds.; Springer Berlin Heidelberg: Berlin, Jermany, 2015; pp. 1–5.

80. El-Sherbiny, I.M.; Yacoub, M.H. Hydrogel scaffolds for tissue engineering: Progress and challenges. *Glob. cardiol. Sci. Pract.* **2013**, *2013*, 316–342.

81. Aszódi, A.; Legate, K.R.; Nakchbandi, I.; Fässler, R. What mouse mutants teach us about extracellular matrix function. *Annu. Rev. Cell. Dev. Biol.* **2006**, *22*, 591–621.

82. Sargeant, T.D.; Gulerb, M.O.; Oppenheimera, S.M.; Mata, A.; Satcher, R.L.; Dunanda, D.C.; Stupp, S.I. Hybrid bone implants: Self-assembly of peptide amphiphile nanofibers within porous titanium. *Biomaterials* **2008**, *29*, 161–171.

83. Hartgerink, J.D.; Beniash, E.; Stupp, S.I. Self-assembly and mineralization of peptide-amphiphile nanofibers. *Science* **2001**, *294*, 1684–1688.

84. Shah, R.M.; Shah, N.A.; Del Rosario Lim, M.M.; Hsieh, C.; Nuber, G.; Stupp, S.I. Supramolecular design of self-assembling nanofibers for cartilage regeneration. *Proc. Natl. Acad. Sci. USA* **2010**, *107*, 3293–3298.

85. Rajangam, K.; Behanna, H.A.; Hui, M.J.; Han, X.; Hulvat, J.F.; Lomasney, J.W.; Stupp, S.I. Heparin binding nanostructures to promote growth of blood vessels. *Nano Lett.* **2006**, *6*, 2086–2090.

86. Stendahl, J.C.; Wang, L.J.; Chow, L.W.; Kaufman, D.B.; Stupp, S.I. Growth factor delivery from self-assembling nanofibers to facilitate islet transplantation. *Transplantation* **2008**, *86*, 478–481.

87. Angeloni, N.L.; Bond, C.W.; Tang, Y.; Harrington, D.A.; Zhang, S.; Stupp, S.I.; McKenna, K.E.; Podlasek, C.A. Regeneration of the cavernous nerve by sonic hedgehog using aligned peptide amphiphile nanofibers. *Biomaterials* **2011**, *32*, 1091–1101.

88. Robinson, L.R.; Fitzgerald, N.C.; Doughty, D.G.; Dawes, N.C.; Berge, C.A.; Bissett, D.L. Topical palmitoyl pentapeptide provides improvement in photoaged human facial skin. *Int. J. Cosmet. Sci.* **2005**, *27*, 155–160.

89. Ferreira, D.S.; Marques, A.P.; Reisa, R.L.; Azevedo, H.S. Hyaluronan and self-assembling peptides as building blocks to reconstruct the extracellular environment in skin tissue. *Biomater. Sci.* **2013**, *1*, 952–964.

90. Germain, L.; Auger, F.A.; Grandbois, E.; Guignard, R.; Giasson, M.; Boisjoly, H.; Guérin, S.L. Reconstructed human cornea produced *in vitro* by tissue engineering. *Pathobiology* **1999**, *67*, 140–147.

91. Griffith, M.; Hakim, M.; Shimmura, S.; Watsky, M.A.; Li, F.; Carlsson, D.; Doillon, C.J.; Nakamura, M.; Suuronen, E.; Shinozaki, N.; *et al.* Artificial human corneas: Scaffolds for transplantation and host regeneration. *Cornea* **2002**, *21*, 54–61.

92. Hashimoto, Y.; Funamoto, S.; Sasaki, S.; Honda, T.; Hattori, S.; Nam, K.; Kimura, T.; Mochizuki, M.; Fujisato, T.; Kobayashi, H.; *et al.* Preparation and characterization of decellularized cornea using high-hydrostatic pressurization for corneal tissue engineering. *Biomaterials* **2010**, *31*, 3941–3948.

93. Chen, B.; Jones, R.R.; Mi, S.; Foster, J.; Alcock, S.; Hamley, I.W.; Connon, C.J. The mechanical properties of amniotic membrane influence its effect as a biomaterial for ocular surface repair. *Soft Matter* **2012**, *8*, 8379–8387.

94. Gomes, J.A.; Dos Santos, M.S.; Cunha, M.C.; Mascaro, V.L.; Barros, J.d.; De Sousa, L.B. Amniotic membrane transplantation for partial and total limbal stem cell deficiency secondary to chemical burn. *Opthalmology* **2003**, *110*, 466–473.

95. Carsson, D.J.; Li, F.; Shimmura, S.; Griffith, M. Bioengineered corneas: How close are we? *Curr. Opin. Opthalmol.* **2003**, *14*, 192–197.

96. Castelletto, V.; Gouveia, R.M.; Connon, C.J.; Hamley, I.W. New rgd-peptide amphiphile mixtures containing a negatively charged diluent. *Faraday Discuss.* **2013**, *166*, 381–397.

97. Castelletto, V.; Gouveia, R.M.; Connon, C.J.; Hamley, I.W.; Seitsonen, J.; Nykänen, A.; Ruokolainen, J. Alanine-rich amphiphilic peptide containing the rgd cell adhesion motif: A coating material for human fibroblast attachment and culture. *Biomater. Sci.* **2014**, *2*, 362–369.

98. Castelletto, V.; Hamley, I.W.; Perez, J.; Abezgauz, L.; Danino, D. Fibrillar superstructure from extended nanotapes formed by a collagen-stimulating peptide. *Chem. Commun.* **2010**, *46*, 9185–9187.

99. Jones, R.R.; Castelletto, V.; Connon, C.J.; Hamley, I.W. Collagen stimulating effect of peptide amphiphile c16-kttks on human fibroblasts. *Mol. Pharm.* **2013**, *10*, 1063–1069.

100. Uzunalli, G.; Soran, Z.; Erkal, T.S.; Dagdas, Y.S.; Dinc, E.; Hondur, A.M.; Bilgihan, K.; Aydin, B.; Guler, M.O.; Tekinay, A.B. Bioactive self-assembled peptide nanofibers for corneal stroma regeneration. *Acta Biomater.* **2014**, *10*, 1156–1166.

101. Hamley, I.W.; Dehsorkhi, A.; Walter, M.N.M.; Connon, C.J.; Reza, M.; Ruokolainen, J. Self-assembly and collagen stimulating activity of a peptide amphiphile incorporating a peptide sequence from lumican. *Langmuir* **2015**, *31*, 4490–4495.

102. Gribova, V.; Crouzier, T.; Picart, C. A material's point of view on recent developments of polymeric biomaterials: Control of mechanical and biochemical properties. *J. Mater. Chem.* **2011**, *21*, 14354–14366.

103. Hersel, U.; Dahmen, C.; Kessler, H. Rgd modified polymers: Biomaterials for stimulated cell adhesion and beyond. *Biomaterials* **2003**, *24*, 4385–4415.

104. Pierschbacher, M.D.; Ruoslahti, E. Cell attachment activity of fibronectin can be duplicated by small synthetic fragments of the molecule. *Nature* **1984**, *309*, 30–33.

105. Gouveia, R.M.; Connon, C.J. The effects of retinoic acid on human corneal stromal keratocytes cultured *in vitro* under serum-free conditions. *Invest. Opthalmol. Vis. Sci.* **2013**, *54*, 7483–7491.

106. Bard, J.B.L.; Hay, E.D. The behavior of fibroblasts from the developing avian cornea: Morphology and movement *in situ* and *in vitro*. *J. Cell Biol.* **1975**, *67*, 400–418.

107. Cintron, C.; Covington, H.; Kublin, C.L. Morphogenesis of rabbit corneal stroma. *Investig. Ophthalmol. Vis. Sci.* **1983**, *24*, 543–556.

108. Hay, E.D. The mesenchymal cell, its role in the embryo, and the remarkable signaling mechanisms that create it. *Dev. Dyn.* **2005**, *233*, 706–720.

109. Maurice, D.M. The structure and transparency of the cornea. *J. Physiol.* **1957**, *136*, 263–286.

110. Trelstad, R.L.; Coulombre, A.J. Morphogenesis of the collagenous stroma in the chick cornea. *J. Cell Biol.* **1971**, *50*, 840–858.

111. Petroll, W.M.; New, K.; Sachdev, M.; Cavanagh, H.D.; Jester, J.V. Radial keratotomy. III: Relationship between wound gape and corneal curvature in primate eyes. *Investig. Ophthalmol. Vis. Sci.* **1992**, *33*, 3283–3291.

112. Moller-Pedersen, T.; Vogel, M.D.; Li, H.; Petroll, W.M.; Cavanagh, H.D.; Jester, J.V. Quantification of stromal thinning, epithelial thickness, and corneal haze following photorefractive keratectomy using *in vivo* confocal microscopy. *Ophthalmology* **1997**, *104*, 360–368.

113. Karamichos, D.; Lakshman, N.; Petroll, W.M. An experimental model for assessing fibroblast migration in 3-d collagen matrices. *Cell Motil. Cytoskelet.* **2009**, *66*, 1–9.

114. Kim, A.; Lakshman, N.; Karamichos, D.; Petroll, W.M. Growth factor regulation of corneal keratocyte differentiation and migration in compressed collagen matrices. *Investig. Ophthalmol. Vis. Sci.* **2010**, *51*, 864–875.

115. Kim, A.; Zhou, C.; Lakshman, N.; Petroll, W.M. Corneal stromal cells use both high- and low-contractility migration mechanisms in 3-D collagen matrices. *Exp. Cell Res.* **2012**, *318*, 741–752.

116. Yamanaka, O.; Yuan, Y.; Coulson-Thomas, V.J.; Gesteira, T.F.; Call, M.K.; Zhang, Y.J.; Zhang, J.H.; Chang, S.H.; Xie, C.C.; Liu, C.Y.; *et al.* Lumican binds alk5 to promote epithelium wound healing. *PLoS ONE* **2013**, *8*.

117. Carlson, E.C.; Lin, M.; Liu, C.Y.; Kao, W.W.; Perez, V.L.; Pearlman, E. Keratocan and lumican regulate neutrophil infiltration and corneal clarity in lipopolysaccharide-induced keratitis by direct interaction with CXCL1. *J. Biol. Chem.* **2007**, *282*, 35502–35509.

118. Saika, S.; Shiraishi, A.; Liu, C.Y.; Funderburgh, J.L.; Kao, C.W.; Converse, R.L.; Kao, W.W. Role of lumican in the corneal epithelium during wound healing. *J. Biol. Chem.* **2000**, *275*, 2607–2612.

119. Aumailley, M.; Smyth, N. The role of laminins in basement membrane function. *J. Anat.* **1998**, *193*, 1–21.

Concise Review: Comparison of Culture Membranes Used for Tissue Engineered Conjunctival Epithelial Equivalents

Jon Roger Eidet, Darlene A. Dartt and Tor Paaske Utheim

Abstract: The conjunctival epithelium plays an important role in ensuring the optical clarity of the cornea by providing lubrication to maintain a smooth, refractive surface, by producing mucins critical for tear film stability and by protecting against mechanical stress and infectious agents. A large number of disorders can lead to scarring of the conjunctiva through chronic conjunctival inflammation. For controlling complications of conjunctival scarring, surgery can be considered. Surgical treatment of symblepharon includes removal of the scar tissue to reestablish the deep fornix. The surgical defect is then covered by the application of a tissue substitute. One obvious limiting factor when using autografts is the size of the defect to be covered, as the amount of healthy conjunctiva is scarce. These limitations have led scientists to develop tissue engineered conjunctival equivalents. A tissue engineered conjunctival epithelial equivalent needs to be easily manipulated surgically, not cause an inflammatory reaction and be biocompatible. This review summarizes the various substrates and membranes that have been used to culture conjunctival epithelial cells during the last three decades. Future avenues for developing tissue engineered conjunctiva are discussed.

Reprinted from *J. Funct. Biomater.* Cite as: Eidet, J.R.; Dartt, D.A.; Utheim, T.P. Concise Review: Comparison of Culture Membranes Used for Tissue Engineered Conjunctival Epithelial Equivalents. *J. Funct. Biomater.* **2015**, *6*, 1064–1084.

1. Conjunctiva

Conjunctival epithelium is non-keratinized and is at least two cell layers thick [1]. The number of cell layers depends on the degree of conjunctival stretching [2]. The conjunctival epithelium consists of two phenotypically distinct cell types, stratified squamous non-goblet cells (90%–95%) and goblet cells (5%–10%) (Figure 1), in addition to occasional lymphocytes [3] and melanocytes. The conjunctival epithelium plays an important role in ensuring the optical clarity of the cornea by providing lubrication to maintain a smooth, refractive surface, and by producing mucins critical for tear film stability [4]. The conjunctiva also protects the eye against mechanical stress and infectious agents. It, furthermore, contributes water and electrolytes to the tear fluid [5]. The squamous cells produce cell membrane-tethered mucins, while the goblet cells secrete the gel-forming mucins, both of which helps to maintain a protective tear film. The superficial surface of the squamous cells are covered by the membrane-tethered mucins mucin-1 (MUC1), mucin-4 (MUC4) and mucin-16 (MUC16) [6], which are essential for tear stability and make up the glycocalyx [6].

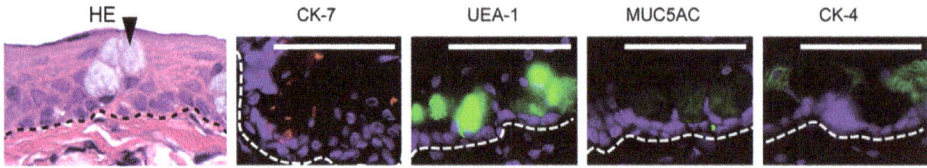

Figure 1. Photomicrographs show hematoxylin and eosin (HE) and immunofluorescently stained sections of rat conjunctiva. The black arrowhead in the HE photomicrograph indicates mucin granules of goblet cells. The black dotted line indicates the basal membrane, which overlies loose vascularized conjunctival forniceal connective tissue. Original magnification of the HE photomicrograph: ×630. Immunofluorescence photomicrographs of forniceal conjunctival sections show conjunctival epithelial cell markers, which include the goblet cell markers anti-cytokeratin 7 (Ck7), Ulex europaeus agglutinin 1 (UEA-1) lectin and anti-mucin 5AC (MUC5AC), as well as the marker for stratified squamous non-goblet cells anti-cytokeratin 4 (Ck4). Nuclei were stained with DAPI (blue). Ck7 is expressed in the goblet cell body, whereas UEA-1 and MUC5AC stain the goblet cell mucin-contents. Ck4 is only detected in squamous cells between goblet cell clusters. The basal membrane is indicated by the white dotted line. Scale bars: 100 µm. Adapted from Fostad *et al.* 2012 [7].

The gel-forming mucin-5AC (MUC5AC) and mucin-2 (MUC2) are secreted by goblet cells into the aqueous layer of the tear film [8,9] (Figure 2). The squamous conjunctival cells also contribute to the hydration of the ocular surface through ion transport across the apical cell membrane with accompanying osmotic water transfer [5]. Goblet cells contain mucin-granules and have traditionally been identified through their secretory product using markers, including the ulex europaeus agglutinin-1 (UEA-1) lectin, anti-mucin-5AC (MUC5AC) and anti-AM3 antibodies, and periodic acid-Schiff (PAS) reagent that target the goblet cell gel-forming mucins [10]. In addition to cytokeratin 4 (Ck4) (Figure 1), squamous conjunctival epithelial cells can be identified by Ck13, a binding pair of Ck4 [11].

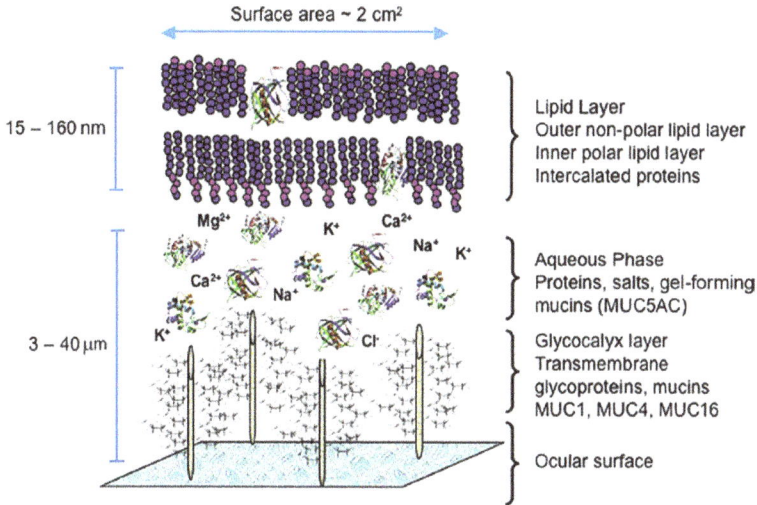

Figure 2. Model of the human tear film. Adapted from Nichols *et al.* 2001 [12].

2. Conjunctival Stem Cells

Conjunctival stem cells continuously regenerate the conjunctiva by giving rise to both stratified squamous non-goblet and goblet cells [13], thereby maintaining a healthy tear film [14]. Disorders that damage these stem cells cause varying extent of keratinization, which disrupts the protective tear film and ultimately leads to limbal stem cell deficiency (LSCD) and visual impairment or blindness. The location of the conjunctival epithelial stem cells has been investigated in several studies on mouse [15–17], rat [18,19], rabbit [20,21] and human [22–24] tissue, yet no real consensus has been reached. The conjunctival stem cells have been suggested to reside in the limbus [18], bulbar conjunctiva [15,22,23], medial canthal [24], forniceal conjunctiva [16,17,20,22,24,25], palpebral conjunctiva [19] and mucocutaneous junction [18,21]. Although the conjunctival stem cells may not solely be located to one single region, their relative number generally appears to be highest in the fornix [26].

Stem cells are surrounded and influenced by a three-dimensional microenvironment known as a niche [27]. The niche comprises of numerous components, including stromal cells, soluble factors, extracellular matrix (ECM), mechanical/spatial cues and signaling molecules that dictates stem cell function [28]. The limbal stem cell niche has been reported to contain specific ECM proteins. Moreover, the specific composition of the ECM shows topographical variations throughout the ocular surface [29]. Thus, the specific composition of the ECM in the substrate may affect the preservation of conjunctival stem cells in culture.

3. Conjunctival Scarring Diseases

A large number of disorders can lead to scarring of the conjunctiva through chronic conjunctival inflammation. Scarring varies in severity and can be self-limited, such as in chemical/thermal burns and infectious diseases due to adeno- and herpes viruses, or progressive, as in cicatrizing conjunctivitis, which consists of several diseases including ocular cicatricial pemphigoid, Stevens-Johnson syndrome, atopic keratoconjunctivitis and Sjögren's syndrome [30]. Cicatrizing conjunctivitis is rare, and in total these disorders have an incidence of 1.2 in 1 million in the United Kingdom [30]. Treatment depends on the disease etiology and severity, but can include various anti-inflammatory, immunomodulatory and immunosuppressive drugs [31].

Surgical treatment of symblepharon includes removal of the scar tissue to reestablish the deep fornix [32]. The surgical defect is then covered with a tissue substitute to prevent re-obliteration. These include mechanical [33], physical [34] or chemical [35] approaches and the grafting of conjunctival or mucous membranes [32]. Surgical techniques for restoration of a diseased conjunctiva have utilized different conjunctival substitutes, including conjunctival autografts [36]. An obvious limiting factor when using autografts is the size of the defect to be covered, as the amount of healthy conjunctiva is limited. These drawbacks have led scientists to develop tissue engineered conjunctival equivalents.

4. Tissue Engineered Conjunctival Equivalents

A tissue engineered conjunctival epithelial equivalent needs to be easily manipulated surgically, not cause an inflammatory reaction, be biocompatible and contain a mix of stratified squamous cells, goblet cells and undifferentiated cells. Unlike tissue engineered corneal equivalents, conjunctival equivalents do not need to be transparent, which increases the range of suitable culture membranes.

In addition to conjunctival epithelial cells (CEC) cultured on amniotic membrane (AM) [4], there is likely a wide range of cell types that can be used for developing a tissue engineered conjunctival equivalent. This assumption is based on multiple studies demonstrating successful restoration of the cornea with cultured non-limbal cells. Tissue engineered corneal equivalents share many of the same prerequisites as conjunctival equivalents, e.g., with regard to barrier function and tear film support. Besides limbal stem cells, corneal equivalents have been developed from oral mucosal epithelial cells [37,38], embryonic stem cells (ESC) [39], bone-marrow-derived mesenchymal stem cells (MSC) [40], immature dental pulp stem cells [41], hair follicle-derived stem cells [42] and umbilical cord lining stem cells [43]. For conjunctival reconstruction, epidermal keratinocytes have been cultured on AM and transplanted to restore the conjunctiva in rhesus monkeys [44]. Although the conjunctival stratified squamous cell markers MUC4 and Ck4 were present in the transplant, goblet cells were absent. In a recent study, a tissue engineered conjunctival equivalent was developed from cultured AM epithelial cells [45]. The conjunctival equivalent contained PAS-positive cells, indicative of goblet cells, and successfully restored the conjunctiva in a rabbit model. Transplants containing goblet cells could also be developed from nasal mucosa, which harbors goblet cells [46]. Thus, there are multiple possible cell sources for developing conjunctival equivalents, though no comparative studies have defined the optimal choice of donor cells.

A number of different substrates and membranes have been attempted for tissue engineering conjunctival epithelial equivalents. These can be categorized into: (1) biological membranes; (2) extracellular matrix protein-containing membranes; and (3) synthetic polymer membranes.

4.1. Biological Membranes

Seventy-six years after it was first used in ophthalmology, AM, which constitutes the innermost layer of the fetal membranes, has a prominent role in ocular surface reconstruction [47]. AM is particularly suited for clinical use as it supports epithelialization [48], reduces scaring [49], suppresses the immune response [50], reduces pain, and decreases inflammation [51]. Prior to AM transplantation (AMT), the AM is cryopreserved, which kills all the AM cells [52]. Hence, AM grafts function primarily as a matrix and not by virtue of transplanted functional cells. The membranes have most commonly been cryopreserved in a basal cell medium at -80 °C [53], but a technique for freeze-drying the AM has also been developed [54]. Freeze-dried AM can be sterilized by gamma-irradiation [54], however, AM treated this way may release a less amount of growth factors than conventionally cryopreserved membranes [55]. In addition, the AM can be sterilized with per-acetic acid/ethanol and air-dried [56]. The latter technique is, on the other hand, reported to yield inferior results compared to cryopreserved AM with respect to rate of cell outgrowth, release of wound-healing factors, and preservation of the AM basement membrane (BM) [57]. In patients with chronic inflammation there is a tendency for recurrent shrinkage and symblepharon formation after restoring the ocular surface with AM [58]. The success of transplanting AM is therefore dependent on the underlying disease [4].

Twelve studies have described culture of CEC on AM, of which eight used denuded AM (dAM) (Table 1). Meller *et al.* first reported the use of dAM for cell culture of CEC since they noticed that the devitalized AM epithelium inhibited adhesion and growth of the CEC [59]. All later studies using intact AM have utilized explant culture.

Eight out of ten studies confirmed the presence of goblet cells on AM (detected either by their mucin content or by Ck7), irrespective of whether dAM or intact AM had been used [59]. Data on actual percentages of goblet cells in CEC cultures on AM are sparse, although one study reported that between 25% and 75% of the cells were MUC5AC positive [60]. Although Ck7 positive goblet cells have been demonstrated under serum-free conditions, addition of 10% FBS improved the preservation of goblet cells [61]. This is in line with a study showing that FBS promotes expression of conjunctival epithelial cytokeratins due to the effect of vitamin A [62]. Development of mucin-containing goblet cells have also been achieved on AM independent of feeder cells, air-lifting or high calcium [60]. Thus, AM generally promotes goblet cell development.

Stratified CEC were obtained in all studies using AM, except one [63]. Culture techniques to induce stratification include the use of explants, air-lifting, feeder layer, and high calcium. Air-lifting promotes cell polarity by increasing the number of microvilli, tight junctions, and hemidesmosomes in CEC cultures [59]. The molecular mechanisms involved in air-lifting include the p38 mitogen-activated protein kinase and Wnt signaling pathways [64]. Stratification was achieved when including a feeder layer [50], air-lifting [59] and/or high calcium [65] in cell cultures.

Stratified CEC cultures were also generated on cadaveric acellular dermis (AlloDerm) coated with collagen type 4 (COL4) [65]. The latter study employed a serum-free culture protocol without feeder cells. Goblet cells, however, were not reported. Hence, except for the latter study on acellular dermis, culture of CEC on biological membranes generally promotes stratified cultures with goblet cells.

Table 1. Conjunctival Epithelial cells cultured on biological membranes.

Substrate (s)	Cell Species	Explant/ Suspension Culture	Culture Time (Days)	Feeder Cells	Air-Lifting	High Calcium	Basal Medium	Serum	Conjunctival Donor Site	Goblet Cells	Comment	Authors
AM	Human	Explant	21	–	–	–	CNT50	FBS/AS	–	Yes (with both serum type)		Rivas et al. 2014
AM	Human	Explant	14	No	No	No	DMEM:F12	5% FBS	Fornix/bulbus	Yes (<25% to 75% MUC5AC+)	Stratified culture	Eidet et al. 2014
AM	Human	Explant	12	No	Yes	No	DMEM:F12	5% FBS	Bulbus	Yes (MUC5AC+, fever than in native conjunctiva)	Stratified culture	Tan et al. 2014
AM	Rabbit	Explant	8–15	3T3/No	Yes	–	–	–	Fornix	No MUC5AC–	Stratified culture (Ck3+/Ck12–)	Cho et al. 2014
AM	Human	Explant	9–11	3T3	Yes	–	DMEM:F12	10% FBS	–	Yes (PAS+, increased by γSI)	Stratified culture	Tian et al. 2014
dAM	Human	Explant	–	No	No	–	DMEM:F12 KM (serum free or DMEM:F12)	FBS	Fornix	Yes (PAS+)	Stratified culture	Silber et al. 2014
dAM	Human	Suspension	5	No	No	–	KM (serum free or DMEM:F12)	0%–20% FBS	Palpebra	Yes (100% Ck7+; best preserved by 10% FBS)		Martinez-Osorio et al. 2009
dAM	Human	Suspension	21	3T3	Yes	Yes	KM (serum free or DMEM:F12)	FBS	–	No (MUC5AC–)	Stratified culture	Tanioka et al. 2006
dAM	Human	Explant	14	No	–	Yes	DMEM:F12	FBS/HS	Bulbus	–	Stratified culture	Ang et al. 2005
dAM	Human	Explant	12–22	No	Yes/No	Yes/No	KGM or DMEM/F12	0 or 10% FBS	Bulbus	Yes (MUC5AC detected by PCR in all groups)	Stratified culture	Ang et al. 2004
dAM	Human	Explant	11–15	No	No	–	KGM:F12	10% FBS	Bulbus	–	Monolayer culture	Sangwan et al. 2003
dAM	Rabbit	Suspension	<28	RCF	Yes/No	No	DMEM:F12	5% FBS	–	Yes (scattered MUC5AC+ cells with/without AL and RCF)	Stratified culture (increased in AL)	Meller et al. 1999
AlloDerm coated with COL4	Human	Suspension	18	No	Yes	Yes	MCDB 153	No	–	–	Stratified culture	Yoshizawa et al. 2004

AL = air-lifting; AM = human amniotic membrane; (–) = not reported; AS = autologous serum; DMEM = Dulbecco's Modified Eagle's Medium; 3T3 = 3T3 feeder cells; γSI = γ-secretase inhibitor; dAM = denuded AM; KM = keratinocyte medium; HS = human serum; KGM = keratinocyte growth medium; RCF = rabbit conjunctival fibroblasts; COL4 = collagen type 4; MUC5AC = mucin 5AC; Ck7 = cytokeratin 7.

4.2. Extracellular Matrix Protein-Containing Membranes

The conjunctival BM is a thin connective tissue membrane, which is composed of collagen type IV (collagen $\alpha1$ and $\alpha2$ chains), laminin ($\alpha5$, $\beta2$ and $\gamma1$ chains), nidogen-1 and -2 and thrombospondin-4 [29]. It is therefore reasonable to assume that a tissue engineered CEC equivalent would benefit from being surfaced by ECM proteins. Nineteen studies have described the culture of CEC on various ECM proteins (Table 2). Collagen type 1 (COL1) was most commonly used, either in the form of a coating [66], a gel [67] or as a compressed gel [68]. The latter two forms offer the mechanical strength to transfer the cultured cells to the surgical site. In addition, fibronectin (FN), laminin (LN), Matrigel, elastin-like polymer (ELP), gelatin-chitosan and poly-l-lysine (PLL) were tried [61,66,69–77].

Goblet cells were seen when CEC were grown inside a collagen gel [78], but not always when grown as a monolayer on top of the collagen gel [78]. Compared to Matrigel, CEC grown on COL1 expressed more MUC5AC RNA than Matrigel cultures [76]. Five percent PAS positive goblet cells were detected when culturing CEC on top of a COL1:COL3 mix in serum-free medium [71]. The latter study also achieved stratification. When cultured without feeder cells, air-lifting or high calcium, the CEC formed monolayer cultures on COL1 [66]. Stratified cultures were achieved with the addition of feeder cells [67], air-lifting [67], or high calcium [76].

Matrigel is composed of LN, COL4, heparan sulfate proteoglycans, entactin, transforming growth factor (TGF), and basic fibroblast growth factor (bFGF) [71]. Cultured CEC generally form aggregates on Matrigel rather than continuous cell sheets [66,76]. In one study the aggregates contained PAS positive goblet cells [66].

Use of fibronectin, either alone or in a mix with COL1, was reported in four studies [69–72]. The CEC formed monolayer cultures [70], but the presence of goblet cells were not reported. Elastin-like polymer has been used to grow Ck7 positive cells of the cell line IOBA-NHC [79]. Gelatin-chitosan yielded stratified cultures with Ck4 positive squamous cells when using explant culture [77]. Of all the ECM protein substrates, collagen gels and compressed collagen appear the most useful for conjunctival tissue engineering due to their mechanical properties and potential promotion of goblet cell formation.

Table 2. Conjunctival epithelial cells cultured on extracellular matrix protein-containing membranes.

Substrate(s)	Cell Species	Normal Cells/Cells Line	Explant/Suspension Culture	Culture Time (Days)	Feeder Cells	Air-Lifting	High Calcium	Basal Medium	Serum	Conjunctival Donor Site	Goblet Cells	Comment	Authors
BSA:COL mix	Rabbit	Normal	Suspension	–	No	No	No	PC-1 (serum free)	No	–	–		Scholz et al. 2002
COL	Rabbit	Normal	Suspension	6	No	Yes/No	No	PC-1 (serum free) or DMEM:F12	No	–	Yes (3% to 4% PAS + in AL group)		Yang et al. 2000
COL:FN mix	Human	Normal	Suspension	–	No	No	No	KGM (serum free)	No	Bulbus	–	Monolayer culture	Cook et al. 1998
COL:FN mix	Human	Normal	Suspension	–	No	No	No	EpiLife	No	–			Gordan et al. 2005
COL1	Bovine	Normal	Suspension	12	No	Yes/No	–	DMEM:F12	10% FBS	Bulbus	Yes (PAS + in both AL and submerged cultures)	Stratified culture (increased by AL)	Civiale et al. 2003
COL1 gel	Rabbit	Normal	Suspension	7–14	No	No	No	DMEM:F12	10% FBS	Bulbus	Yes (PAS + cell within gel, PAS– on the gel surface)	Stratified culture within gel, monolayer on the gel surface	Niiya et al. 1997
COL1 gel with/without 3T3	Rabbit	Normal	Suspension	6	3T3/No	Yes/No	No	DMEM:F12	10% FBS	–	No (PAS–, MUC5AC–)	Stratified culture (increased by AL and 3T3)	Chen et al. 1994
COL1 gel with/without 3T3 or HCF	Human	Normal	Suspension	14	3T3/HCF/no	Yes	No	DMEM:F12	5% FBS	Bulbus	Yes (only with HCF)	Stratified culture (with feeder cells)	Tsai et al. 1994
COL1 or Matrigel	Human	ConjEp-1/p53DD/cdk4R/TERT cell line	Suspension	(3–7 days in high Ca)	3T3/no	No	Yes	KM (serum free) or DMEM:F12	10% FBS	Bulbus	Yes (MUC5AC RNA highest with COL1)	Stratified culture (COL1), aggregates (Matrigel)	Gipson et al. 2003
COL1, COL1:COL3 mix, LN, FN or Matrigel	Rabbit	Normal	Suspension	<14	No	No	No	PC-1 (serum free)	0 or 1% FBS	All conjunctiva	Yes (5% PAS + in serum free cultures on COL1:COL3 mix)	Stratified culture (COL1:COL3 mix)	Saha et al. 1996
COL1, Matrigel or COL1:Matrigel mix	Rabbit	Normal	Suspension	–	No	No	No	DMEM:F12	5% FBS	All conjunctiva	Yes (PAS + cell in cultures on COL1 and in globules on Matrigel)	Monolayer culture (COL1), aggregates (Matrigel)	Tsai et al. 1988
COL1:FN mix	Rabbit	Normal	Suspension	–	No	No	–	PC-1 (serum free)	No	–	–		Basu et al. 1998
COL4	Rat	Normal	Suspension	10	–	Yes	–	KM (serum free) or DMEM:F12	No	Palpebra	–		Yu et al. 2012
Compressed COL	Human	Normal	Suspension	14	No	No	No	DMEM:F12	10% FBS	–	–	Stratified culture	Drechsler et al. 2015

Table 2. *Cont.*

Substrate (s)	Cell Species	Normal Cells/Cells Line	Explant/ Suspension Culture	Culture Time (Days)	Feeder Cells	Air-Lifting	High Calcium	Basal Medium	Serum	Conjunctival Donor Site	Goblet Cells	Comment	Authors
Elastin-like polymer	Human	IOBA-NHC cell line	Suspension	5	No	No	–	DMEM:F12	–	–	Yes (CK7+)	–	Martinez-Osorio et al. 2009
Gelatin-chitosan	Rabbit	Normal	Explant	14	No	No	–	DMEM:F12	10% FBS	–	–	Stratified culture (Ck4+)	Zhu et al. 2006
LN-1, LN-β2 or COL1 gel with BCF	Bovine	Normal	Explant	14	BCF/no	No	No	KBM (serum free) or DMEM (serum)	0 or 10% FBS	Bulbus	–	Stratified culture (DMEM/10% FCS and cultures on COL1 with BCF)	Kurpakus et al. 1999
LN-1, LN-β2 or poly-l-lysine	Bovine	Normal	Suspension	–	No	No	No	KBM (serum free)	No	–	–	–	Lin et al. 1999
LN-10	Human	HC0597 cell line	Suspension	–	No	No	No	KBM (serum free)	No	–	–	–	Lin et al. 2002

AL = air-lifting; BSA = bovine serum albumin; COL = collagen; (–) = not reported; DMEM = Dulbecco's Modified Eagle's Medium; PAS = periodic acid-Schiff; FN = fibronectin; KGM = keratinocyte growth medium; FBS = fetal bovine serum; 3T3 = 3T3 feeder cells; HCF = human conjunctival fibroblasts; KM = keratinocyte medium; KDM = keratinocyte basal medium; LN = laminin; KDM = keratinocyte basal medium; FCS = fetal calf serum; BCF = bovine conjunctival fibroblasts.

4.3. Synthetic Polymer Membranes

Included in this group are polymers of glycolic acid, lactic acid, ε-caprolactone, 1,3-trimethylene carbonate, ethyl acrylate, hydroxyethyl acrylate, and methacrylic acid. One of the benefits of using these polymers is that several of them, including poly(L-lactide-*co*-glycolide) (PLGA) and poly(ε-caprolactone) (PCL), are already approved by the Food and Drug Administration (FDA) for the use in the human body for specific applications. In addition, the biodegradability of these polymers can be adjusted by controlling the ratio and choice of polymers. For instance, PLGA degrades faster than PCL. Furthermore, in contrast to biological membranes, synthetic membranes can be manufactured under sterile conditions, thereby considerably reducing the risk of transferring infectious agents to the patient. Although biodegradable polymers have been investigated at length with various types of cells, only four studies reported biocompatibility with cultured CEC [80–83] (Table 3). Three of these explored growth of CEC on polymer substrates [80–82], whereas one investigated the toxicity of polymer extract on cells cultured on plastic [83]. One of the studies confirmed the presence of MUC5AC positive goblet cells of comparable density to that seen when culturing CEC on AM [80]. The remaining studies did not report presence of goblet cells. The extract study showed lowest to highest viability with 50:50 poly(DL-lactide-*co*-glycolide) (PDLGA); 85:15 PDLGA and Inion GTR^{TM}, respectively [83]. In cell growth studies, substrates with all three polymers demonstrated high viability [82]. Equally high viability was also seen when growing CEC on poly(ethyl acrylate-*co*-hydroxyethyl acrylate) (P(EA-*co*-HEA)) copolymers or 90:10 poly(ethyl acrylate-*co*-methacrylic acid) (P(EA-*co*-MAAc)) copolymers [81]. Interestingly, the latter two polymer substrates showed increased adhesion, proliferation and viability when hydrophobicity was increased. In contrast, Ang, *et al.* demonstrated increased proliferation when decreasing hydrophobicity of their PCL membranes [80]. The latter authors also obtained stratified cultures, which became more stratified by increasing surface hydrophilicity with NaOH. Thus, surface modification of synthetic polymer membranes can affect adhesion, proliferation, viability and stratification. Obvious advantages of synthetic polymer membranes include existing FDA approval for specific uses in the human body, high mechanical strength and biodegradability.

Table 3. Conjunctival Epithelial cells cultured synthetic polymer membranes.

Substrate (s)	Cell Species	Normal Cells/Cells Line	Culture Time (Days)	Explant/ Suspension Culture	Feeder Cells	Air-Lifting	High Calcium	Culture Medium	Serum	Conjunctival Donor Site	Goblet Cells	Comment	Authors
50:50 PDLGA, 85:15 PDLGA or Inion GTR™	Human	IOBA-NHC cell line	–	Suspension	No	No	–	DMEM:F12	10% FBS	–	–	Extract studies showing lowest to highest viability with 50:50 PDLGA; 85:15 PDLGA; Inion GTR™	Huhtala, *et al.* 2008
50:50 PDLGA, 85:15 PDLGA or Inion GTR™	Human	IOBA-NHC cell line	3	Suspension	No	No	–	DMEM:F12	10% FBS	–	–	High viability with all types of polymer	Huhtala, *et al.* 2007
P(EA-co-HEA) or 90:10 P(EA-co-MAAc) copolymers	Human	IOBA-NHC cell line	–	Suspension	No	–	–	DMEM:F12	10% FBS	–	–	All polymers were non-toxic, hydrophobicity increased adhesion, proliferation and viability	Campillo-Fernandez, *et al.* 2007
Ultrathin PCL	Rabbit	Normal	–	Explant/ suspension	No	No	Yes	KGM (serum free)	No	–	Yes (MUC5AC+ comparable to AM)	Stratified culture (increased by NaOH); NaOH surface modification increased hydrophilicity and proliferation	Ang, *et al.* 2006
Temperature-responsive polymer, poly(N-isopropyl-acrylamide; PIPAAm)	Rabbit	Normal	10	Suspension	No	No	No	DMEM:F12	10% FBS	Fornix/ palpebra	Yes (21.5% MUC5AC+, PAS+)	Stratified culture (4–5 cell layers); proliferation rate of 38.4%; high viability; Ck4 mRNA increased with time	Yao, *et al.* 2015

AL = air-lifting; PDLGA = poly(DL-lactide-*co*-glycolide); Inion GTR™ = a blend of 85:15 poly(L-lactide-*co*-glycolide) (PLGA) and 70:30 poly(L-lactide-*co*-1,3-trimethylene carbonate) (PLTMC) copolymers in a major ratio of 70:30; DMEM = Dulbecco's Modified Eagle's Medium; FBS = fetal bovine serum; P(EA-co-HEA) = poly(ethyl acrylate-*co*-hydroxyethyl acrylate); P(EA-co-MAAc) = poly(ethyl acrylate-*co*-methacrylic acid); PCL = poly(ε-caprolactone); (–) = not reported; MUC5AC = mucin 5AC; PAS = periodic acid-Schiff; Ck4 = cytokeratin 4.

5. Future Avenues for Developing Tissue Engineered Conjunctival Epithelial Equivalents

5.1. Comparative Studies of the Effect of Different Substrates on Cultured Conjunctival Epithelial Cells

In 2010, Rama and associates described the importance of the phenotype for clinical success following transplantation of cultured limbal epithelial cells [84]. p63, which is a marker for undifferentiated cells, was a significant predictor of clinical outcome [84]. It is possible that the phenotype of cultured CEC will determine success following transplantation of CEC. Comparative studies on how various substrates affect the cell sheet with regard to the phenotype in particular are, therefore, warranted.

5.2. Storage and Transportation of Cultured Conjunctival Epithelial Cells

With steadily stricter regulations for cell therapy, which lead to centralization of culture units [85], storage technology of cultured CEC has become increasingly important to allow the tissue to be transported to eye clinics worldwide [86]. Keeping in mind the significance of the phenotype for clinical outcome [84], assessment of the phenotype among other parameters prior to surgery should ideally be performed during the storage period. Moreover, storage in a hermetically sealed container enables microbiological assessment [87]. Finally, storage technology has the advantage of offering increased flexibility in scheduling surgery [88]. Comparative studies on how various substrates influence the ability to store cultured CEC with regard to morphology, viability, and phenotype should be performed to enable worldwide access to cultured CEC.

6. Conclusion

Amniotic membrane is the most commonly used substrate for CEC culture. The majority of the studies demonstrated that AM support the growth of goblet cells, in contrast to several alternative substrates. A major weakness in the current literature is the lack of comparative studies, thus such studies should be prioritized to be able to identify the most ideal substrate for ocular surface repair. Considering the disadvantages inherent to the use of a foreign biological material such as AM, clinical studies involving alternative membranes should be carried out as currently only AM has so far been used for transplanting tissue engineered CEC in humans.

Acknowledgments: Publishing costs were covered by Oslo University Hospital.

Author Contributions: Jon Roger Eidet, Darlene A. Dartt and Tor Paaske Utheim defined the topic and wrote the review.

Conflicts of Interest: The authors declare no conflicts of interest.

References

1. Wagoner, M.D. Chemical injuries of the eye: Current concepts in pathophysiology and therapy. *Surv. Ophthalmol.* **1997**, *41*, 275–313.

2. Gipson, I.K.; Joyce, N.; Zieske, J. The Anatomy and Cell Biology of the Human Cornea, Limbus, Conjunctiva, and Adnexa. In *The Cornea*; Foster, C.A.D., Dohlman, C., Eds.; Lippincott Williams & Wilkens: Philadelphia, PA, USA, 2005; pp. 1–35.

3. Steven, P.; Gebert, A. Conjunctiva-associated lymphoid tissue—Current knowledge, animal models and experimental prospects. *Ophthalmic Res.* **2009**, *42*, 2–8.

4. Schrader, S.; Notara, M.; Beaconsfield, M.; Tuft, S.J.; Daniels, J.T.; Geerling, G. Tissue engineering for conjunctival reconstruction: Established methods and future outlooks. *Curr. Eye Res.* **2009**, *34*, 913–924.

5. Yu, D.; Thelin, W.R.; Rogers, T.D.; Stutts, M.J.; Randell, S.H.; Grubb, B.R.; Boucher, R.C. Regional differences in rat conjunctival ion transport activities. *Am. J. Physiol. Cell Physiol.* **2012**, *303*, C767–C780.

6. Gendler, S.J.; Spicer, A.P. Epithelial mucin genes. *Annu. Rev. Physiol.* **1995**, *57*, 607–634.

7. Fostad, I.G.; Eidet, J.R.; Shatos, M.A.; Utheim, T.P.; Utheim, O.A.; Raeder, S.; Dartt, D.A. Biopsy harvesting site and distance from the explant affect conjunctival epithelial phenotype *ex vivo*. *Exp. Eye Res.* **2012**, *104*, 15–25.

8. Dartt, D.A. Control of mucin production by ocular surface epithelial cells. *Exp. Eye Res.* **2004**, *78*, 173–185.

9. Spurr-Michaud, S.; Argueso, P.; Gipson, I. Assay of mucins in human tear fluid. *Exp. Eye Res.* **2007**, *84*, 939–950.

10. Argueso, P.; Gipson, I.K. Epithelial mucins of the ocular surface: Structure, biosynthesis and function. *Exp. Eye Res.* **2001**, *73*, 281–289.

11. Ramirez-Miranda, A.; Nakatsu, M.N.; Zarei-Ghanavati, S.; Nguyen, C.V.; Deng, S.X. Keratin 13 is a more specific marker of conjunctival epithelium than keratin 19. *Mol. Vis.* **2011**, *17*, 1652–1661.

12. Nichols, K.K.; Foulks, G.N.; Bron, A.J.; Glasgow, B.J.; Dogru, M.; Tsubota, K.; Lemp, M.A.; Sullivan, D.A. The international workshop on meibomian gland dysfunction: Executive summary. *Invest. Ophthalmol. Vis. Sci.* **2011**, *52*, 1922–1929.

13. Wei, Z.G.; Lin, T.; Sun, T.T.; Lavker, R.M. Clonal analysis of the *in vivo* differentiation potential of keratinocytes. *Invest. Ophthalmol. Vis. Sci.* **1997**, *38*, 753–761.

14. Mason, S.L.; Stewart, R.M.; Kearns, V.R.; Williams, R.L.; Sheridan, C.M. Ocular epithelial transplantation: Current uses and future potential. *Regen. Med.* **2011**, *6*, 767–782.

15. Nagasaki, T.; Zhao, J. Uniform distribution of epithelial stem cells in the bulbar conjunctiva. *Invest. Ophthalmol. Vis. Sci.* **2005**, *46*, 126–132.

16. Wei, Z.G.; Cotsarelis, G.; Sun, T.T.; Lavker, R.M. Label-retaining cells are preferentially located in fornical epithelium: Implications on conjunctival epithelial homeostasis. *Invest. Ophthalmol. Vis. Sci.* **1995**, *36*, 236–246.

17. Lavker, R.M.; Wei, Z.G.; Sun, T.T. Phorbol ester preferentially stimulates mouse fornical conjunctival and limbal epithelial cells to proliferate *in vivo*. *Invest. Ophthalmol. Vis. Sci.* **1998**, *39*, 301–307.

18. Pe'er, J.; Zajicek, G.; Greifner, H.; Kogan, M. Streaming conjunctiva. *Anat. Rec.* **1996**, *245*, 36–40.

19. Chen, W.; Ishikawa, M.; Yamaki, K.; Sakuragi, S. Wistar rat palpebral conjunctiva contains more slow-cycling stem cells that have larger proliferative capacity: Implication for conjunctival epithelial homeostasis. *Jpn. J. Ophthalmol.* **2003**, *47*, 119–128.

20. Wei, Z.G.; Wu, R.L.; Lavker, R.M.; Sun, T.T. *In vitro* growth and differentiation of rabbit bulbar, fornix, and palpebral conjunctival epithelia. Implications on conjunctival epithelial transdifferentiation and stem cells. *Invest. Ophthalmol. Vis. Sci.* **1993**, *34*, 1814–1828.

21. Wirtschafter, J.D.; Ketcham, J.M.; Weinstock, R.J.; Tabesh, T.; McLoon, L.K. Mucocutaneous junction as the major source of replacement palpebral conjunctival epithelial cells. *Invest. Ophthalmol. Vis. Sci.* **1999**, *40*, 3138–3146.

22. Pellegrini, G.; Golisano, O.; Paterna, P.; Lambiase, A.; Bonini, S.; Rama, P.; de Luca, M. Location and clonal analysis of stem cells and their differentiated progeny in the human ocular surface. *J. Cell Biol.* **1999**, *145*, 769–782.

23. Qi, H.; Zheng, X.; Yuan, X.; Pflugfelder, S.C.; Li, D.Q. Potential localization of putative stem/progenitor cells in human bulbar conjunctival epithelium. *J. Cell Physiol.* **2010**, *225*, 180–185.

24. Stewart, R.M.; Sheridan, C.M.; Hiscott, P.S.; Czanner, G.; Kaye, S.B. Human conjunctival stem cells are predominantly located in the medial canthal and inferior forniceal areas. *Invest. Ophthalmol. Vis. Sci.* **2015**, *56*, 2021–2030.

25. Eidet, J.R.; Fostad, I.G.; Shatos, M.A.; Utheim, T.P.; Utheim, O.A.; Raeder, S.; Dartt, D.A. Effect of biopsy location and size on proliferative capacity of *ex vivo* expanded conjunctival tissue. *Invest. Ophthalmol. Vis. Sci.* **2012**, *53*, 2897–2903.

26. Ramos, T.; Scott, D.; Ahmad, S. An update on ocular surface epithelial stem cells: Cornea and conjunctiva. *Stem Cells Int.* **2015**, *2015*.

27. Schofield, R. The stem cell system. *Biomed. Pharmacother. Biomed. Pharmacother.* **1983**, *37*, 375–380.

28. Watt, F.M.; Hogan, B.L. Out of eden: Stem cells and their niches. *Science* **2000**, *287*, 1427–1430.

29. Schlotzer-Schrehardt, U.; Dietrich, T.; Saito, K.; Sorokin, L.; Sasaki, T.; Paulsson, M.; Kruse, F.E. Characterization of extracellular matrix components in the limbal epithelial stem cell compartment. *Exp. Eye Res.* **2007**, *85*, 845–860.

30. Radford, C.F.; Rauz, S.; Williams, G.P.; Saw, V.P.; Dart, J.K. Incidence, presenting features, and diagnosis of cicatrising conjunctivitis in the United Kingdom. *Eye* **2012**, *26*, 1199–1208.

31. Sobolewska, B.; Deuter, C.; Zierhut, M. Current medical treatment of ocular mucous membrane pemphigoid. *Ocul. Surf.* **2013**, *11*, 259–266.

32. Solomon, A.; Espana, E.M.; Tseng, S.C. Amniotic membrane transplantation for reconstruction of the conjunctival fornices. *Ophthalmology* **2003**, *110*, 93–100.

33. Patel, B.C.; Sapp, N.A.; Collin, R. Standardized range of conformers and symblepharon rings. *Ophthalmic Plast. Reconstr. Surg.* **1998**, *14*, 144–145.

34. Fein, W. Repair of total and subtotal symblepharons. *Ophthalmic Surg.* **1979**, *10*, 44–47.

35. Donnenfeld, E.D.; Perry, H.D.; Wallerstein, A.; Caronia, R.M.; Kanellopoulos, A.J.; Sforza, P.D.; D'Aversa, G. Subconjunctival mitomycin C for the treatment of ocular cicatricial pemphigoid. *Ophthalmology* **1999**, *106*, 72–79.

36. Thoft, R.A. Conjunctival transplantation. *Arch. Ophthalmol.* **1977**, *95*, 1425–1427.

37. Nakamura, T.; Inatomi, T.; Cooper, L.J.; Rigby, H.; Fullwood, N.J.; Kinoshita, S. Phenotypic investigation of human eyes with transplanted autologous cultivated oral mucosal epithelial sheets for severe ocular surface diseases. *Ophthalmology* **2007**, *114*, 1080–1088.

38. Burillon, C.; Huot, L.; Justin, V.; Nataf, S.; Chapuis, F.; Decullier, E.; Damour, O. Cultured autologous oral mucosal epithelial cell sheet (caomecs) transplantation for the treatment of corneal limbal epithelial stem cell deficiency. *Invest. Ophthalmol. Vis. Sci.* **2012**, *53*, 1325–1331.

39. Homma, R.; Yoshikawa, H.; Takeno, M.; Kurokawa, M.S.; Masuda, C.; Takada, E.; Tsubota, K.; Ueno, S.; Suzuki, N. Induction of epithelial progenitors *in vitro* from mouse embryonic stem cells and application for reconstruction of damaged cornea in mice. *Invest. Ophthalmol. Vis. Sci.* **2004**, *45*, 4320–4326.

40. Reinshagen, H.; Auw-Haedrich, C.; Sorg, R.V.; Boehringer, D.; Eberwein, P.; Schwartzkopff, J.; Sundmacher, R.; Reinhard, T. Corneal surface reconstruction using adult mesenchymal stem cells in experimental limbal stem cell deficiency in rabbits. *Acta Ophthalmol.* **2011**, *89*, 741–748.

41. Gomes, J.A.; Geraldes Monteiro, B.; Melo, G.B.; Smith, R.L.; Cavenaghi Pereira da Silva, M.; Lizier, N.F.; Kerkis, A.; Cerruti, H.; Kerkis, I. Corneal reconstruction with tissue-engineered cell sheets composed of human immature dental pulp stem cells. *Invest. Ophthalmol. Vis. Sci.* **2010**, *51*, 1408–1414.

42. Meyer-Blazejewska, E.A.; Call, M.K.; Yamanaka, O.; Liu, H.; Schlotzer-Schrehardt, U.; Kruse, F.E.; Kao, W.W. From hair to cornea: Toward the therapeutic use of hair follicle-derived stem cells in the treatment of limbal stem cell deficiency. *Stem Cells* **2011**, *29*, 57–66.

43. Reza, H.M.; Ng, B.Y.; Gimeno, F.L.; Phan, T.T.; Ang, L.P. Umbilical cord lining stem cells as a novel and promising source for ocular surface regeneration. *Stem Cell Rev.* **2011**, *7*, 935–947.

44. Lu, R.; Zhang, X.; Huang, D.; Huang, B.; Gao, N.; Wang, Z.; Ge, J. Conjunctival reconstruction with progenitor cell-derived autologous epidermal sheets in rhesus monkey. *PLoS One* **2011**, *6*.

45. Yang, S.P.; Yang, X.Z.; Cao, G.P. Conjunctiva reconstruction by induced differentiation of human amniotic epithelial cells. *Genet. Mol. Res.* **2015**, *14*, 13823–13834.

46. Wenkel, H.; Rummelt, V.; Naumann, G.O. Long term results after autologous nasal mucosal transplantation in severe mucus deficiency syndromes. *Br. J. Ophthalmol.* **2000**, *84*, 279–284.

47. Tseng, S.C. Amniotic membrane transplantation for ocular surface reconstruction. *Biosci. Rep.* **2001**, *21*, 481–489.

48. Touhami, A.; Grueterich, M.; Tseng, S.C. The role of NGF signaling in human limbal epithelium expanded by amniotic membrane culture. *Invest. Ophthalmol. Vis. Sci.* **2002**, *43*, 987–994.

49. Lee, S.B.; Li, D.Q.; Tan, D.T.; Meller, D.C.; Tseng, S.C. Suppression of TGF-β signaling in both normal conjunctival fibroblasts and pterygial body fibroblasts by amniotic membrane. *Curr. Eye Res.* **2000**, *20*, 325–334.

50. Ueta, M.; Kweon, M.N.; Sano, Y.; Sotozono, C.; Yamada, J.; Koizumi, N.; Kiyono, H.; Kinoshita, S. Immunosuppressive properties of human amniotic membrane for mixed lymphocyte reaction. *Clin. Exp. Immunol.* **2002**, *129*, 464–470.

51. Solomon, A.; Rosenblatt, M.; Monroy, D.; Ji, Z.; Pflugfelder, S.C.; Tseng, S.C. Suppression of interleukin 1α and interleukin 1β in human limbal epithelial cells cultured on the amniotic membrane stromal matrix. *Br. J. Ophthalmol.* **2001**, *85*, 444–449.

52. Kruse, F.E.; Joussen, A.M.; Rohrschneider, K.; You, L.; Sinn, B.; Baumann, J.; Volcker, H.E. Cryopreserved human amniotic membrane for ocular surface reconstruction. *Graefe's Arch. Clin. Exp. Ophthalmol.* **2000**, *238*, 68–75.

53. Lee, S.H.; Tseng, S.C. Amniotic membrane transplantation for persistent epithelial defects with ulceration. *Am. J. Ophthalmol.* **1997**, *123*, 303–312.

54. Nakamura, T.; Yoshitani, M.; Rigby, H.; Fullwood, N.J.; Ito, W.; Inatomi, T.; Sotozono, C.; Nakamura, T.; Shimizu, Y.; Kinoshita, S. Sterilized, freeze-dried amniotic membrane: A useful substrate for ocular surface reconstruction. *Invest. Ophthalmol. Vis. Sci.* **2004**, *45*, 93–99.

55. Russo, A.; Bonci, P.; Bonci, P. The effects of different preservation processes on the total protein and growth factor content in a new biological product developed from human amniotic membrane. *Cell. Tissue Banking* **2012**, *13*, 353–361.

56. Von Versen-Hoeynck, F.; Steinfeld, A.P.; Becker, J.; Hermel, M.; Rath, W.; Hesselbarth, U. Sterilization and preservation influence the biophysical properties of human amnion grafts. *Biol. J. Int. Assoc. Biol. Standard.* **2008**, *36*, 248–255.

57. Thomasen, H.; Pauklin, M.; Steuhl, K.P.; Meller, D. Comparison of cryopreserved and air-dried human amniotic membrane for ophthalmologic applications. *Graefe's Arch. Clin. Exp. Ophthalmol.* **2009**, *247*, 1691–1700.

58. Henderson, H.W.; Collin, J.R. Mucous membrane grafting. *Dev. Ophthalmol.* **2008**, *41*, 230–242.

59. Meller, D.; Tseng, S.C. Conjunctival epithelial cell differentiation on amniotic membrane. *Invest. Ophthalmol. Vis. Sci.* **1999**, *40*, 878–886.

60. Eidet, J.R.; Utheim, O.A.; Raeder, S.; Dartt, D.A.; Lyberg, T.; Carreras, E.; Huynh, T.T.; Messelt, E.B.; Louch, W.E.; Roald, B.; *et al.* Effects of serum-free storage on morphology, phenotype, and viability of *ex vivo* cultured human conjunctival epithelium. *Exp. Eye Res.* **2012**, *94*, 109–116.

61. Martinez-Osorio, H.; Calonge, M.; Corell, A.; Reinoso, R.; Lopez, A.; Fernandez, I.; San Jose, E.G.; Diebold, Y. Characterization and short-term culture of cells recovered from human conjunctival epithelium by minimally invasive means. *Mol. Vis.* **2009**, *15*, 2185–2195.

62. Fuchs, E.; Green, H. Regulation of terminal differentiation of cultured human keratinocytes by vitamin A. *Cell* **1981**, *25*, 617–625.

63. Sangwan, V.S.; Vemuganti, G.K.; Iftekhar, G.; Bansal, A.K.; Rao, G.N. Use of autologous cultured limbal and conjunctival epithelium in a patient with severe bilateral ocular surface disease induced by acid injury: A case report of unique application. *Cornea* **2003**, *22*, 478–481.

64. Tan, Y.; Qiu, F.; Qu, Y.L.; Li, C.; Shao, Y.; Xiao, Q.; Liu, Z.; Li, W. Amniotic membrane inhibits squamous metaplasia of human conjunctival epithelium. *Am. J. Physiol. Cell Physiol.* **2011**, *301*, C115–C125.

65. Yoshizawa, M.; Feinberg, S.E.; Marcelo, C.L.; Elner, V.M. *Ex vivo* produced human conjunctiva and oral mucosa equivalents grown in a serum-free culture system. *J. Oral Maxillofac. Surg.* **2004**, *62*, 980–988.

66. Tsai, R.J.; Tseng, S.C. Substrate modulation of cultured rabbit conjunctival epithelial cell differentiation and morphology. *Invest. Ophthalmol. Vis. Sci.* **1988**, *29*, 1565–1576.

67. Chen, W.Y.; Mui, M.M.; Kao, W.W.; Liu, C.Y.; Tseng, S.C. Conjunctival epithelial cells do not transdifferentiate in organotypic cultures: Expression of K12 keratin is restricted to corneal epithelium. *Curr. Eye Res.* **1994**, *13*, 765–778.

68. Drechsler, C.C.; Kunze, A.; Kureshi, A.; Grobe, G.; Reichl, S.; Geerling, G.; Daniels, J.T.; Schrader, S. Development of a conjunctival tissue substitute on the basis of plastic compressed collagen. *J. Tissue Eng. Regen. Med.* **2015**.

69. Gordon, Y.J.; Huang, L.C.; Romanowski, E.G.; Yates, K.A.; Proske, R.J.; McDermott, A.M. Human cathelicidin (LL-37), a multifunctional peptide, is expressed by ocular surface epithelia and has potent antibacterial and antiviral activity. *Curr. Eye Res.* **2005**, *30*, 385–394.

70. Cook, E.B.; Stahl, J.L.; Miller, S.T.; Gern, J.E.; Sukow, K.A.; Graziano, F.M.; Barney, N.P. Isolation of human conjunctival mast cells and epithelial cells: Tumor necrosis factor-alpha from mast cells affects intercellular adhesion molecule 1 expression on epithelial cells. *Invest. Ophthalmol. Vis. Sci.* **1998**, *39*, 336–343.

71. Saha, P.; Kim, K.J.; Lee, V.H. A primary culture model of rabbit conjunctival epithelial cells exhibiting tight barrier properties. *Curr. Eye Res.* **1996**, *15*, 1163–1169.

72. Basu, S.K.; Haworth, I.S.; Bolger, M.B.; Lee, V.H. Proton-driven dipeptide uptake in primary cultured rabbit conjunctival epithelial cells. *Invest. Ophthalmol. Vis. Sci.* **1998**, *39*, 2365–2373.

73. Kurpakus, M.A.; Lin, L. The lack of extracellular laminin β2 chain deposition correlates to the loss of conjunctival epithelial keratin K4 localization in culture. *Curr. Eye Res.* **1999**, *18*, 28–38.

74. Lin, L.; Kurpakus Wheater, M. Differential rapid adhesion of bovine ocular surface epithelial cells to laminin isoforms. *Curr. Eye Res.* **1999**, *19*, 293–299.

75. Lin, L.; Kurpakus-Wheater, M. Laminin α5 chain adhesion and signaling in conjunctival epithelial cells. *Invest. Ophthalmol. Vis. Sci.* **2002**, *43*, 2615–2621.

76. Gipson, I.K.; Spurr-Michaud, S.; Argueso, P.; Tisdale, A.; Ng, T.F.; Russo, C.L. Mucin gene expression in immortalized human corneal-limbal and conjunctival epithelial cell lines. *Invest. Ophthalmol. Vis. Sci.* **2003**, *44*, 2496–2506.

77. Zhu, X.; Beuerman, R.W.; Chan-Park, M.B.; Cheng, Z.; Ang, L.P.; Tan, D.T. Enhancement of the mechanical and biological properties of a biomembrane for tissue engineering the ocular surface. *Ann. Acad. Med. Singapore* **2006**, *35*, 210–214.

78. Niiya, A.; Matsumoto, Y.; Ishibashi, T.; Matsumoto, K.; Kinoshita, S. Collagen gel-embedding culture of conjunctival epithelial cells. *Graefe's Arch. Clin. Exp. Ophthalmol.* **1997**, *235*, 32–40.

79. Martinez-Osorio, H.; Juarez-Campo, M.; Diebold, Y.; Girotti, A.; Alonso, M.; Arias, F.J.; Rodriguez-Cabello, J.C.; Garcia-Vazquez, C.; Calonge, M. Genetically engineered elastin-like polymer as a substratum to culture cells from the ocular surface. *Curr. Eye Res.* **2009**, *34*, 48–56.

80. Ang, L.P.; Cheng, Z.Y.; Beuerman, R.W.; Teoh, S.H.; Zhu, X.; Tan, D.T. The development of a serum-free derived bioengineered conjunctival epithelial equivalent using an ultrathin poly(epsilon-caprolactone) membrane substrate. *Invest. Ophthalmol. Vis. Sci.* **2006**, *47*, 105–112.

81. Campillo-Fernandez, A.J.; Pastor, S.; Abad-Collado, M.; Bataille, L.; Gomez-Ribelles, J.L.; Meseguer-Duenas, J.M.; Monleon-Pradas, M.; Artola, A.; Alio, J.L.; Ruiz-Moreno, J.M. Future design of a new keratoprosthesis. Physical and biological analysis of polymeric substrates for epithelial cell growth. *Biomacromolecules* **2007**, *8*, 2429–2436.

82. Huhtala, A.; Pohjonen, T.; Salminen, L.; Salminen, A.; Kaarniranta, K.; Uusitalo, H. *In vitro* biocompatibility of degradable biopolymers in cell line cultures from various ocular tissues: Direct contact studies. *J. Biomed. Mater. Res. Part A* **2007**, *83*, 407–413.

83. Huhtala, A.; Pohjonen, T.; Salminen, L.; Salminen, A.; Kaarniranta, K.; Uusitalo, H. *In vitro* biocompatibility of degradable biopolymers in cell line cultures from various ocular tissues: Extraction studies. *J. Mater. Sci. Mater. Med.* **2008**, *19*, 645–649.

84. Rama, P.; Matuska, S.; Paganoni, G.; Spinelli, A.; de Luca, M.; Pellegrini, G. Limbal stem-cell therapy and long-term corneal regeneration. *N Engl. J. Med.* **2010**, *363*, 147–155.

85. Daniels, J.T.; Secker, G.A.; Shortt, A.J.; Tuft, S.J.; Seetharaman, S. Stem cell therapy delivery: Treading the regulatory tightrope. *Regen. Med.* **2006**, *1*, 715–719.

86. Ahmad, S.; Osei-Bempong, C.; Dana, R.; Jurkunas, U. The culture and transplantation of human limbal stem cells. *J. Cell Physiol.* **2010**, *225*, 15–19.

87. Utheim, T.P.; Raeder, S.; Utheim, O.A.; de la Paz, M.; Roald, B.; Lyberg, T. Sterility control and long-term eye-bank storage of cultured human limbal epithelial cells for transplantation. *Br. J. Ophthalmol.* **2009**, *93*, 980–983.

88. O'Callaghan, A.R.; Daniels, J.T. Concise review: Limbal epithelial stem cell therapy: Controversies and challenges. *Stem Cells* **2011**, *29*, 1923–1932.

Pre-Clinical Cell-Based Therapy for Limbal Stem Cell Deficiency

Amer Sehic, Øygunn Aass Utheim, Kristoffer Ommundsen and
Tor Paaske Utheim

Abstract: The cornea is essential for normal vision by maintaining transparency for light transmission. Limbal stem cells, which reside in the corneal periphery, contribute to the homeostasis of the corneal epithelium. Any damage or disease affecting the function of these cells may result in limbal stem cell deficiency (LSCD). The condition may result in both severe pain and blindness. Transplantation of *ex vivo* cultured cells onto the cornea is most often an effective therapeutic strategy for LSCD. The use of *ex vivo* cultured limbal epithelial cells (LEC), oral mucosal epithelial cells, and conjunctival epithelial cells to treat LSCD has been explored in humans. The present review focuses on the current state of knowledge of the many other cell-based therapies of LSCD that have so far exclusively been explored in animal models as there is currently no consensus on the best cell type for treating LSCD. Major findings of all these studies with special emphasis on substrates for culture and transplantation are systematically presented and discussed. Among the many potential cell types that still have not been used clinically, we conclude that two easily accessible autologous sources, epidermal stem cells and hair follicle-derived stem cells, are particularly strong candidates for future clinical trials.

Reprinted from *J. Funct. Biomater.* Cite as: Sehic, A.; Utheim, Ø.A.; Ommundsen, K.; Utheim, T.P. Pre-Clinical Cell-Based Therapy for Limbal Stem Cell Deficiency. *J. Funct. Biomater.* **2015**, *6*, 863–888.

1. Cornea and Limbal Stem Cells

The cornea is the anterior, transparent, and avascular tissue with high refractive power that directs light bundles to the retina [1]. The highly specialized structure of the cornea is essential for normal vision. From anterior to posterior, the cornea is composed of five layers, *i.e.*, epithelium, Bowman's membrane, stroma, Descemet's membrane, and endothelium. The corneal epithelium is composed of a basal layer of column-shaped cells, a suprabasal layer of cuboid wing cells, and a superficial layer of flat squamous cells [2]. The thickness of the corneal epithelium in different species, e.g., human, mouse, and rabbit, is conspicuously perpetual, ranging from 45 to 50 μm [3–5]. The renewal of corneal epithelium differs between species and is renewed every 9–12 months in rabbits [6]. The corneal epithelium plays an essential role in maintaining the cornea's avascularity and transparency [7].

The self-renewing properties of the corneal epithelium are an important requirement for corneal integrity and function [8]. This process is dependent on a small population of limbal stem cells that are situated in the basal region of the limbus [9,10]. Limbal stem cells are presented in the basal layer of the limbal epithelium and give rise to fast-dividing, transient amplifying cells [11]. Transient amplifying cells go through a restricted number of divisions before becoming terminally differentiated cells [12]. It has been hypothesized that corneal epithelial maintenance can be defined by the equation $X + Y = Z$, in which X refers to proliferation of basal cells; Y is the centripetal movement of peripheral cells; and Z is the epithelial cell loss from the corneal surface [13].

2. Limbal Stem Cell Deficiency

Any process or disease that results in dysfunction or loss of the limbal epithelial cells (LEC) may result in limbal stem cell deficiency (LSCD) [7]. In LSCD, the conjunctival epithelium migrates across the limbus, resulting in loss of corneal clarity and visual impairment. The condition is painful and potentially blinding [14]. Normal and well-functioning LEC act as an important barrier, preventing invasion of the cornea by conjunctival tissue. Limbal stem cell deficiency typically worsens over time since chronic inflammation not only results in the death of LEC, but also negatively affects the remaining stem cells and their function [14].

The prevalence and incidence of LSCD worldwide are not known. In India, the prevalence is estimated to be approximately 1.5 million [15], and the incidence in North America is estimated to be "thousands" [16]. The etiology of many cases of LSCD is known; however, idiopathic cases also exist [17,18]. Acquired causes of LSCD include thermal and chemical burns of the ocular surface, contact lens wear, ultraviolet radiation, extensive cryotherapy, or surgery to the limbus [7]. There are also numerous hereditary causes of LSCD, including aniridia, where the anterior segment of the eye including the limbus is imperfectly developed. Furthermore, autoimmune diseases involving the ocular surface, e.g., Stevens-Johnson syndrome and ocular cicatricial pemphigoid, are examples of nonhereditary causes of LSCD.

Limbal stem cell deficiency is classified as either partial or total, depending on the extent of the disorder. Conjunctivalization is pathognomonic for LSCD. Other signs are persistent epithelial defects, superficial and deep corneal vascularization, and fibrovascular pannus. Limbal stem cell deficiency in patients with significantly dry eyes results in a partial or total keratinized epithelium [19]. The diagnosis can be corroborated by detection of conjunctival cells on the corneal surface by cytological analysis [20] or *in vivo* confocal microscopy [21], but is seldom performed as the diagnosis is often obvious.

3. Treatment Approaches for Limbal Stem Cell Deficiency

The core of conservative treatment for LSCD lies in the improvement of epithelial healing. A range of clinical procedures, with distinctive benefits and limitations, are currently available for treating LSCD. However, variations in both the severity and causes of LSCD explain why the application of one treatment approach will not be adequate for all. A great variety of cell-based therapeutic strategies have been suggested for LSCD over the past 10 years. In cases of partial LSCD, amniotic membrane (AM) can be applied to the affected eye and aids in repopulating the ocular surface with corneal epithelium [22]. With increased understanding of the origin of the stem cells in the limbus [10], the transplantation of limbal grafts was introduced in 1989 [23], a promising treatment strategy for restoring the ocular surface following LSCD. This procedure, however, carried a risk of inducing LSCD in the healthy eye due to the need of large limbal biopsy, making the therapy impossible in cases of bilateral LSCD.

In 1997, a groundbreaking therapeutic strategy involving *ex vivo* expansion of LEC was introduced [24]. The principle of this method is to culture LEC harvested from the patient, a living relative, or a cadaver on a substrate in the laboratory and then transfer the cultured tissue onto the eyes of patients suffering from LSCD. This therapy has gained popularity in ophthalmology as it increases cell numbers before transplantation without the need for a large limbal biopsy. It is suggested that the mechanism underlying the improvement in the ocular surface after LEC allograft transplantation is due to the stimulation of a small number of residual dormant host cells, rather than transplanted cells, permanently replacing the ocular surface [25]. Another possibility is that the transplanted graft somehow attends to stimulate progenitor cells in the blood stream to repopulate the ocular surface [25].

Recently, the use of induced pluripotent stem cells (iPSCs) has attracted great attention [26,27]. Following culture for two weeks on an amniotic membrane, limbal iPSCs developed substantially higher expression of several putative limbal stem cell markers, including ABCG2 and ΔNp63α, than did fibroblast iPSCs [27]. The successful generation of iPSCs from human primary LEC, and subsequent re-differentiation back to the limbal corneal epithelium, has been demonstrated *in vitro* [27]. However, IPSCs have so far not been used in clinical studies or experimental animals for ocular surface reconstruction, despite the great promise this treatment holds.

Since 1997, several research groups have shown favorable effects of *ex vivo* cultured cell therapy for LSCD in both clinical studies and experimental animals. There is currently a strong trend toward applying autologous sources as there is no risk for immunological reactions and, therefore, no requirement for immunosuppressive therapy with all known side effects [28]. Since 2003, several non-limbal cells have been successfully used to reconstruct the corneal epithelium in bilateral LSCD, in which limbal tissue is not recommended for harvest. Among non-limbal cell types, oral mucosal epithelial cells and conjuctival epithelial cells are the only laboratory cultured cell sources that have been explored in humans. Oral mucosal epithelial cells were the first non-limbal cell type to be identified as a potential source for LSCD. So far, 242 patients have been reported to be treated with a success rate of 72% [29]. Since 2009, conjunctival epithelial cells have also been used with the purpose of reverting LSCD in clinical trials, but the number of patients treated is small [30]. Since 2010, there have been two clinical studies including 17 eyes that have used nasal mucosal epithelial cells to treat LSCD with promising results [31,32]. In contrast to most of the other cell types that have been used for LSCD therapy, nasal mucosa was transplanted to the eyes without prior *ex vivo* cultivation, which substantially simplifies the procedure.

A number of other non-ocular cells have been investigated as alternative stem cell sources for treating LSCD; however, they have only been studied in animal experiments. As none of the cell types used in clinical trials have proved to be successful in more than about three of four cases [7,29], there has been a constant search for novel cell types that potentially could be more effective in reverting LSCD. The present review focuses on these cell types. The review was prepared by searching the National Library of Medicine database using the broad search term "limbal" in an attempt not to leave out any relevant publications. In total, the search resulted in 3634 studies, whereof 19 studies, published from 2004 to 2014, were related directly to the core topic of the present review. These studies include the following cultured cell types: (1) bone marrow-derived mesenchymal stem cells (Table 1) [33–40]; (2) embryonic stem cells (Table 2) [41–44]; (3) epidermal stem cells (Table 3) [45–47]; (4) hair follicle-derived stem cells (Table 4) [48]; (5) immature dental pulp stem cells (Table 4) [49,50]; (6) orbital fat-derived stem cells (Table 4) [51]; and (7) umbilical cord stem cells (Table 4) [52]. Various substrates and methods have been applied to culture and transplant these cell sources onto damaged corneas of mice, rats, rabbits, pigs, and goats (Figure 1, Table 5). In the present review, the ability of all these cell sources to treat LSCD is discussed.

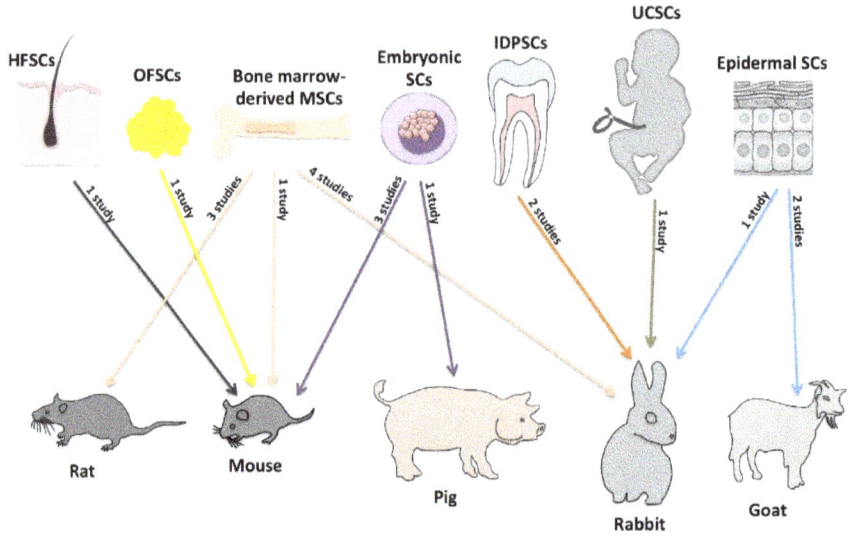

Figure 1. Overview of stem cell sources used in animal experiments. Arrows, including number of studies, indicate the connection between different stem cell sources and LSCD animal models that they have been transplanted to. HFSCs, hair follicle-derived stem cells; MSCs, mesenchymal stem cells; SCs, stem cells; IDPDSCs, immature dental pulp stem cells; OFSCs, orbital fat-derived stem cells; UCSCs, umbilical cord stem cells.

Table 1. Reconstruction of ocular surface using cultured bone marrow-derived mesenchymal stem cells.

Author, Year, (Reference)	Cell Source	Methods	LSCD Model	Follow-up Time	Evaluation	Results
Ma et al. 2006 [35]	Bone Marrow-Derived MSCs; Human	Cultured on AM carrier; Transplanted ($n = 16$); Control groups: 1) transplanted with fibroblast cells on AM ($n = 8$) and 2) transplanted with only AM ($n = 7$)	Rats; Disc paper saturated with 1 N NaOH onto cornea	4 weeks	Slit lamp evaluation; Histology; IH	Reconstruction in 100% (16/16) of animals; Cornea completely transparent in 56.3% (9/16) of animals; Neovascularization detected within 2 mm and over 2 mm in 37.5% (6/16) and 12.5% (2/16) of animals, respectively; No complications
Ye et al. 2006 [39]	Bone Marrow-Derived MSCs; Rabbit	Cultured in α-MEM; IV injection; Four groups: 1) normal BM function, without MSCs injection ($n = 15$), 2) normal BM function, with MSCs injection ($n = 15$), 3) BM suppressed by CP, without MSCs injection ($n = 15$), 4) BM suppressed by CP, with MSCs injection ($n = 15$)	Rabbits; Filter paper saturated with 1 N NaOH onto cornea	1 month	Slit lamp evaluation; IH	Reconstruction in 100% (15/15) of animals in Group 2; Cornea more clear in group 2 compared with other groups; Neovascularization appeared on day 14 in Group 2; No complications
Gu et al. 2009 [33]	Bone Marrow-Derived MSCs; Rabbit	Cultured on fibrin carrier; Transplanted ($n = 10$); Control: eyes transplanted with only fibrin graft gel ($n = 10$)	Rabbits; Cornea treated with n-heptanol	4 weeks	Slit lamp evaluation; Histology; FC; IF	Reconstruction in 100% (10/10) of animals; Iris partially clear in 30% (3/10) and completely obscure in 70% (7/10) of animals; Neovascularization detected over 3 mm from the limbus in 80% (8/10) of animals; No complications
Omoto et al. 2009 [36]	Bone Marrow-Derived MSCs; Human	Cultured in α-MEM; Carrier-free sheets transplanted; Control: no transplantation; Number of animals not reported	Rabbits; Cornea treated with n-heptanol	4 weeks	Slit lamp evaluation; Histology; IH; RT-PCR	Reconstruction of corneal epithelium successful; Corneal clarity: no data; Neovascularization: no data; No complications
Jiang et al. 2010 [34]	Bone Marrow-Derived MSCs; Rat	Cultured on AM carrier. Three groups: 1) transplanted with only AM ($n = 12$); 2) MSCs on AM ($n = 12$); 3) MSCs induced by CSCs on AM ($n = 12$); Control: no transplantation ($n = 12$)	Rats; Filter paper saturated with 1 N NaOH onto cornea	10 weeks	Slit lamp evaluation; Histology; CLCM; SEM; FC; IF; IH	Reconstruction in 75% (9/12) of animals in group 3; Cornea completely transparent in 75% (9/12) of animals; Neovascularization limited within 2 mm of the limbus; No complications

269

Table 1. *Cont.*

Author, Year, (Reference)	Cell Source	Methods	LSCD Model	Follow-up Time	Evaluation	Results
Zajicova *et al.* 2010 [40]	Bone Marrow-Derived MSCs; Mouse	Cultured on nanofiber scaffold carrier; Co-transplantation of LSC and MSCs; Control: normal eyes; Number of animals not reported	Mice; Epithelial debridement with a needle	2 weeks	Slit lamp evaluation; CLCM; FC; RT-PCR	Significantly inhibited local inflammatory reactions and supported healing process; Corneal clarity: no data; Neovascularization: no data; No complications
Reinshagen *et al.* 2011 [37]	Bone Marrow-Derived MSCs; Rabbit	Cultured in DMEM; Three groups: 1) MSCs injected under transplanted AM (*n* = 6); 2) transplanted with only AM (*n* = 5); 3) transplanted with AM and autologous LEC (*n* = 4). Control: no transplantation (*n* = 6)	Rabbits; Cornea treated with n-heptanol	6 months	Slit lamp evaluation; Histology; IH	Reconstruction in 100% (6/6) of animals in Group 1; Improved corneal clarity; Neovascularization of the entire cornea in all animals; No complications
Rohaina *et al.* 2014 [38]	Bone Marrow-Derived MSCs; Human	Cultured on AM carrier; Transplanted (*n* = 4); Control groups: 1) transplanted with only AM (*n* = 5); 2) no transplantation (*n* = 6)	Rats; Disc paper saturated with 1 N NaOH onto cornea	8 weeks	Slit lamp evaluation; Histology; IH; OCT; RT-PCR	Reconstruction in 100% (4/4) of animals; Moderate corneal clarity; Minimal vascularization; No complications

AM, amniotic membrane; BM, bone marrow; CFE, colony-forming efficiency; CLCM, confocal laser corneal microscopy; CP, cyclophosphamide; CSCs, corneal stromal cells; FC, flow cytometry; IH, immunohistochemistry; IF, immunofluorescence; IV, intravenous; LEC, limbal epithelial cells; LSC, limbal stem cells; LSCD, limbal stem cell deficiency; MSCs, mesenchymal stem cells; OCT, optical coherence tomography; RT-PCR, reverse transcriptase polymerase chain reaction; SEM, scanning electron microscopy.

Table 2. Reconstruction of ocular surface using cultured embryonic stem cells.

Author, Year, (Reference)	Cell Source	Methods	LSCD Model	Follow-up Time	Evaluation	Results
Homma et al. 2004 [41]	Embryonic SCs; Mouse	Cultured on collagen IV-coated plates; Carrier-free sheets transplanted ($n = 10$); Control: no transplantation ($n = 10$)	Mice; Cornea treated with n-heptanol	24 h	FC; Histology; RT-PCR; WB	Reconstruction in 100% (10/10) of animals; Corneal clarity: no data; Neovascularization: no data; No complications
Ueno et al. 2007 [44]	Embryonic SCs; Mouse	Cultured on gelatin-coated plates; Transfected with Pax6; Carrier-free sheets transplanted ($n = 5$); Control groups: 1) normal eyes ($n = 5$); 2) no transplantation ($n = 5$)	Mice; Cornea treated with n-heptanol	24 h	Histology; IF; RT-PCR	Reconstruction in 100% (5/5) of animals 12 h after transplantation; Corneal clarity: no data; Neovascularization: no data; No complications
Kumagai et al. 2010 [42]	Embryonic SCs; Monkey	Cultured on collagen IV-coated plates; Carrier-free sheets transplanted ($n = 10$); Control groups: 1) normal eyes ($n = 10$); 2) no transplantation ($n = 10$)	Mice; Cornea treated with n-heptanol	6 h	CLCM; IF; RT-PCR	Transplanted cells adhered to the corneal stroma and formed multiple cell layers in 100% (10/10) of animals; Corneal clarity: no data; Neovascularization: no data; No complications
Notara et al. 2013 [43]	Embryonic SCs; Mouse	Cultured on collagen IV-coated plates; Carrier-free sheets transplanted; Control: no transplantation; Number of animals not reported	Pigs; Epithelial debridement with a blade	5 weeks	Histology; IH; RT-PCR; WB	Reconstruction after 1 week; Corneal clarity: no data; Neovascularization: no data; Mild immune reaction

CLCM, confocal laser corneal microscopy; FC, flow cytometry; IF, immunofluorescence; IH, immunohistochemistry; RT-PCR, reverse transcriptase polymerase chain reaction; SCs, stem cells; SEM, scanning electron microscopy; WB, western blotting.

Table 3. Reconstruction of ocular surface using cultured epidermal stem cells.

Author, Year, (Reference)	Cell Source	Methods	LSCD Model	Follow-up Time	Evaluation	Results
Yang et al. 2007 [47]	Epidermal SCs; Goat	Cultured on AM carrier; Transplanted ($n = 7$); Control groups: 1) transplantation with AM ($n = 4$); 2) no transplantation ($n = 4$)	Goats; Excision of the cornea and limbus	24 months	IH; SEM; TEM	Reconstruction in 100% (7/7) of animals; Two or three quadrants of clear cornea in 71.4% (5/7) of animals at follow-up time to 24 months; Minimal neovascularization; Perforation through the pupil during operation in one eye
Yang et al. 2008 [46]	Epidermal SCs; Goat	Cultured on AM carrier; Transplanted ($n = 10$); Control groups: 1) transplanted with only AM ($n = 8$); 2) no transplantation ($n = 8$)	Goats; Excision of the cornea and limbus; Burned with 1 N NaOH	30 months	Digital camera; Histology; IH	Reconstruction in 100% (10/10) of animals; Three or four quadrants of clear cornea in 80% (8/100) of animals at follow-up time to 30 months; Minimal neovascularization; No complications
Ouyang et al. 2014 [45]	Epidermal SCs; Human	Cultured on fibrin carrier; Transduction of Pax6 converted these cells into LSC-like cells; Transplanted and covered with AM ($n = 5$); Control: transplanted with only AM ($n = 4$)	Rabbits; Excision of the cornea and limbus	3 months	CLCM; IF; Microarrays; Quantitative PCR; RNA-sequencing; WB	Reconstruction in 100% (5/5) of animals; Transparent cornea in 100% (5/5) of animals for over 3 months; Minimal neovascularization; No complications

AM, amniotic membrane; CLCM, confocal laser corneal microscopy; IF, immunofluorescence; IH, immunohistochemistry; LEC, limbal epithelial cells; LSC, limbal stem cells; LSCD, limbal stem cell deficiency; PCR, polymerase chain reaction; SCs, stem cells; SEM, scanning electron microscopy; TEM, transmission electron microscopy; WB, western blotting.

272

Table 4. Reconstruction of ocular surface using cultured immature dental pulp stem cells, hair follicle-derived stem cells, umbilical cord stem cells, and orbital fat-derived stem cells.

Author, Year, (Reference)	Cell Source	Methods	LSCD Model	Follow-up Time	Evaluation	Results
Monteiro et al. 2009 [50]	IDPSCs; Human	Cultured on AM carrier; Transplanted (n = 5); Control: transplanted with only AM (n = 5)	Rabbits; Chemical burn of the cornea	3 months	Slit lamp evaluation; CLCM; IF; RT-PCR	Reconstruction in 100% (5/5) of animals; Gradual improvement in corneal transparency in 100% (5/5) of animals during follow-up time of 3 months; Neovascularization: no data; No complications
Gomes et al. 2010 [49]	IDPSCs; Human	Cultured on AM carrier; MCB (n = 5), SCB (n = 4); Transplanted and covered with AM; Control: transplanted with only AM (n = 6)	Rabbits; Filter paper saturated with 0.5 M NaOH for 25 s (MCB), and for 45 s (SCB)	3 months	Slit lamp evaluation; EM; Histology; IH	Reconstruction in 100% (5/5) of animals; Less organized and loose corneal epithelium in 75% (3/4) of SCB animals; Improved corneal clarity in 100% (5/5) of MCB animals; Superficial neovascularization in one animal No complications
Meyer-Blazejewska et al. 2011 [48]	HFSCs; Mouse	Cultured on fibrin carrier; Transplanted (n = 31); Control: no transplantation (n = 31)	Mice; Cornea and limbus removed	5 weeks	Slit lamp evaluation; Histology; IF	Reconstruction in 87.5% (7/8) of animals after two weeks Improved corneal clarity; Neovascularization in 12.5% (1/8) of animals; No complications
Reza et al. 2011 [52]	UCSCs; Human	Cultured on AM carrier; Three groups: 1) transplanted cell sheets on AM (n = 6); 2) transplanted with only AM; 3) no transplantation	Rabbits; Cornea and limbus removed	4 weeks	Slit lamp evaluation; Histology; IC; IH; RT-PCR	Reconstruction in 66.7% (4/6) of animals; Corneal clarity: no data; Severe neovascularization in one eye; Mild superficial inflammation in one other
Lin et al. 2013 [51]	OFSCs; Human	Cultured in MesenPro medium; Topical application of cells (n = 9), Intra-limbal injection of cells (n = 3); Control: Topical application of PBS (n = 6), Injection of PBS (n = 3), no treatment (n = 3)	Mice; Filter paper saturated with 0.5 N NaOH onto cornea	1 week	Digital camera; Histology; IH; IF; WB	Reconstruction of corneal epithelium after 1 week; Improved corneal clarity; No neovascularization; No complications

AM, amniotic membrane; CLCM, Confocal laser corneal microscopy; EM, electron microscopy; HFSCs, hair follicle-derived stem cells; IC, immunocytochemistry; IF, immunofluorescence; IH, immunohistochemistry; IDPSCs, immature dental pulp stem cells; LSC, limbal stem cells; LSCD, limbal stem cell deficiency; MCB, mild chemical burn; OFSCs, orbital fat-derived stem cells; PBS, phosphate buffered saline; RT-PCR, reverse transcriptase polymerase chain reaction; SC, stem cell; SCB, severe chemical burn; UCSCs, umbilical cord stem cells.

Table 5. Different culture and carrier biomaterials and methods used in cell-based therapies of LSCD, explored in animal models.

Methods	Materials	References
Transplantation	Carrier-free cell sheets	[36,41–44]
Transplantation	Amniotic membrane	[34,38,46,47,49,50,52,53]
Intravenous injection	–	[39]
Transplantation	Fibrin scaffold	[33,45,48]
Transplantation	Nanofiber scaffold	[40]
Injection under amniotic membrane	–	[37]
Topical application/Intra-limbal injection	–	[51]

4. Substrates for Corneal Reconstruction

To what extent biomechanical properties of the underlying substrate determine the success of *ex vivo* expansion of stem cells in treatment of LSCD is unknown. It is reasonable to assume that the optimal substrate will at least in some way resemble the limbal niche, in which limbal stem cells reside. The most common culture substrate for corneal reconstruction has so far been human AM. However, a number of alternative biological, biosynthetic, or synthetic substrates have been suggested as potential materials for ocular surface reconstruction (Table 6). The fundamental characteristics of an appropriate scaffold include cell attachment and cell proliferation both in culture and after transplantation, transparency, mechanical stability, and biocompatibility. In the studies on cell-based therapies for LSCD that have only been investigated in animal experiments, three substrates have so far been used: AM [34,38,46,47,49,50,52,53], nanofiber scaffold [40], and fibrin scaffold [33,45,48]. In addition, carrier-free methods [36,41–44], transplanting intact cell sheets without an underlying supportive membrane, injection of cells under transplanted AM [37], topical application of cells [51], intra-limbal injection of cells [51], and intravenous injection through an ear vein [39] have been applied (Table 5).

Amniotic membrane promotes cellular growth and exhibits anti-angiogenic and anti-inflammatory characteristics [54]. However, AM exhibits some significant disadvantages, including limited transparency and mechanical strength, poor standardization of preparation, risk for disease transmission, and biological variability (Table 7) [55]. There are extensive similarities between the basement membrane composition of AM and limbal niche, but AM lacks limbus-specific environmental factors, making it unsuitable as a surrogate niche for limbal stem cells [56]. In the studies on cell-based therapies of LSCD that have only been investigated in animal experiments, AM, with favorable results (Tables 1, 3 and 4), has been used as a substrate for culture and transplantation of bone marrow-derived mesenchymal stem cells (MSCs) [34,35,38], epidermal stem cells (SCs) [46,47], immature dental pulp stem cells (IDPSCs) [49,50], and umbilical cord stem cells (UCSCs) [52].

Table 6. Potential biomaterials and carriers for ocular surface reconstruction.

Biological/Biosynthetic	Synthetic
Amniotic membrane [57]	Contact lenses [58]
Chemically cross-linked hyaluronic acid-based hydrogels [59]	Mebiol Gel (thermo-reversible polymer gel) [53]
Chitosan matrix/silver matrix/gold matrix [60]	Nanofiber scaffolds [40]
Collagen IV-coated plates [61]	Petrolatum gauze [24]
Collagen membranes [62]	Plastic [25]
Corneal stroma [63]	Poly(lactide-co-glycolide) electrospun scaffolds [64]
Fibrin [65]	Poly-ε-caprolactone electrospun scaffolds [66]
Human keratoplasty lenticules [67]	
Laminin-coated compressed collagen gel [68]	
Matrigel (reconstituted basement membrane extract) [69]	
Plastic compressed collagen [70]	
Recombinant human cross-linked collagen scaffold [71]	
Silk fibroin [72]	
Silk fibroin mixed with polyethylene glycol [72]	

The list of possibilities is not complete.

As a substitution for natural extracellular matrix, investigators have attempted to produce synthetic nanofiber scaffolds, primarily using electrospinning [66], with the purpose of supporting cellular growth in corneal engineering. Nanofibers are three-dimensional (3D) and exhibit an enormous surface area. Polycaprolactone, which is a degradable polyester, has been found to have sufficient mechanical strength, high biocompatibility, low production costs, and ease of use (Table 7) [73]. Polycaprolactone has proved to be a suitable substrate for culture of corneal [66], limbal [66], and conjuntival cells [35]. Zajiceva *et al.* cultured bone marrow-derived MSCs on 3D nanofiber scaffolds fabricated from polyamide and transplanted the sheets onto the cornea of LSCD mice models [40]. The viability and morphology of cells grown on these nanofibers were comparable with those grown on plastic. Recently, a protocol for the use of nanofiber scaffolds for the growth of MSCs and limbal stem cells, and for their transplantation onto a damaged ocular surface in a mouse model, has been described, demonstrating the potential for nanofibers in clinical studies [74]. There are no studies, however, that have used nanofiber scaffolds for ocular surface reconstruction in humans.

Fibrin, a degradable natural substrate, has been used as a culture membrane in the treatment of LSCD in humans [75,76]. Fibrin substrates provide several advantages, such as relatively high mechanical strength, a high degree of transparency, and rapid bioadsorbence (Table 7) [54]. Fibrin, compared to, for example, collagen, has been shown to promote growth, survival, and an undifferentiated phenotype of cultured LEC [77]. The value of this membrane in ocular surface reconstruction has been further supported in LSCD rabbit models, using bone marrow-derived MSCs [33] and epidermal SCs [45], and in mice with hair follicle-derived stem cells (HFDSCs) [48].

Most of the cell-based therapeutic strategies entail the use of underlying substrate scaffolds. However, carrier-free methods, without a supportive membrane, have also been applied. Polymers that are responsive to temperature can detach adherent cells by reducing the temperature from 37 °C to 20 °C [78]. Carrier-free techniques take advantage of adhesive properties of the basement membranes. It was demonstrated that the presence of β_1 integrin in the carrier-free group is important for the attachment of cell sheets to the ocular surface [79]. Promising results with carrier-free transplantation in animal studies are reported using bone marrow-derived MSCs in rabbits [36] and embryonic SCs in pigs [43] and mice [41,42,44].

Table 7. Properties, advantages, and disadvantages of different carrier biomaterials and methods used in cell-based therapies of LSCD, explored in animal models.

Carriers/Methods	Transparency	Mechanical Strength	Elasticity	Advantages	Disadvantages
AM	+	++	+++	Stimulates cell growth, anti-inflammation, anti-angiogenesis, proper elasticity	Limited transparency, variable quality, risk of disease transmission, limited mechanical strength, poor standardization
Carrier-free method	N/A	N/A	N/A	Rapid adhesion, does not require preparation and standardization of membranes, does not require sutures	Possibility for detachment from the ocular surface in the early period after surgery
Fibrin gel	++	+++	+++	Proper transparency, good bioadsorbence, easy manipulation, good mechanical strength, elasticity, degradable	Possibility for immune response, risk for disease transmission
Nanofiber	++	++++	++	Good transparency, high mechanical strength, highly flexible, proper biocompatibility, easy to use, controlled shape and pore size, low cost	Limited elasticity, high cost

N/A indicates not applicable.

5. Cultured Bone Marrow-Derived Mesenchymal Stem Cells

Mesenchymal stem cells have multi-lineage potential [80]. Previous studies have reported that bone marrow-derived MSCs have a beneficial effect on the survival, growth, and proliferation of various types of cells, such as cardiac progenitor cells [81], neural stem cells [82], neurons [83], and Schwann cells [84]. Studies have demonstrated that *in vivo* administration of MSCs decreases the incidence of graft-*versus*-host disease in humans and mice [85,86], inhibits the manifestation of autoimmune diseases [87], impairs septic complications [88], and considerably counteracts rejection of allogeneic corneal allografts [89]. After *in vivo* application of MSCs, these cells migrate into the damaged area, thus supporting tissue healing [90].

The role of bone marrow-derived MSCs has also been investigated in corneal tissue regeneration. To date, as many as eight animal studies have been performed using this cell source for corneal repair following induced LSCD (Table 1). Various substrates and methods have been applied to transplant cultured MSC cells to damaged cornea of mice, rats, and rabbits, including AM [34,35,38], nanofiber scaffold [40], fibrin scaffold [33], carrier-free sheets [36], injection under transplanted AM [37], and intravenous injection through an ear vein [39].

Overall, the results obtained from animal experiments show that bone marrow-derived MSCs have a favorable effect with regard to cell differentiation into a corneal epithelial phenotype, improved corneal clarity, and reduced vascularization (Table 1). In one mouse study, with the short follow-up time of two weeks, the authors reported that transplantation of bone marrow-derived MSCs on nanofiber scaffold carriers supported the epithelial healing and inhibited local inflammatory reactions [40]. The other studies, with follow-up times ranging from one to six months, reported that the reconstruction of corneal epithelium after transplantation of bone-marrow derived MSCs was achieved in 90.6% (29/32) of the experimental rats [34,35,38] and 100% (31/31) of the experimental rabbits [33,36,37,39]. In rats with induced LSCD, where cultured cells were transplanted on AM, the improved corneal clarity was achieved in 87.5% (28/32) of the transplanted animals, and the cornea was completely transparent in 78.6% (22/28) of the animals [34,35,38]. However, no studies reported that the cornea was completely transparent after transplantation in rabbit LSCD models [33,36,37,39]. In one of these studies where MSCs were transplanted on a fibrin carrier, the iris was partially clear in 30% (3/10) and completely obscure in 70% (7/10) of the transplanted animals [33]. The studies in both rats and rabbits have also revealed that some neovascularization was observed in all transplanted eyes, with the best outcome being neovascularization limited to 2 mm central to the limbus 10 weeks after the transplantation [34].

It is speculated that the favorable effect of bone marrow-derived MSCs may be mediated by the intercellular signaling of epidermal growth factor (EGF) [91]. It has been suggested that EGF may be one of the most important mitogens of corneal epithelial cells [33,34]. Furthermore, bone marrow-derived MSCs induced to corneal lineage exhibited up-regulation of the putative limbal epithelial stem cell-specific genes p63 and β_1-integrin, and protein levels of p63 and CK3 were increased [38]. Other investigators have reported similar findings with the up-regulation of key putative stem cell markers [33,34,36,37]. This may be particularly important in the light of the recent finding by Rama *et al.* that the phenotype of cultured LEC is critical to ensure successful reconstruction of the ocular surface following LSCD [76]. The authors found that cell cultures in which p63-bright cells constituted more than

3% of the total number of cells were associated with successful transplantation in 78% of patients. In contrast, cultures in which p63-bright cells made up 3% or less of the total number of cells, successful transplantation was only seen in 11% of patients. In conclusion, the investigations performed in animal experiments suggest that bone marrow-derived MSCs may serve as a possible stem cell source for corneal reconstruction in humans, however, neovascularization was a consistent feature following transplantation.

6. Cultured Embryonic Stem Cells

Embryonic SCs are widely accepted as a significant cell source in tissue regeneration due to their great plasticity. A number of cell types have been induced from embryonic SCs *in vitro*, e.g., lung alveolar epithelial cells [92] and epithelial cells of the thymus [93]. It has also been demonstrated that embryonic SCs are capable of differentiating into corneal epithelial-like cells [94,95]. There are hitherto four studies that have investigated the potential of embryonic SCs for regeneration of the cornea in animal LSCD models (Table 2). In these studies, embryonic SCs were either cultured on collagen IV [41–43] or gelatin coated plates [44]. After culture, the carrier-free cell sheets were transplanted onto the corneas of mice [41,42,44] and pigs [43] (Table 2).

Following transplantation of cultured embryonic SCs onto corneas of LSCD animal models, re-epithelialization of the corneal surface with monolayer [41] and multilayer [42–44] epithelial-like cells was observed. The restored epithelium exhibited high levels of expression of CD44 and E-cadherin, which are important in corneal epithelial wound healing [41,42,44]. Furthermore, it has been demonstrated that embryonic SCs induced into epithelial-like cells expressed the basal limbal epithelial marker p63 [42,43] and the mature corneal epithelial marker CK12 [41–44].

Disadvantages of using embryonic SCs include difficulty of access, ethical concerns, high costs, immunogenicity, and risk of tumor formation [96]. None of the studies using embryonic SCs in animals have reported the degree of success in terms of number of animals with corneal reconstruction, or the effect on corneal transparency and neovascularization. Moreover, the follow-up time is very short (from one day to five weeks). Taken together, more studies with longer follow-up times, which also inform on the degree of success, are warranted prior to clinical trials.

7. Cultured Epidermal Stem Cells

Epidermal SCs have the remarkable ability to differentiate into other types of tissues [97]. Three studies have so far demonstrated the potential of epidermal SCs to regenerate the corneal surface following LSCD (Table 3). Two of the studies used AM for the culture and transplantation of epidermal SCs onto the cornea of goats [46,47], whereas the other used fibrin scaffold in rabbits [45]. These studies demonstrated that culture and transplantation of epidermal SCs onto damaged cornea successfully restored the corneal epithelium in 100% (22/22) of the animals. Moreover, the cornea became completely transparent with only mild neovascularization [45–47]. In one study, the corneal surface was intact with normal transparency for over three months [45]. In a study by Yang and colleagues, with a follow-up time to 30 months, the cornea was clear in three or four quadrants in 80% (8/10) of animals [46]. In a third study, with a follow-up time to 24 months, 71.4% (5/7) of the eyes of the treated animals had two or three quadrants of clear cornea [47]. Corneal perforation during the operation was reported in one animal [47]. No other complications were noted in any of the animals.

Following transplantation of the epidermal SCs onto the cornea of goats, the epidermal markers CK1/10 were down-regulated in the corneal stroma at 12 months, whereas the expression of the CK3, CK12, and PAX6 was up-regulated in the reconstructed epithelium [46]. The authors suggested that a possible mechanism of epidermal SCs in reconstruction of the damaged corneal epithelium involves the down-regulation of CK1/10 and up-regulation of PAX6. The PAX6 gene is involved in controlling eye formation during embryonic development [45,98,99], and recently the transduction of PAX6 in skin epithelial stem cells has been demonstrated to be adequate to transform epidermal SCs to limbal stem cell-like cells [45].

In conclusion, the results obtained with epidermal SCs in animal studies are very promising, with a high degree of success following transplantation in many animals, even with a follow-up period of 2.5 years [46,47]. Since epidermal SCs are also exceptionally easy to access, they may prove to be an excellent cell type for treating LSCD in humans.

8. Cultured Hair Follicle-Derived Stem Cells

The hair follicle harbors mesenchymal stem cells in the dermal papilla and connective tissue sheath that have large plasticity and can differentiate—given appropriate conditions *in vitro* and *in vivo*—into several cell lineages. These include chondrogenic, osteogenic, adipogenic, myogenic, neurogenic, and hematopoietic cell lineages [100–102]. In addition, the hair follicle comprises stem cells of epithelial origin, residing in the bulge region of the outer root sheath. The cells possess the ability to differentiate into hair follicles and sebaceous glands under physiological conditions. Following injury, however, these stem cells differentiated into epidermis [103–105].

By means of conditioned media harvested from corneal and limbal stromal fibroblasts, Meyer-Blazejewska *et al.* found that hair follicle-derived stem cells (HFSCs) were able to be reprogrammed *in vitro* into cells with a corneal epithelial phenotype [106]. In a follow-up study, the same research group performed *in vivo* experiments using a transgenic mouse model that allows HFSCs to change color upon differentiation to corneal epithelial cells, in which CK12 is expressed [48]. Hair follicle-derived stem cells were cultured on fibrin scaffolds and transplanted onto the cornea of mice with induced LSCD. The achieved results were promising, with cell differentiation into a corneal epithelial phenotype and suppression of vascularization and conjunctival ingrowth with reconstruction of the ocular surface in 87.5% (7/8) of the transplanted animals two weeks following transplantation.

Due to promising results in an animal study comprising as many as 31 mice and extremely easy access, HFSCs clearly warrant further investigations.

9. Cultured Immature Dental Pulp Stem Cells

Human immature dental pulp cells (IDPSCs) are capable of differentiation into a multitude of cell types, including neurons, smooth and skeletal muscle, cartilage, and bone [107]. There are two animal studies using human IDPSCs to treat LSCD in which the cells were cultured on AM and transplanted onto the damaged cornea of rabbits [49,50]. Human immature dental pulp cells expressed markers in common with LEC/corneal cells, such as ABCG2, β_1-integrin, p63, and CK3/12 [50]. In 2009, Monteiro et al. [50] demonstrated that transplantation of IDPSCs resulted in reconstruction of the ocular surface in 100% (5/5) of experimental animals. The authors also reported gradual improvement in corneal transparency during a follow-up time of three months [50]. One year later, Gomes and colleagues showed that rabbit eyes after transplantation of IDPSCs exhibited well-organized corneal epithelium and improved corneal transparency in 100% (5/5) of animals with mild chemical burn damage, while control corneas developed total conjunctivalization and opacification [49]. In the animals with severe chemical burns, 75% (3/4) of transplanted eyes showed less organized and loose corneal epithelium and inflammatory cells within the superficial and stromal layers. Furthermore, one animal exhibited a thin corneal epithelium and superficial neovascularization [49].

Overall, these two studies using IDPSC have shown that the transplantation of tissue engineered IDPSC sheets could successfully restore the ocular surface in animal models of LSCD. Human IDPSC are relatively easy to access from the dental pulp; however, the need for extraction of the tooth is a clear disadvantage with this technology.

10. Cultured Umbilical Cord Stem Cells

There is only one study on the potential use of umbilical cord stem cells (UCSCs) to reverse LSCD in animals [52]. The UCSCs were cultured on AM and then transplanted onto the cornea of a LSCD rabbit model, resulting in regeneration of a clear corneal epithelium with a smooth surface and minimal corneal neovascularization in 66.7% (4/6) of the animals. Mild superficial inflammation was reported in one eye, whereas severe neovascularization was observed in the other. Furthermore, it was demonstrated that this new corneal smooth surface exhibited expression of normal corneal-specific markers CK3 and CK12, but not CK4 or CK1/10. Compared to embryonic SCs, umbilical cord stem cells have the advantage of being less immunogenic [108], non-tumorigenic [108], and ethically acceptable [52]. Compared to hair follicles and epidermal cells, the disadvantages of UCSCs include more complicated accessibility and allogeneic transplantation.

11. Cultured Orbital Fat-Derived Stem Cells

Multipotent stem cells have recently been successfully isolated and purified from human orbital fat tissue [109]. It has been demonstrated that the growth kinetics of orbital fat-derived stem cells (OFSCs) resemble those of bone marrow-derived MSCs, and that they share several surface markers [110]. Low immunogenicity of OFSC transplantation has been demonstrated in a xenotransplant model [110]. Furthermore, OFSCs possess adipogenic, chondrogenic, and osteogenic differentiation capacity, and are capable of differentiating into corneal epithelial cells *in vitro* [109]. So far, there is only one study on the potential use of OFSCs to treat damaged ocular surfaces in mice [51]. The authors reported that the topical administration and intra-limbal injection of OFSCs resulted in the reconstruction of clear corneal epithelium one week after treatment. It is suggested that inflammatory inhibition and corneal epithelial differentiation by OFSCs are responsible for corneal wound healing in the first few days, and that corneal stroma engraftment of OFSCs at a late stage is associated with corneal transparency [51]. The possibility of a topical approach to deliver OFSCs to reconstruct the ocular surface is particularly promising as it represents a non-invasive method. So far, few other non-invasive strategies have been suggested for the treatment of LSCD, and currently include the use of amniotic membrane extract [111], limbal fibroblast conditioned medium [112], and autologous serum [113], "a tonic for the ailing epithelium" [114].

12. Challenges and Future Perspectives

Over the past 10 years, a number of stem cell sources have been suggested for the treatment of ocular surface disorders. The clinical decision as to the optimal approach to treat LSCD has become challenging due to a precipitous increase in treatment options coupled with an almost absence of comparative studies. Comparisons between animal experiments of cell-based therapies of LSCD are difficult due to the following factors: (a) various methods for inducing LSCD in animals, (b) assorted culture techniques, (c) various transplantation methods, (d) differences in postoperative treatment, (e) disparities in follow-up time, and (f) huge differences in the presentation of experimental data. Increased standardization of these parameters will simplify the comparisons between animal experiments involving different stem cell sources, thereby encouraging corneal regenerative medicine.

Mechanisms through which cell-based therapies reconstruct the ocular surface are still elusive. The transplanted cells may substitute the progenitor/stem cells of the host for a period of time and/or revitalize the stem cells of the host, e.g., by secreting growth factors. There are several lines of evidence supporting the hypothesis that cultured cells transplanted onto the cornea primarily work by providing a favorable environment. The fact that LSCD can be successfully treated by a number of cell types implies that factors other than the choice of cell type may govern clinical success. The identification of factors secreted from cultured non-limbal epithelial cells that may be involved in the revitalization of limbal stem cells is an exciting future avenue for research.

It is likely that the phenotype of cultured non-limbal cells affects success following transplantation [76]. Studies on how various culture parameters affect the cell sheet, with particular emphasis on the phenotype, are warranted.

13. Conclusions

Animal experiments with epidermal SCs, HFSCs, IDPSCs, and bone marrow-derived MSCs have all shown promising results for the treatment of LSCD (Table 8). They represent an autologous source of cells in contrast to embryonic SCs and UCSCs. The long-term effects using embryonic SCs and UCSCs are unknown as none of the cell types have a follow-up time longer than five weeks. This contrasts sharply with the 2.5 year follow-up time for transplanted cultured epidermal SCs. Epidermal SCs and HFSCs both have the distinct benefit of exceptional ease of access. Coupled with promising results in many animals, these two types are particularly strong candidates for future clinical trials. Future research on these cells could include the development of a xenobiotic culture and storage [115–120] system that can keep the cells in a relatively undifferentiated state [76], while maintaining sufficient strength to be suitable for transplantation. Such a system would increase the safety [121], flexibility [122], global impact [123], and, most likely, the clinical results of the transplants [76].

Table 8. Overall success in ocular surface reconstruction using different stem cell sources.

Types of Stem Cells	Success	Complications (Number of Animals)	Ease of Access	Number of Animals (Number of Studies)	Autologous Source	Ethical Concerns
Bone Marrow-Derived MSCs	+++	–	++	63 (8) [1]	Yes	No
Embryonic SCs	+	Mild immune reaction *	+	25 (4) [2]	No	Yes
Epidermal SCs	++++	Perforation (1)	++++	22 (3)	Yes	No
HFSCs	+++	–	++++	31 (1)	Yes	No
IDPSCs	+++	–	++	14 (2)	Yes	No
OFSCs	++	–	++	12 (1)	Yes	No
UCSCs	++	Mild superficial inflammation (1)	++	6 (1)	No	No

[1] number of animals not reported in two studies; [2] number of animals not reported in one study; * number of animals not reported; HFSCs, hair follicle-derived stem cells; MSCs, mesenchymal stem cells; SCs, stem cells; OFSCs, orbital fat-derived stem cells; UCSCs, umbilical cord stem cells; +: low degree; ++: moderate degree; +++: high degree; ++++: very high degree.

Acknowledgments: Funding received from Department of Oral Biology, Faculty of Dentistry, University of Oslo and Department of Medical Biochemistry, Oslo University Hospital, Oslo, Norway.

Conflicts of Interest: The authors declare no conflict of interest.

References

1. Land, M.F.; Fernald, R.D. The evolution of eyes. *Annu. Rev. Neurosci.* **1992**, *15*, 1–29.
2. Beuerman, R.W.; Pedroza, L. Ultrastructure of the human cornea. *Microsc. Res. Tech.* **1996**, *33*, 320–335.
3. Robertson, D.M.; Ladage, P.M.; Yamamoto, N.; Jester, J.V.; Petroll, W.M.; Cavanagh, H.D. Bcl-2 and bax regulation of corneal homeostasis in genetically altered mice. *Eye Contact Lens* **2006**, *32*, 3–7.
4. Moller-Pedersen, T.; Li, H.F.; Petroll, W.M.; Cavanagh, H.D.; Jester, J.V. Confocal microscopic characterization of wound repair after photorefractive keratectomy. *Invest. Ophthalmol. Vis. Sci.* **1998**, *39*, 487–501.
5. Cavanagh, H.D.; Ladage, P.M.; Li, S.L.; Yamamoto, K.; Molai, M.; Ren, D.H.; Petroll, W.M.; Jester, J.V. Effects of daily and overnight wear of a novel hyper oxygen-transmissible soft contact lens on bacterial binding and corneal epithelium: A 13-month clinical trial. *Ophthalmology* **2002**, *109*, 1957–1969.
6. Sharma, A.; Coles, W.H. Kinetics of corneal epithelial maintenance and graft loss. A population balance model. *Invest. Ophthalmol. Vis. Sci.* **1989**, *30*, 1962–1971.
7. Utheim, T.P. Limbal epithelial cell therapy: Past, present, and future. *Methods Mol. Biol.* **2013**, *1014*, 3–43.
8. Nishida, T. Fundamentals, diagnosis and management. In *Cornea*, 2nd ed.; Krachmer, J.H., Mannis, M.J., Holland, E., Eds.; Elsevier: Philadelphia, PA, USA, 2005; pp. 3–26.
9. Cotsarelis, G.; Cheng, S.Z.; Dong, G.; Sun, T.T.; Lavker, R.M. Existence of slow-cycling limbal epithelial basal cells that can be preferentially stimulated to proliferate: Implications on epithelial stem cells. *Cell* **1989**, *57*, 201–209.

10. Davanger, M.; Evensen, A. Role of the pericorneal papillary structure in renewal of corneal epithelium. *Nature* **1971**, *229*, 560–561.

11. Zieske, J.D. Perpetuation of stem cells in the eye. *Eye* **1994**, *8*, 163–169.

12. Schermer, A.; Galvin, S.; Sun, T.T. Differentiation-related expression of a major 64K corneal keratin *in vivo* and in culture suggests limbal location of corneal epithelial stem cells. *J. Cell Biol.* **1986**, *103*, 49–62.

13. Pellegrini, G.; Rama, P.; Mavilio, F.; De Luca, M. Epithelial stem cells in corneal regeneration and epidermal gene therapy. *J. Pathol.* **2009**, *217*, 217–228.

14. Schwartz, G.; Holland, E.J. Cornea: Fundamentals, diagnosis and management. In *Classification and Staging of Ocular Surface Disease*, 2nd ed.; Krachmer, J.H., Mannis, M.J., Holland, E., Eds.; Elsevier: Philadelphia, PA, USA, 2005; pp. 1785–1797.

15. Vemuganti, G.K.; Sangwan, V.S. Interview: Affordability at the cutting edge: Stem cell therapy for ocular surface reconstruction. *Regen. Med.* **2010**, *5*, 337–340.

16. Schwab, I.R.; Isseroff, R.R. Bioengineered corneas—The promise and the challenge. *N. Engl. J. Med.* **2000**, *343*, 136–138.

17. Puangsricharern, V.; Tseng, S.C. Cytologic evidence of corneal diseases with limbal stem cell deficiency. *Ophthalmology* **1995**, *102*, 1476–1485.

18. Espana, E.M.; Grueterich, M.; Romano, A.C.; Touhami, A.; Tseng, S.C. Idiopathic limbal stem cell deficiency. *Ophthalmology* **2002**, *109*, 2004–2010.

19. Dua, H.S.; Joseph, A.; Shanmuganathan, V.A.; Jones, R.E. Stem cell differentiation and the effects of deficiency. *Eye* **2003**, *17*, 877–885.

20. Di Iorio, E.; Ferrari, S.; Fasolo, A.; Bohm, E.; Ponzin, D.; Barbaro, V. Techniques for culture and assessment of limbal stem cell grafts. *Ocul. Surf.* **2010**, *8*, 146–153.

21. Shortt, A.J.; Secker, G.A.; Rajan, M.S.; Meligonis, G.; Dart, J.K.; Tuft, S.J.; Daniels, J.T. *Ex vivo* expansion and transplantation of limbal epithelial stem cells. *Ophthalmology* **2008**, *115*, 1989–1997.

22. Rauz, S.; Saw, V.P. Serum eye drops, amniotic membrane and limbal epithelial stem cells—tools in the treatment of ocular surface disease. *Cell Tissue Bank.* **2010**, *11*, 13–27.

23. Kenyon, K.R.; Tseng, S.C. Limbal autograft transplantation for ocular surface disorders. *Ophthalmology* **1989**, *96*, 709–722.

24. Pellegrini, G.; Traverso, C.E.; Franzi, A.T.; Zingirian, M.; Cancedda, R.; De Luca, M. Long-term restoration of damaged corneal surfaces with autologous cultivated corneal epithelium. *Lancet* **1997**, *349*, 990–993.

25. Daya, S.M.; Watson, A.; Sharpe, J.R.; Giledi, O.; Rowe, A.; Martin, R.; James, S.E. Outcomes and DNA analysis of *ex vivo* expanded stem cell allograft for ocular surface reconstruction. *Ophthalmology* **2005**, *112*, 470–477.

26. Hayashi, R.; Ishikawa, Y.; Ito, M.; Kageyama, T.; Takashiba, K.; Fujioka, T.; Tsujikawa, M.; Miyoshi, H.; Yamato, M.; Nakamura, Y.; *et al.* Generation of corneal epithelial cells from induced pluripotent stem cells derived from human dermal fibroblast and corneal limbal epithelium. *PloS ONE* **2012**, *7*.

27. Sareen, D.; Saghizadeh, M.; Ornelas, L.; Winkler, M.A.; Narwani, K.; Sahabian, A.; Funari, V.A.; Tang, J.; Spurka, L.; Punj, V.; *et al.* Differentiation of human limbal-derived induced pluripotent stem cells into limbal-like epithelium. *Stem Cells Transl. Med.* **2014**, *3*, 1002–1012.

28. Niethammer, D.; Kummerle-Deschner, J.; Dannecker, G.E. Side-effects of long-term immunosuppression versus morbidity in autologous stem cell rescue: Striking the balance. *Rheumatology* **1999**, *38*, 747–750.

29. Utheim, T.P. Concise review: Transplantation of cultured oral mucosal epithelial cells for treating limbal stem cell deficiency-current status and future perspectives. *Stem Cells* **2015**, *33*, 1685–1695.

30. Di Girolamo, N.; Bosch, M.; Zamora, K.; Coroneo, M.T.; Wakefield, D.; Watson, S.L. A contact lens-based technique for expansion and transplantation of autologous epithelial progenitors for ocular surface reconstruction. *Transplantation* **2009**, *87*, 1571–1578.

31. Chun, Y.S.; Park, I.K.; Kim, J.C. Technique for autologous nasal mucosa transplantation in severe ocular surface disease. *Eur. J. Ophthalmol.* **2011**, *21*, 545–551.

32. Kim, J.H.; Chun, Y.S.; Lee, S.H.; Mun, S.K.; Jung, H.S.; Lee, S.H.; Son, Y.; Kim, J.C. Ocular surface reconstruction with autologous nasal mucosa in cicatricial ocular surface disease. *Am. J. Ophthalmol.* **2010**, *149*, 45–53.

33. Gu, S.; Xing, C.; Han, J.; Tso, M.O.; Hong, J. Differentiation of rabbit bone marrow mesenchymal stem cells into corneal epithelial cells *in vivo* and *ex vivo*. *Molecular vision* **2009**, *15*, 99–107.

34. Jiang, T.S.; Cai, L.; Ji, W.Y.; Hui, Y.N.; Wang, Y.S.; Hu, D.; Zhu, J. Reconstruction of the corneal epithelium with induced marrow mesenchymal stem cells in rats. *Mol. Vis.* **2010**, *16*, 1304–1316.

35. Ma, Y.; Xu, Y.; Xiao, Z.; Yang, W.; Zhang, C.; Song, E.; Du, Y.; Li, L. Reconstruction of chemically burned rat corneal surface by bone marrow-derived human mesenchymal stem cells. *Stem Cells* **2006**, *24*, 315–321.

36. Omoto, M.; Miyashita, H.; Shimmura, S.; Higa, K.; Kawakita, T.; Yoshida, S.; McGrogan, M.; Shimazaki, J.; Tsubota, K. The use of human mesenchymal stem cell-derived feeder cells for the cultivation of transplantable epithelial sheets. *Invest. Ophthalmol. Vis. Sci.* **2009**, *50*, 2109–2115.

37. Reinshagen, H.; Auw-Haedrich, C.; Sorg, R.V.; Boehringer, D.; Eberwein, P.; Schwartzkopff, J.; Sundmacher, R.; Reinhard, T. Corneal surface reconstruction using adult mesenchymal stem cells in experimental limbal stem cell deficiency in rabbits. *Acta Ophthalmol.* **2011**, *89*, 741–748.

38. Rohaina, C.M.; Then, K.Y.; Ng, A.M.; Wan Abdul Halim, W.H.; Zahidin, A.Z.; Saim, A.; Idrus, R.B. Reconstruction of limbal stem cell deficient corneal surface with induced human bone marrow mesenchymal stem cells on amniotic membrane. *Transl. Res.* **2014**, *163*, 200–210.

39. Ye, J.; Yao, K.; Kim, J.C. Mesenchymal stem cell transplantation in a rabbit corneal alkali burn model: Engraftment and involvement in wound healing. *Eye* **2006**, *20*, 482–490.

40. Zajicova, A.; Pokorna, K.; Lencova, A.; Krulova, M.; Svobodova, E.; Kubinova, S.; Sykova, E.; Pradny, M.; Michalek, J.; Svobodova, J.; *et al.* Treatment of ocular surface injuries by limbal and mesenchymal stem cells growing on nanofiber scaffolds. *Cell Transpl.* **2010**, *19*, 1281–1290.

41. Homma, R.; Yoshikawa, H.; Takeno, M.; Kurokawa, M.S.; Masuda, C.; Takada, E.; Tsubota, K.; Ueno, S.; Suzuki, N. Induction of epithelial progenitors *in vitro* from mouse embryonic stem cells and application for reconstruction of damaged cornea in mice. *Invest. Ophthalmol. Vis. Sci.* **2004**, *45*, 4320–4326.

42. Kumagai, Y.; Kurokawa, M.S.; Ueno, H.; Kayama, M.; Tsubota, K.; Nakatsuji, N.; Kondo, Y.; Ueno, S.; Suzuki, N. Induction of corneal epithelium-like cells from cynomolgus monkey embryonic stem cells and their experimental transplantation to damaged cornea. *Cornea* **2010**, *29*, 432–438.

43. Notara, M.; Hernandez, D.; Mason, C.; Daniels, J.T. Characterization of the phenotype and functionality of corneal epithelial cells derived from mouse embryonic stem cells. *Regen. Med.* **2012**, *7*, 167–178.

44. Ueno, H.; Kurokawa, M.S.; Kayama, M.; Homma, R.; Kumagai, Y.; Masuda, C.; Takada, E.; Tsubota, K.; Ueno, S.; Suzuki, N. Experimental transplantation of corneal epithelium-like cells induced by PAX6 gene transfection of mouse embryonic stem cells. *Cornea* **2007**, *26*, 1220–1227.

45. Ouyang, H.; Xue, Y.; Lin, Y.; Zhang, X.; Xi, L.; Patel, S.; Cai, H.; Luo, J.; Zhang, M.; Zhang, M.; *et al.* WNT7a and PAX6 define corneal epithelium homeostasis and pathogenesis. *Nature* **2014**, *511*, 358–361.

46. Yang, X.; Moldovan, N.I.; Zhao, Q.; Mi, S.; Zhou, Z.; Chen, D.; Gao, Z.; Tong, D.; Dou, Z. Reconstruction of damaged cornea by autologous transplantation of epidermal adult stem cells. *Mol. Vis.* **2008**, *14*, 1064–1070.

47. Yang, X.; Qu, L.; Wang, X.; Zhao, M.; Li, W.; Hua, J.; Shi, M.; Moldovan, N.; Wang, H.; Dou, Z. Plasticity of epidermal adult stem cells derived from adult goat ear skin. *Mol. Reprod. Dev.* **2007**, *74*, 386–396.

48. Meyer-Blazejewska, E.A.; Call, M.K.; Yamanaka, O.; Liu, H.; Schlotzer-Schrehardt, U.; Kruse, F.E.; Kao, W.W. From hair to cornea: Toward the therapeutic use of hair follicle-derived stem cells in the treatment of limbal stem cell deficiency. *Stem Cells* **2011**, *29*, 57–66.

49. Gomes, J.A.; Geraldes Monteiro, B.; Melo, G.B.; Smith, R.L.; Cavenaghi Pereira da Silva, M.; Lizier, N.F.; Kerkis, A.; Cerruti, H.; Kerkis, I. Corneal reconstruction with tissue-engineered cell sheets composed of human immature dental pulp stem cells. *Invest. Ophthalmol. Vis. Sci.* **2010**, *51*, 1408–1414.

50. Monteiro, B.G.; Serafim, R.C.; Melo, G.B.; Silva, M.C.; Lizier, N.F.; Maranduba, C.M.; Smith, R.L.; Kerkis, A.; Cerruti, H.; Gomes, J.A.; *et al.* Human immature dental pulp stem cells share key characteristic features with limbal stem cells. *Cell Prolif.* **2009**, *42*, 587–594.

51. Lin, K.J.; Loi, M.X.; Lien, G.S.; Cheng, C.F.; Pao, H.Y.; Chang, Y.C.; Ji, A.T.; Ho, J.H. Topical administration of orbital fat-derived stem cells promotes corneal tissue regeneration. *Stem Cell Res. Ther.* **2013**, *4*.

52. Reza, H.M.; Ng, B.Y.; Gimeno, F.L.; Phan, T.T.; Ang, L.P. Umbilical cord lining stem cells as a novel and promising source for ocular surface regeneration. *Stem Cell Rev.* **2011**, *7*, 935–947.

53. Sudha, B.; Madhavan, H.N.; Sitalakshmi, G.; Malathi, J.; Krishnakumar, S.; Mori, Y.; Yoshioka, H.; Abraham, S. Cultivation of human corneal limbal stem cells in mebiol gel—A thermo-reversible gelation polymer. *Indian J. Med. Res.* **2006**, *124*, 655–664.

54. Feng, Y.; Borrelli, M.; Reichl, S.; Schrader, S.; Geerling, G. Review of alternative carrier materials for ocular surface reconstruction. *Curr. Eye Res.* **2014**, *39*, 541–552.

55. Levis, H.; Daniels, J.T. New technologies in limbal epithelial stem cell transplantation. *Curr. Opin. Biotechnol.* **2009**, *20*, 593–597.

56. Dietrich-Ntoukas, T.; Hofmann-Rummelt, C.; Kruse, F.E.; Schlotzer-Schrehardt, U. Comparative analysis of the basement membrane composition of the human limbus epithelium and amniotic membrane epithelium. *Cornea* **2012**, *31*, 564–569.

57. Meller, D.; Tseng, S.C. Conjunctival epithelial cell differentiation on amniotic membrane. *Investig. Ophthalmol. Vis. Sci.* **1999**, *40*, 878–886.

58. Di Girolamo, N.; Chui, J.; Wakefield, D.; Coroneo, M.T. Cultured human ocular surface epithelium on therapeutic contact lenses. *Br. J. Ophthalmol.* **2007**, *91*, 459–464.

59. Fiorica, C.; Senior, R.A.; Pitarresi, G.; Palumbo, F.S.; Giammona, G.; Deshpande, P.; MacNeil, S. Biocompatible hydrogels based on hyaluronic acid cross-linked with a polyaspartamide derivative as delivery systems for epithelial limbal cells. *Int. J. Pharm.* **2011**, *414*, 104–111.

60. Sudha, B.; Jasty, S.; Krishnan, S.; Krishnakumar, S. Signal transduction pathway involved in the *ex vivo* expansion of limbal epithelial cells cultured on various substrates. *Indian J. Med. Res.* **2009**, *129*, 382–389.

61. Li, D.Q.; Chen, Z.; Song, X.J.; de Paiva, C.S.; Kim, H.S.; Pflugfelder, S.C. Partial enrichment of a population of human limbal epithelial cells with putative stem cell properties based on collagen type IV adhesiveness. *Exp. Eye Res.* **2005**, *80*, 581–590.

62. Kito, K.; Kagami, H.; Kobayashi, C.; Ueda, M.; Terasaki, H. Effects of cryopreservation on histology and viability of cultured corneal epithelial cell sheets in rabbit. *Cornea* **2005**, *24*, 735–741.

63. Friend, J.; Kinoshita, S.; Thoft, R.A.; Eliason, J.A. Corneal epithelial cell cultures on stromal carriers. *Investig. Ophthalmol. Vis. Sci.* **1982**, *23*, 41–49.

64. Deshpande, P.; McKean, R.; Blackwood, K.A.; Senior, R.A.; Ogunbanjo, A.; Ryan, A.J.; MacNeil, S. Using poly(lactide-co-glycolide) electrospun scaffolds to deliver cultured epithelial cells to the cornea. *Regen Med.* **2010**, *5*, 395–401.

65. Talbot, M.; Carrier, P.; Giasson, C.J.; Deschambeault, A.; Guerin, S.L.; Auger, F.A.; Bazin, R.; Germain, L. Autologous transplantation of rabbit limbal epithelia cultured on fibrin gels for ocular surface reconstruction. *Mol. Vis.* **2006**, *12*, 65–75.

66. Sharma, S.; Mohanty, S.; Gupta, D.; Jassal, M.; Agrawal, A.K.; Tandon, R. Cellular response of limbal epithelial cells on electrospun poly-epsilon-caprolactone nanofibrous scaffolds for ocular surface bioengineering: A preliminary *in vitro* study. *Mol. Vis.* **2011**, *17*, 2898–2910.

67. Barbaro, V.; Ferrari, S.; Fasolo, A.; Ponzin, D.; Di Iorio, E. Reconstruction of a human hemicornea through natural scaffolds compatible with the growth of corneal epithelial stem cells and stromal keratocytes. *Mol. Vis.* **2009**, *15*, 2084–2093.

68. Mi, S.; Chen, B.; Wright, B.; Connon, C.J. *Ex vivo* construction of an artificial ocular surface by combination of corneal limbal epithelial cells and a compressed collagen scaffold containing keratocytes. *Tissue Eng. Part A* **2010**, *16*, 2091–2100.

69. Ahmadiankia, N.; Ebrahimi, M.; Hosseini, A.; Baharvand, H. Effects of different extracellular matrices and co-cultures on human limbal stem cell expansion *in vitro*. *Cell Biol. Int.* **2009**, *33*, 978–987.

70. Levis, H.J.; Brown, R.A.; Daniels, J.T. Plastic compressed collagen as a biomimetic substrate for human limbal epithelial cell culture. *Biomaterials* **2010**, *31*, 7726–7737.

71. Dravida, S.; Gaddipati, S.; Griffith, M.; Merrett, K.; Lakshmi Madhira, S.; Sangwan, V.S.; Vemuganti, G.K. A biomimetic scaffold for culturing limbal stem cells: A promising alternative for clinical transplantation. *J. Tissue Eng. Regen Med.* **2008**, *2*, 263–271.

72. Higa, K.; Takeshima, N.; Moro, F.; Kawakita, T.; Kawashima, M.; Demura, M.; Shimazaki, J.; Asakura, T.; Tsubota, K.; Shimmura, S. Porous silk fibroin film ass a transparent carrier for cultivated corneal epithelial sheets. *J. Biomater Sci. Polym. Ed.* **2010**, *22*.

73. Redenti, S.; Tao, S.; Yang, J.; Gu, P.; Klassen, H.; Saigal, S.; Desai, T.; Young, M.J. Retinal tissue engineering using mouse retinal progenitor cells and a novel biodegradable, thin-film poly(e-caprolactone) nanowire scaffold. *J. Ocul. Biol. Dis. Infor.* **2008**, *1*, 19–29.

74. Holan, V.; Javorkova, E.; Trosan, P. The growth and delivery of mesenchymal and limbal stem cells using copolymer polyamide 6/12 nanofiber scaffolds. *Methods Mol. Biol.* **2013**, *1014*, 187–199.

75. Rama, P.; Bonini, S.; Lambiase, A.; Golisano, O.; Paterna, P.; De Luca, M.; Pellegrini, G. Autologous fibrin-cultured limbal stem cells permanently restore the corneal surface of patients with total limbal stem cell deficiency. *Transplantation* **2001**, *72*, 1478–1485.

76. Rama, P.; Matuska, S.; Paganoni, G.; Spinelli, A.; De Luca, M.; Pellegrini, G. Limbal stem-cell therapy and long-term corneal regeneration. *N. Engl. J. Med.* **2010**, *363*, 147–155.

77. Forni, M.F.; Loureiro, R.R.; Cristovam, P.C.; Bonatti, J.A.; Sogayar, M.C.; Gomes, J.A. Comparison between different biomaterial scaffolds for limbal-derived stem cells growth and enrichment. *Curr. Eye Res.* **2013**, *38*, 27–34.

78. Nishida, K.; Yamato, M.; Hayashida, Y.; Watanabe, K.; Maeda, N.; Watanabe, H.; Yamamoto, K.; Nagai, S.; Kikuchi, A.; Tano, Y.; *et al.* Functional bioengineered corneal epithelial sheet grafts from corneal stem cells expanded *ex vivo* on a temperature-responsive cell culture surface. *Transplantation* **2004**, *77*, 379–385.

79. Higa, K.; Shimmura, S.; Kato, N.; Kawakita, T.; Miyashita, H.; Itabashi, Y.; Fukuda, K.; Shimazaki, J.; Tsubota, K. Proliferation and differentiation of transplantable rabbit epithelial sheets engineered with or without an amniotic membrane carrier. *Invest. Ophthalmol. Vis. Sci.* **2007**, *48*, 597–604.

80. Pittenger, M.F.; Mackay, A.M.; Beck, S.C.; Jaiswal, R.K.; Douglas, R.; Mosca, J.D.; Moorman, M.A.; Simonetti, D.W.; Craig, S.; Marshak, D.R. Multilineage potential of adult human mesenchymal stem cells. *Science* **1999**, *284*, 143–147.

81. Nakanishi, C.; Yamagishi, M.; Yamahara, K.; Hagino, I.; Mori, H.; Sawa, Y.; Yagihara, T.; Kitamura, S.; Nagaya, N. Activation of cardiac progenitor cells through paracrine effects of mesenchymal stem cells. *Biochem. Biophys. Res. Commun.* **2008**, *374*, 11–16.

82. Yoo, S.W.; Kim, S.S.; Lee, S.Y.; Lee, H.S.; Kim, H.S.; Lee, Y.D.; Suh-Kim, H. Mesenchymal stem cells promote proliferation of endogenous neural stem cells and survival of newborn cells in a rat stroke model. *Exp. Mol. Med.* **2008**, *40*, 387–397.

83. Deng, Y.B.; Ye, W.B.; Hu, Z.Z.; Yan, Y.; Wang, Y.; Takon, B.F.; Zhou, G.Q.; Zhou, Y.F. Intravenously administered BMSCs reduce neuronal apoptosis and promote neuronal proliferation through the release of VEGF after stroke in rats. *Neurol. Res.* **2010**, *32*, 148–156.

84. Wang, J.; Ding, F.; Gu, Y.; Liu, J.; Gu, X. Bone marrow mesenchymal stem cells promote cell proliferation and neurotrophic function of Schwann cells *in vitro* and *in vivo*. *Brain Res.* **2009**, *1262*, 7–15.

85. Lazarus, H.M.; Koc, O.N.; Devine, S.M.; Curtin, P.; Maziarz, R.T.; Holland, H.K.; Shpall, E.J.; McCarthy, P.; Atkinson, K.; Cooper, B.W.; *et al.* Cotransplantation of HLA-identical sibling culture-expanded mesenchymal stem cells and hematopoietic stem cells in hematologic malignancy patients. *Biol. Blood Marrow Transpl.* **2005**, *11*, 389–398.

86. Le Blanc, K.; Rasmusson, I.; Sundberg, B.; Gotherstrom, C.; Hassan, M.; Uzunel, M.; Ringden, O. Treatment of severe acute graft-*versus*-host disease with third party haploidentical mesenchymal stem cells. *Lancet* **2004**, *363*, 1439–1441.

87. Zappia, E.; Casazza, S.; Pedemonte, E.; Benvenuto, F.; Bonanni, I.; Gerdoni, E.; Giunti, D.; Ceravolo, A.; Cazzanti, F.; Frassoni, F.; *et al.* Mesenchymal stem cells ameliorate experimental autoimmune encephalomyelitis inducing T-cell anergy. *Blood* **2005**, *106*, 1755–1761.

88. Wu, Y.; Chen, L.; Scott, P.G.; Tredget, E.E. Mesenchymal stem cells enhance wound healing through differentiation and angiogenesis. *Stem Cells* **2007**, *25*, 2648–2659.

89. Jia, Z.; Jiao, C.; Zhao, S.; Li, X.; Ren, X.; Zhang, L.; Han, Z.C.; Zhang, X. Immunomodulatory effects of mesenchymal stem cells in a rat corneal allograft rejection model. *Exp. Eye Res.* **2012**, *102*, 44–49.

90. Sasaki, M.; Abe, R.; Fujita, Y.; Ando, S.; Inokuma, D.; Shimizu, H. Mesenchymal stem cells are recruited into wounded skin and contribute to wound repair by transdifferentiation into multiple skin cell type. *J. Immunol.* **2008**, *180*, 2581–2587.

91. Hu, N.; Zhang, Y.Y.; Gu, H.W.; Guan, H.J. Effects of bone marrow mesenchymal stem cells on cell proliferation and growth factor expression of limbal epithelial cells *in vitro*. *Ophthalmic Res.* **2012**, *48*, 82–88.

92. Ali, N.N.; Edgar, A.J.; Samadikuchaksaraei, A.; Timson, C.M.; Romanska, H.M.; Polak, J.M.; Bishop, A.E. Derivation of type II alveolar epithelial cells from murine embryonic stem cells. *Tissue Eng.* **2002**, *8*, 541–550.

93. Rodewald, H.R.; Paul, S.; Haller, C.; Bluethmann, H.; Blum, C. Thymus medulla consisting of epithelial islets each derived from a single progenitor. *Nature* **2001**, *414*, 763–768.

94. Ahmad, S.; Stewart, R.; Yung, S.; Kolli, S.; Armstrong, L.; Stojkovic, M.; Figueiredo, F.; Lako, M. Differentiation of human embryonic stem cells into corneal epithelial-like cells by *in vitro* replication of the corneal epithelial stem cell niche. *Stem Cells* **2007**, *25*, 1145–1155.

95. Brzeszczynska, J.; Samuel, K.; Greenhough, S.; Ramaesh, K.; Dhillon, B.; Hay, D.C.; Ross, J.A. Differentiation and molecular profiling of human embryonic stem cell-derived corneal epithelial cells. *Int. J. Mol. Med.* **2014**, *33*, 1597–1606.

96. Cui, L.; Guan, Y.; Qu, Z.; Zhang, J.; Liao, B.; Ma, B.; Qian, J.; Li, D.; Li, W.; Xu, G.T.; *et al.* WNT signaling determines tumorigenicity and function of ESC-derived retinal progenitors. *J. Clin. Investg.* **2013**, *123*, 1647–1661.

97. Liang, L.; Bickenbach, J.R. Somatic epidermal stem cells can produce multiple cell lineages during development. *Stem Cells* **2002**, *20*, 21–31.

98. Tsonis, P.A.; Fuentes, E.J. Focus on molecules: PAX-6, the eye master. *Exp. Eye Res.* **2006**, *83*, 233–234.

99. Treisman, J.E. How to make an eye. *Development* **2004**, *131*, 3823–3827.

100. Jahoda, C.A.; Whitehouse, J.; Reynolds, A.J.; Hole, N. Hair follicle dermal cells differentiate into adipogenic and osteogenic lineages. *Exp. Dermatol.* **2003**, *12*, 849–859.

101. Lako, M.; Armstrong, L.; Cairns, P.M.; Harris, S.; Hole, N.; Jahoda, C.A. Hair follicle dermal cells repopulate the mouse haematopoietic system. *J. Cell Sci.* **2002**, *115*, 3967–3974.

102. Richardson, G.D.; Arnott, E.C.; Whitehouse, C.J.; Lawrence, C.M.; Reynolds, A.J.; Hole, N.; Jahoda, C.A. Plasticity of rodent and human hair follicle dermal cells: Implications for cell therapy and tissue engineering. *J. Investig. Dermatol. Symp. Proc.* **2005**, *10*, 180–183.

103. Cotsarelis, G. Epithelial stem cells: A folliculocentric view. *J. Investig. Dermatol.* **2006**, *126*, 1459–1468.

104. Cotsarelis, G.; Sun, T.T.; Lavker, R.M. Label-retaining cells reside in the bulge area of pilosebaceous unit: Implications for follicular stem cells, hair cycle, and skin carcinogenesis. *Cell* **1990**, *61*, 1329–1337.

105. Taylor, G.; Lehrer, M.S.; Jensen, P.J.; Sun, T.T.; Lavker, R.M. Involvement of follicular stem cells in forming not only the follicle but also the epidermis. *Cell* **2000**, *102*, 451–461.

106. Blazejewska, E.A.; Schlotzer-Schrehardt, U.; Zenkel, M.; Bachmann, B.; Chankiewitz, E.; Jacobi, C.; Kruse, F.E. Corneal limbal microenvironment can induce transdifferentiation of hair follicle stem cells into corneal epithelial-like cells. *Stem Cells* **2009**, *27*, 642–652.

107. Kerkis, I.; Kerkis, A.; Dozortsev, D.; Stukart-Parsons, G.C.; Gomes Massironi, S.M.; Pereira, L.V.; Caplan, A.I.; Cerruti, H.F. Isolation and characterization of a population of immature dental pulp stem cells expressing OCT-4 and other embryonic stem cell markers. *Cells Tissues Organs* **2006**, *184*, 105–116.

108. Xiong, N.; Cao, X.; Zhang, Z.; Huang, J.; Chen, C.; Zhang, Z.; Jia, M.; Xiong, J.; Liang, Z.; Sun, S.; *et al.* Long-term efficacy and safety of human umbilical cord mesenchymal stromal cells in rotenone-induced hemiparkinsonian rats. *Biol. Blood Marrow Transpl.* **2010**, *16*, 1519–1529.

109. Ho, J.H.; Ma, W.H.; Tseng, T.C.; Chen, Y.F.; Chen, M.H.; Lee, O.K. Isolation and characterization of multi-potent stem cells from human orbital fat tissues. *Tissue Eng. Part A* **2011**, *17*, 255–266.

110. Chien, M.H.; Bien, M.Y.; Ku, C.C.; Chang, Y.C.; Pao, H.Y.; Yang, Y.L.; Hsiao, M.; Chen, C.L.; Ho, J.H. Systemic human orbital fat-derived stem/stromal cell transplantation ameliorates acute inflammation in lipopolysaccharide-induced acute lung injury. *Crit Care Med.* **2012**, *40*, 1245–1253.

111. Liang, L.; Li, W.; Ling, S.; Sheha, H.; Qiu, W.; Li, C.; Liu, Z. Amniotic membrane extraction solution for ocular chemical burns. *Clin. Exp. Ophthalmol.* **2009**, *37*, 855–863.

112. Amirjamshidi, H.; Milani, B.Y.; Sagha, H.M.; Movahedan, A.; Shafiq, M.A.; Lavker, R.M.; Yue, B.Y.; Djalilian, A.R. Limbal fibroblast conditioned media: A non-invasive treatment for limbal stem cell deficiency. *Mol. Vis.* **2011**, *17*, 658–666.

113. Geerling, G.; Maclennan, S.; Hartwig, D. Autologous serum eye drops for ocular surface disorders. *Br. J. Ophthalmol.* **2004**, *88*, 1467–1474.

114. Pflugfelder, S.C. Is autologous serum a tonic for the ailing corneal epithelium? *Am. J. Ophthalmol.* **2006**, *142*, 316–317.

115. Utheim, T.P.; Raeder, S.; Utheim, O.A.; Cai, Y.; Roald, B.; Drolsum, L.; Lyberg, T.; Nicolaissen, B. A novel method for preserving cultured limbal epithelial cells. *Br. J. Ophthalmol.* **2007**, *91*, 797–800.

116. Pasovic, L.; Utheim, T.P.; Maria, R.; Lyberg, T.; Messelt, E.B.; Aabel, P.; Chen, D.F.; Chen, X.; Eidet, J.R. Optimization of Storage Temperature for Cultured ARPE-19 Cells. *J. Ophthalmol.* **2013**, *2013*.

117. Pasovic, L.; Eidet, J.R.; Lyberg, T.; Messelt, E.B.; Aabel, P.; Utheim, T.P. Antioxidants Improve the Viability of Stored Adult Retinal Pigment Epithelial-19 Cultures. *Ophthalmol. Ther.* **2014**, *3*, 49–61.

118. Jackson, C.; Aabel, P.; Eidet, J.R.; Messelt, E.B.; Lyberg, T.; von Unge, M.; Utheim, T.P. Effect of storage temperature on cultured epidermal cell sheets stored in xenobiotic-free medium. *PLoS ONE* **2014**, *9*.

119. Islam, R.; Jackson, C.; Eidet, J.R.; Messelt, E.B.; Corraya, R.M.; Lyberg, T.; Griffith, M.; Dartt, D.A.; Utheim, T.P. Effect of storage temperature on structure and function of cultured human oral keratinocytes. *PLoS ONE* **2015**, *10*.

120. Eidet, J.R.; Utheim, O.A.; Raeder, S.; Dartt, D.A.; Lyberg, T.; Carreras, E.; Huynh, T.T.; Messelt, E.B.; Louch, W.E.; Roald, B.; *et al.* Effects of serum-free storage on morphology, phenotype, and viability of *ex vivo* cultured human conjunctival epithelium. *Exp. Eye Res.* **2012**, *94*, 109–116.

121. Utheim, T.P.; Raeder, S.; Utheim, O.A.; de la Paz, M.; Roald, B.; Lyberg, T. Sterility control and long-term eye-bank storage of cultured human limbal epithelial cells for transplantation. *Br. J. Ophthalmol.* **2009**, *93*, 980–983.

122. O'Callaghan, A.R.; Daniels, J.T. Concise review: Limbal epithelial stem cell therapy: Controversies and challenges. *Stem Cells* **2011**, *29*, 1923–1932.

123. Ahmad, S.; Osei-Bempong, C.; Dana, R.; Jurkunas, U. The culture and transplantation of human limbal stem cells. *J. Cell Physiol.* **2010**, *225*, 15–19.

MDPI AG

Klybeckstrasse 64

4057 Basel, Switzerland

Tel. +41 61 683 77 34

Fax +41 61 302 89 18

http://www.mdpi.com/

JFB Editorial Office

E-mail: jfb@mdpi.com

http://www.mdpi.com/journal/jfb

www.ingramcontent.com/pod-product-compliance
Lightning Source LLC
Chambersburg PA
CBHW051923190326
41458CB00026B/6391